THE ESSENTIAL EINSTEIN
PUBLIC WRITINGS

THE ESSENTIAL EINSTEIN
PUBLIC WRITINGS

ALBERT EINSTEIN

EDITED BY

DIANA KORMOS BUCHWALD

&

TILMAN SAUER

PRINCETON UNIVERSITY PRESS
PRINCETON & OXFORD

Copyright © 2025 by Princeton University Press

Princeton University Press is committed to the protection of copyright and the intellectual property our authors entrust to us. Copyright promotes the progress and integrity of knowledge created by humans. By engaging with an authorized copy of this work, you are supporting creators and the global exchange of ideas. As this work is protected by copyright, any reproduction or distribution of it in any form for any purpose requires permission; permission requests should be sent to permissions@press.princeton.edu. Ingestion of any PUP IP for any AI purposes is strictly prohibited.

Published by Princeton University Press
41 William Street, Princeton, New Jersey 08540
99 Banbury Road, Oxford OX2 6JX

press.princeton.edu

GPSR Authorized Representative: Easy Access System Europe - Mustamäe tee 50, 10621 Tallinn, Estonia, gpsr.requests@easproject.com

All Rights Reserved

ISBN 9780691131108
ISBN (e-book) 9780691272191

Library of Congress Control Number: 2025936081

British Library Cataloging-in-Publication Data is available

Editorial: Eric Crahan and Rebecca Binnie
Production Editorial: Terri O'Prey
Text Design: Wanda España
Jacket Design: Chris Ferrante
Production: Erin Suydam
Publicity: Julia Haav and Kate Farquhar-Thomson
Copyeditor: Madeleine Adams

Cover image: Philippe Halsman, Albert Einstein in his study in Princeton, New Jersey, 1947, © Philippe Halsman Estate 2025.

This book has been composed in Proxima Nova and Minion Pro

Printed in the United States of America

10 9 8 7 6 5 4 3 2 1

Contents

	Introduction to the Public Writings	xi
1	On the Principle of Relativity (1914)	1
2	Inaugural Lecture at the Prussian Academy of Sciences (1914)	4
3	Manifesto to the Europeans (1914) with G. F. Nicolai and F. W. Förster	7
4	Ernst Mach (1916)	10
5	My Opinion on the War (1916)	17
6	The Nightmare (1917)	20
7	Motives for Research (1918)	21
8	Dialogue about Objections to the Theory of Relativity (1921)	24
9	Time, Space, and Gravitation (1919)	34
10	Induction and Deduction in Physics (1919)	39
11	Immigration from the East (1919)	41
12	Uproar in the Lecture Hall/An Explanation (1920)	43
13	A Confession (1920)	44
14	Ether and the Theory of Relativity (1920)	46
15	To the General Association for Popular Technical Education (1920)	57
16	On New Sources of Energy (1920)	58
17	My Response: On the Anti-Relativity Company (1920)	60

18 On the Contribution of Intellectuals to International Reconciliation (1920) 63

19 The Common Element in Artistic and Scientific Experience (1921) 65

20 Geometry and Experience (1921) 65

21 How I Became a Zionist (1921) 78

22 The Development and Present Position of the Theory of Relativity (1921) 83

23 On a Jewish Palestine (1921) 86

24 The Impact of Science on the Development of Pacifism (1921) 89

25 The Plight of German Science: A Danger for the Nation (1921) 90

26 Preface to Bertrand Russell's *Political Ideals* (1922) 92

27 Review of Wolfgang Pauli, *The Theory of Relativity* (1922) 93

28 In Memoriam Walther Rathenau (1922) 94

29 On the Present Crisis of Theoretical Physics (1922) 96

30 Musings on My Impressions in Japan (1923) 102

31 My Impressions of Palestine (1923) 107

32 Antisemitism and Academic Youth (1923) 110

33 Review of Josef Winternitz's *Relativity Theory and Epistemology* (1923) 112

34 Sound Recording for the Prussian State Library (1924) 115

35 The Compton Experiment: Does Science Exist for Its Own Sake? (1924) 116

36 On the League of Nations (1924) 121

37	Review of Alfred C. Elsbach's *Kant and Einstein* (1924)	123
38	Non-Euclidean Geometry and Physics (1925)	130
39	Introductory Letter to *Letters from Russian Prisons* (1925)	135
40	Why Zionism/A Message (1925)	136
41	Pan-Europe (1925)	138
42	The Mission of the Hebrew University/A Word for the Journey (1925)	141
43	On Ideals (1925)	143
44	Space-Time, *Encyclopædia Britannica* (1926)	146
45	New Experiments on the Effect of the Earth's Motion on Light Velocity with Respect to the Earth (1927)	157
46	Speech at Rally for the Keren Ha-Yesod in Berlin (1926)	160
47	Newton's Mechanics and Its Influence on the Formation of Theoretical Physics (1927)	161
48	Review of Émile Meyerson's *La Déduction Relativiste* (1928)	169
49	The New Field Theory (1929)	174
50	Einstein Believes in Spinoza's God (1929)	183
51	The Palestine Troubles (1929)	185
52	To a Young Scholar (1929)	192
53	What I Believe: Living Philosophies XIII (1930)	193
54	Religion and Science (1930)	197
55	Militant Pacifism/The Two Percent Speech (1930)	201
56	Some Remarks Concerning My American Impressions (1931)	203

57	The 1932 Disarmament Conference (1931)	206
58	To American Negroes (1932)	208
59	Why War? (1932)	210
60	Is There a Jewish Philosophy? (1932)	213
61	Statement on Hitler upon Leaving Pasadena (1933)	215
62	Letter of Resignation from Prussian Academy of Sciences (1933)	216
63	Militant Pacifism No More (1933)	217
64	On the Method of Theoretical Physics/The Herbert Spencer Lecture (1933)	218
65	Science and Civilization/The Albert Hall Speech (1933)	225
66	Foreword to "The Contribution of the Jews of Germany to German Civilization" (1933)	228
67	On Germany and Hitler (1935)	230
68	Some Thoughts Concerning Education (1936)	231
69	The Calling of the Jews (1936)	236
70	Why Do They Hate the Jews? (1938)	238
71	Our Debt to Zionism (1938)	245
72	Ten Fateful Years: Living Philosophies, Revised (1938)	247
73	Letter to Franklin D. Roosevelt (1939)	250
74	Freedom and Science (1940)	252
75	The Common Language of Science (1942)	255
76	Newton's 300th Birthday (1942)	258
77	Remarks on Bertrand Russell's Theory of Knowledge (1944)	262

78	To the Heroes of the Battle of the Warsaw Ghetto (1944)	268
79	On the Atomic Bomb (1945)	269
80	On the American Council for Judaism (1945)	274
81	Commemorative Words for Franklin D. Roosevelt (1945)	275
82	The Way Out (1946)	277
83	Foreword to *Spinoza: Portrait of a Spiritual Hero* (1946)	281
84	A Message to My Adopted Country (1946)	283
85	The Military Mentality (1947)	285
86	In the Shadow of the Atomic Bomb (1947)	287
87	A Plea for International Understanding (1947)	289
88	Quantum Mechanics and Reality (1948)	291
89	A Reply to the Soviet Scientists (1948)	296
90	Religion and Science: Irreconcilable? (1948)	302
91	Why Socialism? (1949)	305
92	Autobiographical Notes (1949)	312
93	Statement to the Society for Social Responsibility in Science (1950)	353
94	Letter Declining the Presidency of Israel (1952)	354
95	Elementary Considerations on the Interpretation of the Foundations of Quantum Mechanics (1953)	355
96	Recollections–Souvenirs (1955)	361
Bibliography		369
Index		375

Introduction to the Public Writings

In late 1921, Michele Besso, one of Einstein's oldest friends, wrote to him: "A complete collection of all your papers until 1919... would be essential for insight into the historical development of the problems... Many would procure the collection at any price... it would be of immense material and intellectual worth."

One hundred years later, *The Collected Papers of Albert Einstein*, the annotated edition of Einstein's immense written legacy, is still a work in progress. Seventeen volumes covering the first fifty-two years of Einstein's life have been published since 1987 by Princeton University Press. Some thirteen further volumes are expected to appear in the future. The extant volumes contain as full text or in abstract more than ten thousand unique documents. Among them are more than five hundred writings.

Before 1919, Einstein did not publish any works that were not strictly scientific. His papers appeared in specialized journals and conference proceedings and were addressed to his professional colleagues.

The situation gradually began to change during the First World War. By then a full professor at the University of Berlin and a permanent member of the Prussian Academy of Sciences, Einstein started to respond to the political, social, and cultural imperatives of the day. The breakdown in scientific cooperation and exchanges during the war, the rising antisemitism at German universities, the atrocities committed in the trenches became topics of concern among some—but by no means the majority—of his academic colleagues. It will surprise even a careful analyst of Einstein's published and unpublished oeuvre to find that during the years 1916 to 1929, Einstein wrote some four hundred papers, books, newspaper articles, appeals, addresses, poems, calculations, and aphorisms. Of these, a third were on political and social issues, on pacifism, Jewish matters, Zionism, and antisemitism, containing also general addresses, obituaries, and prefaces. Another third comprised papers on relativity, gravitational waves, cosmology, quantum theory, and various other scientific topics. And the last third comprised popularizations of science and technology, patents and expert opinions, and foundational issues in science, philosophy, and religion.

From this multifarious treasure trove, the second volume of *The Essential Einstein* presents almost exclusively those general writings that Einstein himself approved for publication during his lifetime. The volume distinguishes itself from previous compilations by properly dating and presenting in full text writings that have previously been presented in excerpt or as interlineations in collected volumes. All texts have been checked against Einstein's manuscripts.

For all documents published before 1930, we have essentially used the English text published in the companion translation volumes of the *Collected Papers of Albert Einstein* (CPAE, Volumes 1–17). We have occasionally tacitly revised or corrected existing translations, and at times provided explanatory notes. The translations have been conformed to U.S. English spelling. Documents 13, 27, and 29 in the *Scientific Writings*, and several documents in the *Public Writings* (such as Documents 57, 95, and 96) were translated by us.

We here take the opportunity to thank all past and present collaborators at the Einstein Papers Project. We thank especially Jackie Bussone, Barbara and Michael Ferrara, Jennifer Nollar James, and Natalia Poleacova, who, at various stages, have assisted in the technical preparation of the volumes. We also express our thanks to the patient and dedicated colleagues at Princeton University Press: Anne Savarese, Rebecca Binnie, Peter Dougherty, Eric Crahan, Madeleine B. Adams, Max McMaster, and especially Terri O'Prey, for her impeccable production standards and her unstinting devotion to all things Einstein.

The editors have selected some one hundred writings, presented in chronological order, that we deem essential for a comprehensive understanding of Einstein's ethical worldview.

THE ESSENTIAL
EINSTEIN
PUBLIC WRITINGS

1. On the Principle of Relativity (1914)

Vossische Zeitung (26 April 1914), Morgen-Ausgabe, no. 2, pp. [1–2] of 8. [*Collected Papers* Vol. 6, Doc. 1].

By April 1914, Einstein had published two non-technical expositions of special relativity, and one journal article in which he touched on general relativity as well. This document is the first exposition to appear in a daily newspaper.

The editorial staff of the *Vossische Zeitung* has asked me to relate something about my field of work to readers. I gladly honor this request. Although a deeper understanding of the theory of relativity is hardly possible without considerable effort, it may still be appealing for the non-scientist to hear something about the methods and results of this new branch of theoretical research.

Even a cursory analysis of the processes we call motion already teaches us that we can perceive only the relative motion of things with respect to one another. We sit in a railway carriage and see (on the adjacent track) another carriage pass by. If we ignore the vibration of our carriage, we have no immediate means by which to decide whether the two carriages are moving "in reality." We find only that the relative position of the carriages changes over time. Even if we look at the telegraph poles alongside the track, nothing essential changes in this situation. For when we usually refer to telegraph poles (and the earth's surface) as "at rest" and every object moving relative to them as "in motion," we merely use a customary and handy expression without deeper meaning. An observer in a "moving" railway carriage will not come into conflict with his perceptions if he states that the carriage is at rest and the ground and telegraph poles are in motion.

Physicists have found over time that this characteristic of motion, to appear purely relative, is not merely attributable to primitive perception, but rather that one is justified to call any single thing "at rest" among a multitude of things that are in relative (uniform) motion with respect to one another. Let's think again of a uniformly moving carriage on a straight track. Let the windows be closed airtight, with no light coming in; wheels and tracks are completely smooth. Inside the carriage is a physicist with all kinds of apparatuses imaginable. We do know all experiments done by this physicist would come out exactly the same if the carriage were not moving, or if it were moving at a different velocity, for that matter. This statement is essentially what physicists call the "principle of relativity." One can generally phrase this principle as: "The laws of nature perceived by an observer are *independent* of his state of motion."

This statement sounds harmless and self-evident. It would not have excited anybody were it not for the fact that the laws of the propagation of light, which have emerged from the recent development in electrodynamics, seem incompatible with this principle. The phenomena of the optics of moving bodies have led to the interpretation that light always propagates with the same velocity in empty space, irrespective of the state of motion of the light source. Yet, this result seems to contradict the aforementioned principle of relativity. After all, when a beam of light travels with a stated velocity relative to one observer, then a second observer who is himself traveling in the direction of the propagation of the light beam, so it seems, should find the light beam propagating at a lesser velocity than the first observer would. If this were indeed true, then the law of light propagation in a vacuum would not be the same for two observers in relative uniform motion to each other, in contradiction to the principle of relativity stated above.

This is where the theory of relativity comes in. This theory shows that the law of constancy of light propagation in a vacuum can be satisfied simultaneously for two observers in relative motion to each other, such that the same beam of light shows the same velocity to both.

At first glance, the possibility for such a paradoxical interpretation can be understood through a more detailed analysis of the physical meaning of spatial and temporal statements. Recognizing the relativity of the concept of simultaneity is of special importance to this question. Before the theory of relativity, it was believed that the statement "two events happening at two different places are simultaneous" had a clear meaning—clear without a special need to define the concept of simultaneity. A more detailed investigation that did not skirt the issue of defining simultaneity showed, however, that the simultaneity of two events is not absolute, but instead can only be defined relative to one observer of a given state of motion. It turns out that two events simultaneous with respect to one observer are, in general, not simultaneous with respect to a second observer moving relative to the first one. This signifies a fundamental change in our concept of time. (This is the most important and most controversial theorem of the new theory of relativity. It is impossible to enter here into an in-depth discussion of the epistemological and scientific philosophical assumptions and consequences evolving from this basic principle).[1]

By combining the principle of relativity with the results of the constancy of the speed of light in vacuum, one arrives at what is today called "relativity theory" in a purely deductive manner. This theory has already proven itself as an aid for the theoretical deduction of the laws of nature. Its significance lies in the fact it provides conditions every general law of nature must satisfy, for the theory teaches that events in nature are such that the laws do not depend on the state of motion of the observer to whom the events are related spatially and temporally.

Two of the major results of the theory of relativity should be mentioned here, as they are also of interest to the layman. First: the hypothesis of the existence of a space-filling medium for light propagation, the so-called light-ether, must be abandoned. According to this theory, light no longer appears as a state of motion of an unknown carrier, but rather as a physical structure with a physical existence of its own. Second: the theory establishes that the inertia of a body is not an absolutely unchanging constant, but instead grows with the energy content. The important conservation theorems of mass and energy melt into a single theorem; the energy of a body also determines its mass.

Is the theory of relativity sketched above basically complete, or is it only a first step in a continuing development? Physicists who value the theory of relativity still hold differing opinions on this question. Nevertheless, weighty arguments speak for the second alternative. We have stated above that the laws of nature are the same for a "uniformly moving" observer as they are for one "at rest." This means an observer cannot find criteria that would allow him to decide if he is at rest or in a state of uniform motion. "At rest" and "in uniform motion" are physically

---

[1] Those who want to familiarize themselves with a more detailed substantiation and justification can find sufficient instruction—without difficult mathematical derivations—in E. Cohn's pamphlet, "Physikalisches über Raum und Zeit,"[a] and in an essay by Jos. Petzoldt, "Die Relativitästheorie der Physik,"[b] in the most recent issue of the *Zeitschrift für positivistische Philosophie*.

equivalent. This raises the question whether the principle of relativity is limited to uniform motion. Could the laws of nature not be such that they are the same for two observers who are in nonuniform motion relative to each other? Recently, it has turned out such an extension of relativity theory can be carried out and leads to a general theory of gravitation that contains the Newtonian theory as a first approximation. According to this theory, light rays suffer a curvature in a gravitational field; though minute, it is just within the range of astronomical measurement. The future will teach us whether this generalized relativity theory, which is very satisfying from an epistemological aspect, conforms to reality.

Editorial Notes

[a]   Cohn 1913.   [b]   Petzolt 1914.

2. Inaugural Lecture at the Prussian Academy of Sciences (1914)

Königlich Preußische Akademie der Wissenschaften. Berlin: Sitzungsberichte (1914), pp. 739–742. [*Collected Papers* Vol. 6, Doc. 3].

> This lecture was delivered on 2 July 1914 to the Prussian Academy of Sciences in Berlin. Einstein had been appointed a member of the Academy in November 1913 and had moved from Zurich to Berlin only a few weeks before the lecture. The text contains an early discussion of the deductive character of relativity, which he contrasts with other physical theories that seek to find foundational principles inductively.

Most honored colleagues!

Please accept, first, my heartfelt gratitude for the greatest favor you could have bestowed upon a person of my kind. By appointing me to your Academy, you have allowed me to devote myself wholly to scientific studies and freed me from the tribulations and worries of a practical profession. I ask you to believe in my feelings of gratitude and the diligence of my striving, even if the fruits of my efforts may appear to you as meager ones.

Allow me to append a few general remarks about the position that my field of work, theoretical physics, takes concerning experimental physics. Recently, a mathematician friend half-jokingly said to me: "The mathematician knows some things, no doubt, but of course not what one wants to have from him in each case." The theoretical physicist is in a very similar position when the experimental physicist asks him for advice. So what are the reasons for this strange lack of adaptability?

The method of the theoretician implicitly mandates that he uses, as his basis, general assumptions—so-called principles—from which he can then deduce conclusions. His activity, therefore, has two parts: first, he must ferret out these principles; and second, he has to develop the conclusions that can be deduced from these principles. His school provides him with excellent tools with which to fulfill the second-named task. Consequently, when the first one of his tasks is already solved for some area or a complex of connections, then sufficient diligence and insight ensure success will not be denied to him. But the former task—namely, to establish these principles that can serve as the basis of his deductions—is one of a wholly different kind. There is no learnable, systematically applicable method here that would lead him to the objective. The researcher must instead eavesdrop on nature to become privy to these general principles, recognizing in larger sets of experiential facts that certain general traits can then be strictly and precisely formulated.

A chain of conclusions sets in once this formulation is achieved, often with unforeseen connections far transcending the domain of facts from which the principle has been wrested. However, if the principles that must serve as the basis for the deduction remain undiscovered, the individual experimental fact is of no help to the theoretician. In fact, he cannot even do much with individual empirically established general laws. Instead, he must remain in a state of helplessness regarding the individual results of empirical research until principles that he can transform into the basis of deductive developments have become clear to him.

It is in such a position that theory finds itself at present regarding the laws of heat radiation and molecular movement at low temperatures. Just fifteen years ago, nobody doubted that Galilean-Newtonian mechanics applied to molecular motions and the Maxwellian theory of the electromagnetic field would lead to a correct representation of the electrical, optical, and thermal properties of bodies. Then Planck showed that establishing a law of heat radiation consistent with experience required a method of calculation whose incompatibility with the principles of classical mechanics was becoming ever more apparent. With this method of calculation, Planck introduced into physics the so-called quantum hypothesis, which has since found outstanding confirmations. With this quantum hypothesis, he toppled classical mechanics for cases of sufficiently small masses with sufficiently small velocities and sufficiently large accelerations, and so we today can accept the laws of motion established by Galileo and Newton only as limit-laws. But despite the most strenuous efforts of theoreticians, we have not yet succeeded in replacing the principles of mechanics with those that are adequate for Planck's law of heat radiation or the quantum hypothesis. As incontrovertibly as it has been proven that heat derives from molecular movement, we still must concede today that the basic laws of this motion similarly confront us as the planetary motions confronted pre-Newtonian astronomers.

I have just pointed out a set of facts for whose theoretical treatment we lack the principles. But it can happen as well that clearly formulated principles lead to consequences that are completely, or almost completely, outside the factual domain currently accessible to our experience. In this case, protracted experimental research may be needed to find out whether the theoretical principles correspond to reality. Such a case offers itself with the theory of relativity.

An analysis of the basic temporal and spatial concepts has shown that the theorem of a constant light velocity in a vacuum, which follows from the optics of moving matter, does not at all force us toward a theory of a light-ether at rest. Instead, it has been possible to establish a general theory that takes cognizance of the fact the translational motion of the Earth is never noticeable in the experiments conducted on Earth. Here, we use the relativity principle, which states that the laws of nature do not change their form when one changes from the original (admissible) coordinate system to a new one that is in uniform translational motion relative to the former. This theory found noteworthy confirmations in experience and has led to a simplification of the theoretical representation of whole complexes of facts that were already connected.

From a theoretical perspective, however, this theory does not grant full satisfaction, as the above-stated relativity principle favors *uniform* motion. If it is indeed true that uniform motion cannot have an absolute meaning from the physical perspective, we then have the obvious question of whether this statement can be extended to non-uniform motions. It turns out that one arrives at a very distinct extension of relativity theory when they use a relativity principle as a basis, in this sense, extended. One is thereby led to a general theory of gravitation, including its dynamics. As of now, we do not yet have the factual material needed to test the justification of introducing this principle.[a]

We have determined that inductive physics has questions for deductive physics and vice versa; eliciting the answers will require the application of our utmost efforts. May we, using united labor, soon succeed in advancing toward conclusive progress.[b]

Editorial Notes

[a] See Planck 1914 for Planck's comments about the need to generalize special relativity, made in response to Einstein's lecture.

[b] Einstein's theory proved unable to account for the anomalous motion of Mercury's perihelion. Both his attempts to measure the predicted gravitational redshift of spectral lines and gravitational light deflection proved inconclusive.

3. Manifesto to the Europeans (1914)

A. Einstein, G. F. Nicolai, and F. W. Förster

> G. F. Nicolai. *Die Biologie des Krieges: Betrachtungen eines deutschen Naturforschers.* Zurich: Füssli (1917), pp. 9–11.[a] [*Collected Papers* Vol. 6, Doc. 8].

This document was drafted in collaboration with G. F. Nicolai and F. W. Förster as a response to a 1914 manifesto drawn up by ninety-three German artists and intellectuals denying German culpability in the World

> War, titled "To the Civilized World."[b] Though informally circulated among academics since 1915, its text was not published until 1917 in Nicolai's *Die Biologie des Krieges* as the book's introduction.

While technology and traffic clearly drive us toward a factual recognition of international relations, and thus toward a common world civilization, it is also true that no war has ever interrupted the cultural communalism of cooperative work as intensively as this present war does. Perhaps we have come to such a salient awareness only due to the numerous erstwhile common bonds whose interruption we now sense so painfully.

Even if this state of affairs should not surprise us—those whose hearts are at all concerned about common world civilization[c]—would have a doubled obligation to fight to uphold those principles. However, those of whom one should expect such convictions—that is, principally, scientists and artists—have thus far almost exclusively uttered statements that would suggest that their desire to maintain these relations has evaporated concurrently with the interruption of the relations. They have spoken with explainable martial spirit, but have spoken of peace least of all.[d]

Such a mood cannot be excused by any national passion; it is unworthy of all the world has understood by the name of culture to date. Should this mood achieve a certain universality among the educated, it would be a disaster. It would not only be a disaster for civilization, but—and we are firmly convinced of this—a disaster for the national survival of individual states, the very cause for which all this barbarity ultimately has been unleashed.

Through technology, the world has become *smaller*; the *states* of the large peninsula of Europe appear today as close to each other as the cities of each small Mediterranean peninsula appeared in ancient times. In the needs and experiences of every individual, based on his awareness of a manifold of relations, Europe—one could almost say the world—already outlines itself as an element of unity.

It would consequently be a duty of educated and well-meaning Europeans to at least attempt to prevent Europe, on account of its deficient organization as a whole, from suffering the same tragic fate as ancient Greece once did. Should Europe also gradually exhaust itself and thus perish from fratricidal war?

The struggle raging today will likely produce no victor; it will probably leave only the vanquished. Therefore, it seems not only *good*, but rather bitterly *necessary* that educated men of all nations marshal their influence such that, whatever the still-uncertain end of the war may be, *the terms of peace shall not become the wellspring of future wars*. The evident fact that all European relational conditions slipped into an *unstable and plasticized* state through this war should rather be used to create an organic European whole. The technological and intellectual conditions for this are extant.

It need not be deliberated herein by which manner this (new) ordering in Europe is possible. We want merely to emphasize very fundamentally that we are firmly convinced the time has come when *Europe must act as one to protect her soil, her inhabitants, and her culture*.

To this end, it seems first of all to be a necessity that all those who have a place in their hearts for European culture and civilization—in other words, those who can be called in *Goethe's*[e] prescient words *"good Europeans"*[f]—come together. For we must not, after all, give up the hope that their raised and collective voices, even beneath the din of arms, will not resound unheard, especially if we find among these "good Europeans" of tomorrow all those who enjoy esteem and authority among their educated peers.

But it is necessary that Europeans first come together, and if, as we hope, enough *Europeans in Europe* can be found—that it is to say, people to whom Europe is not merely a geographical concept, but rather a dear affair of the heart—then we shall try to call together such a union of Europeans, thereupon such a union shall speak and decide.

To this end, we only want to urge and appeal; and if you feel as we do, if you are like-mindedly determined to *provide the European will the farthest-reaching possible resonance*, then we ask you to please send your (supporting) signature to us.

Editorial Notes

[a] An alternative English translation can be found in Lief 1933.
[b] Fulda et al. 1914, Bayer et al. 1919.
[c] One may surmise that in 1914 all European nations understood "world civilization" as "western civilization."

[d] The German original is here syntactically faulty and not clear.
[e] Johann Wolfgang von Goethe (1749–1832), the most influential German-language writer, poet, essayist, and critic.
[f] The term was actually coined by Friedrich Nietzsche, citing Goethe as the model for such a "good European" (Nietzsche 1886).

4. Ernst Mach (1916)

Physikalische Zeitschrift 17, no. 7 (1916), pp. 101–104. [*Collected Papers* Vol. 6, Doc. 29].

> Ernst Mach was Professor Emeritus of History and Theory of the Inductive Sciences at the University of Vienna. He died on 19 February 1916 at the age of 78. Einstein would reflect negatively on this obituary in a letter to Katja Alder, claiming, "It is not really good; my style is laborious and wooden, my knowledge of the literature is meager."

Recently, Ernst Mach left us—a man who had the greatest influence upon the epistemological orientation of natural scientists in our times, a man of rare independence of judgment. In him, the direct joy of seeing and understanding—Spinoza's *amor dei intellectualis*[a]—was so prevalent, he could look into the world with the curious eyes of a child even in his old age and thus be perfectly happy to enjoy himself by understanding interwoven connections.

But why does a properly gifted natural scientist care for epistemology at all? Isn't there more important work in his field? That's what I hear many a colleague say, and I sense that even more feel that way. I cannot share this conviction. When I think of the most able students I met in my teaching,—and I mean those who distinguished themselves not just by mere skills, but by their independence in judgment—I must state that they all had a lively interest in epistemology. They liked to start discussions about the final goals and methods of the sciences and undoubtedly showed, by the hardheadedness with which they defended their views, that the topic was important to them, and this is not surprising.

If I turn toward a science for external reasons such as ambition or earning an income rather than, or at least not exclusively, the mere sportive joy and fun of brain-acrobatics, then I must ask myself the question: what is the final goal that the science I am devoted to will and can reach? To what extent are its general results "true"? What is essential, and what is based only on accident in its development?

To appreciate Mach's merits, one must not ask: what has Mach found, in all these general questions, that no man before him saw or thought of? Again and again, the truth in these things must be chiseled out by energetic personalities and must always be commensurate with the needs of the times in which the sculptor works. If this truth is not always created anew, we will finally lose it altogether. Hence, it is difficult, albeit also not so essential, to answer the question; what did Mach teach us that would, in principle, be new when compared to Bacon and Hume? Or, as far as the general epistemological perspective toward individual sciences is concerned, what distinguishes him basically from Stuart Mill, Kirchhoff, Hertz, or Helmholtz? It is a fact that Mach has had a tremendous impact on our generation of natural scientists, in particular with his historical-critical writings, wherein he follows the evolution of individual sciences with so much love, probes practically the most remote brain cells of researchers who broke new paths in their fields. I even believe that those who consider themselves to be opponents of Mach barely know how many of his views they absorbed, in a manner of speaking, with their mother's milk.[b]

Science is, according to Mach, nothing but the comparison and orderly arrangement of factually given contents of our consciousness in accord with certain gradually acquired points of view and methods. Therefore, physics and psychology differ from each other not so much in the subject matter, but rather only in the points of view of the arrangement and connection of the various topics. It appears that Mach sought to demonstrate how this order evolved in the particular sciences he knew as his most important task. As a result of this activity of orderly arrangement, one obtains the abstract concepts and laws (rules) of their connection. Both are chosen such that they together form a scheme of order wherein the facts can be securely and comprehensively arranged. From what has been said, concepts make sense only insofar as they can be pointed out in things, as well as the points of view according to which concepts are associated with things (analysis of concepts).

The significance of personalities like Mach by no means lies only in the fact that they satisfy the philosophical needs of their times, an endeavor the hard-nosed specialist may dismiss as a luxury. Concepts that have proven useful in ordering things can easily attain authority over us, such that we forget their worldly origin and take them as immutably given. Then, they are rather rubber-stamped as a "*sine qua non*" of thought, as an "*a priori* given," etc. Such errors then render the road of scientific progress impassable for long stretches of time. Therefore, it is not at all idle play when we are trained to analyze the entrenched concepts and point out the circumstances that promoted their justification and usefulness and how they evolved from the experience at hand. This breaks their all-too-powerful authority. They are removed when they cannot properly legitimize themselves; they are corrected when their association with given things is too sloppy; they are replaced by others when a new system can be established that, for various reasons, we prefer.

The specialized scientist who concentrates more on details often views such analysis as superfluous, as far-out, and occasionally as outright ridiculous. But the situation changes when one of the traditionally used concepts is replaced by a more precise one because the development of the science in question demands it. Then those who do not cleanly use their own concepts raise vigorous protest and complain of a revolutionary endangerment of their most sacred treasures. Mixed into the outcry are the voices of those philosophers who believe they cannot dispense with the disputed concepts because they have incorporated them into their treasure chest of "absolutes" and "*a prioris*," or, in short, because they have already proclaimed their fundamental immutability.

The reader can already guess that I am touching on certain concepts of the philosophy of space and time, as well as mechanics, that have undergone a modification by the theory of relativity. Nobody can deny that epistemologists paved the road for progress, and I know for myself, at least, that Hume and Mach have helped me a lot, directly and indirectly. I ask the reader to look at Mach's opus, "*Die Mechanik in ihrer Entwicklung*,"[c] especially at the deliberations under 6. and 7. of chapter 2 ("Newton's Views of Time, Space, and Motion" and "Synoptical Critique of the Newtonian Enunciations"). There, one finds superbly explained thoughts that, until now, have by no means become commonplace among physicists. These portions are especially attractive because they are tied to Newton with verbatim quotations. Here are just some of these nuggets:

Newton:

Absolute, true, and mathematical time, of itself and by its own nature, flows uniformly on, without regard to anything external. It is also called *duration*. Relative, apparent, and common time is some sensible and external measure of absolute duration, whether accurate or inequable, commonly employed in place of true time; as an hour, a day, a month, a year.

Mach:

When we say a thing A changes with the time, we mean simply the conditions that determine a thing A depend on the conditions that determine another thing B. The vibrations of a pendulum take place in time when its excursion depends on the position of the earth. Since, however, in the observation of the pendulum, we do not need to consider its dependence on the position of the earth, but may compare it with any other thing, ... the opinion[d] easily arises that all the things with which we compare it are unessential.... It is beyond[e] our power to measure the changes of things by time. Quite the contrary, time is an abstraction at which we arrive by means of the changes of things, made because we are not restricted to any one definite measure, all being interconnected.

Newton:

Absolute space, in its own nature and without regard to anything external, always remains similar and immovable. Relative space is some movable dimension or measure of absolute space, which our senses determine by its position with respect to other bodies, and which is commonly taken for immovable (absolute) space.

This is followed by the corresponding definitions of "absolute motion" and "relative motion," which is then succeeded by the following:

The effects by which absolute and relative motions are distinguished from one another are centrifugal forces, which produce a tendency of recession from the axis. For in a circular motion which is purely relative, no such forces exist; but in a true and absolute circular motion, they do exist and are greater or less according to the quantity of the (absolute) motion.

Next follows a description of the well-known experiment with the pail,[f] intended to illustrate the foregoing reasoning.

It is interesting to look at Mach's critique of this position, and I shall quote a few particularly characteristic portions from it.

> If we say that a body K alters its direction and velocity solely through the influence of another body K', we cannot conclude such unless other bodies $A, B, C \ldots$ are present against which we judge the movement of body K. In reality, we are simply cognizant of a relation of the body K to $A, B, C \ldots$ If now we suddenly neglect $A, B, C \ldots$ and attempt to speak of the deportment of the body K in absolute space, we implicate ourselves in a twofold error. In the first place, we cannot know how K would act in the absence of $A, B, C \ldots$; and in the second place, we would have no means to judge the behavior of K and put to the test what we had predicated,—which, therefore, would be bereft of all scientific significance.
>
> The motion of a body K can only be estimated by reference to other bodies $A, B, C \ldots$ But since we always have at our disposal a sufficient number of bodies that are, as respects each other, relatively fixed or only slowly change their positions, we are restricted to no one definite body and can[g] leave out of account now this one, and now that one. In this way, the conviction arose that these bodies are indifferent generally.
>
> Newton's experiment with the rotating vessel of water simply informs us that the relative rotation of the water with respect to the sides of the vessel produces no noticeable centrifugal forces, but that such forces are produced by its relative rotation with respect to the mass of the earth and the other celestial bodies. No one is competent to say how the experiment would turn out if the sides of the vessel increased in thickness and mass till they were ultimately several leagues thick.

These quotations show that Mach clearly recognized the weak points of classical mechanics and thus came close to demanding a general theory of relativity—and this was almost half a century ago! It is not improbable that Mach would have come upon relativity theory if, at the time when he was in fresh and youthful spirit, physicists had already been intrigued by the question of the significance of the constancy of the speed of light. Absent this stimulation, which flows from Maxwell-Lorentz electrodynamics, even Mach's critical urge did not sufficiently arouse a feeling for the need for a definition of simultaneity for spatially distant events.

The contemplation on Newton's bucket experiment demonstrates how close his thinking was to the demands of relativity in a wider sense (relativity of accelerations). Admittedly, the vivid awareness that the equivalence of inertial and gravitational mass elicits a postulate of relativity in a wider sense was missing, as we are unable to decide through experiments whether the falling of a body relative to a coordinate system is caused by the presence of a gravitational field or by a state of acceleration of the coordinate system.

In his intellectual development, Mach was not a philosopher who chose natural science as an object of speculation, but rather an ardent researcher of nature with many interests, one who obviously found pleasure in researching detailed questions off the trodden path of general interest. The almost innumerable individual investigations he published—partly by himself, partly together with students—in the fields of physics and empirical psychology are testimony. Among his experimental investigations in physics, those of the acoustic waves generated by bullets are best known. While the basic idea was not new, the investigations revealed an extraordinary experimental talent. He succeeded in photographing the density distribution of air around a supersonic bullet and thus shed new light upon a group of acoustic processes about which nothing was known before him. Everyone who likes the subjects of physics will enjoy his popular lecture about it.[h]

All of Mach's philosophical studies sprang from his desire to find a perspective from which the various branches of science to which he dedicated his lifelong labor can be seen as an integrated endeavor. He comprehended all science as striving for order among individual elementary experiences, which he called "sensations" (*Empfindungen*). This choice of word is probably the reason those who are less familiar with his works often mistook the sober, careful thinker for a philosophical idealist and solipsist.

When reading Mach's works, one truly feels the cozy ease the author must have felt when he effortlessly wrote his concise, striking sentences. But it is not merely the enjoyment of good style and mere intellectual joy that make reading his books so attractive; it is also the kind, humanitarian, and hopeful attitude that gleams from between the lines when he talks about human things in general. These sentiments and convictions also made him immune to another disease of our time, from which few are spared today—national fanaticism. In his popular essay "On Phenomena of Flying Projectiles," he could not refrain from expressing his hope for an understanding among nations in the last paragraph.[i]

(Received on March 14, 1916)

Editorial Notes

[a] *Amor dei intellectualis* (the intellectual love of God) is a concept Spinoza introduced in his philosophical treatise *Ethics* (Spinoza 1677), in which he defines it as "pleasure accompanied by the idea of God as cause, that is, the love of God; not in so far as we imagine him as present, but in so far as we understand him to be eternal."

[b] Einstein identifies Max Planck as one such opponent in a letter sent to Mach on 9 August 1909, which makes similar comments.

[c] Mach 1897. Einstein's quoted sections differ slightly from editions of *Die Mechanik in ihrer Entwicklung* that would have been available to him. Differences apart from changes in spelling convention will be noted. The English translation here lines up mostly with the original 1893 T. J. McCormack translation, though slightly altered to account for Einstein's misquotations and McCormack's liberties, which do not match what Einstein himself cites.

[d] Einstein uses "opinion" ("Meinung") here, while the original text uses "illusion" ("Täuschung").

[e] Einstein removes "utterly" ("ganz") from this sentence.

[f] Newton describes the thought experiment in *Principia* (Newton 1846):

> The effects that distinguish absolute from relative motion are the forces of receding from the axis of circular motion. For there are no such forces in a circular motion purely relative, but in a true and absolute circular motion, they are greater or less, according to the quantity of the motion. If a vessel hung by a long cord is so often turned about that the cord is strongly twisted, then filled with water and held at rest together with the water; after it is whirled about the contrary way by the sudden action of another force, and while the cord is untwisting itself, the vessel continues for some time in this motion; the surface of the water will at first be plain, as before the vessel began to move: but the vessel, by gradually communicating its motion to the water, will make it begin sensibly to revolve and recede little by little from the middle and ascend to the sides of the vessel, forming itself into a concave figure (as I have experienced); and the swifter the motion becomes, the higher will the water rise, till at last performing its revolutions in the same times with the vessel, it becomes relatively at rest in it. This ascent of the water shows its endeavor to recede from the axis of its motion; and the true and absolute circular motion of the water, which is here directly contrary to the relative, discovers itself and may be measured by this endeavor. At first, when the relative motion of the water in the vessel was greatest, it produced no endeavor to recede from the axis; the water showed no tendency to the circumference nor any ascent toward the sides of the vessel, but remained of a plain surface, and therefore its true circular motion had not yet begun. But afterward, when the relative motion of the water had decreased, the ascent thereof towards the sides of the vessel proved its endeavor to recede from the axis; and this endeavour showed the real circular motion of the water perpetually increasing till it had acquired its greatest quantity, when the water rested relatively in the vessel. And therefore this endeavor does not depend upon any translation of the water in respect of the ambient bodies, nor can true circular motion be defined by such translation.

[g] Einstein removes "alternatively" ("abwechselnd") from this sentence.
[h] Mach 1903.
[i] The final passage of Mach 1898. (republished in German in Mach 1903) reads as follows:
> Let the hate of races and nationalities run riot as it may; the intercourse of nations will still increase and grow more intimate. By the side of the problems that separate nations, the great and common ideals claiming the exclusive powers of the men of the future appear, one after another, in greater distinctness and greater might.

5. My Opinion on the War (1916)

Berliner Goethebund. *Das Land Goethes 1914–1916: Ein vaterländisches Gedenkbuch.* Stuttgart, Berlin: Deutsche Verlags-Anstalt (1916), p. 30.[a] [*Collected Papers* Vol. 6, Doc. 20].

> The Berlin chapter of the Goethebund, founded in 1900 to broaden the public understanding of the arts and sciences, solicited this manuscript from Einstein on 23 October 1915. The organization planned a "patriotic commemorative volume" titled *Das Land Goethes* 1914–1916, in which the heirs of Goethe were requested to defend German culture against criticism of the country by other nations in the wake of World War I. Somewhere between 23 October and 11 November 1915, Einstein submitted the manuscript, which he later hesitantly revised, at the Goethebund's request, by deleting its fourth and fifth paragraphs.

The psychological roots of war are, in my opinion, biologically founded in the aggressive characteristics of the male creature. We "jewels of the creation" are not the only ones who can boast this distinction; some animals outdo us on this point, e.g., the bull and the rooster. This aggressive tendency comes to the fore whenever individual males are placed side by side, and even more so when relatively close-knit societies must deal with each other. Almost without fail, they will end up in disputes that escalate into quarrels and murder unless special precautions are taken to prevent such occurrences. I will never forget the[b] honest hatred the schoolmates of my age felt for years against the first-graders of a school on a neighboring street. Innumerable fistfights occurred, resulting in many a hole in the heads of those little striplings. Who could doubt that vendetta and dueling spring from such feelings? I even believe that the *honor* [*Ehre*] so carefully groomed by us gains its major nourishment from such sources.

Understandably, the more modern organized states had to vigorously push those manifestations of primitive virile characteristics into the background. But wherever two nation-states are next to each other without a joint superpower above them, those feelings, at times, generate tensions in temperament [*Gemüt*] that lead to catastrophes of war. In saying so, I consider the so-called aims and causes of war as rather meaningless, as they are always found when passion needs them.

The more subtle intellects of all times have agreed that war is one of the worst enemies of human development and that everything must be done to prevent it. Notwithstanding the unspeakably sad conditions of the present times, I have the conviction it is possible to form a state-like organization in Europe that makes European wars impossible in the near future, just as war between Bavaria and Württemberg is now impossible in the German Reich. No friend of spiritual evolution should fail to stand up for this most important political aim of our time.[c]

But how can a powerless individual creature contribute to reaching this goal? Should, perhaps, everybody devote a considerable portion of his time and resources to politics? I really believe that the intellectually more mature people of Europe have sinned when they neglected to care for general political questions, yet I do not see in political commitment the most important effectiveness of an individual in this matter. I instead believe that everybody should act individually in the sense that those feelings I have elaborately discussed before are steered on a course that can no longer become a curse for the public.

Every man who knows he has acted to his best knowledge and ability should feel honorable, without regard to the words and deeds of others. Words and actions of others or other groups cannot violate one's personal honor or that of one's group. Power-hunger and greed shall, as in the past, be treated as despicable vices; the same applies to hatred and quarrelsomeness. I do not suffer from an overestimation of the past, but, in my opinion, we have not made progress on this important point;

we instead fell back. Every well-meaning person should work hard on himself and in his personal circle to improve in these respects. The heavy burdens presently plaguing us in such a horrible way will then vanish too.

But why so many words when I can say it all in one sentence, and a sentence very appropriate for a Jew? Honor your Master, Jesus Christ, not only in words and songs, but rather, foremost, by your deeds.

Editorial Notes

[a] The manuscript submitted to Goethebund consists of three unnumbered pages. Page numbers provided are of those from the manuscript rather than the final publication.

[b] The words "bloody and" were excised from the manuscript.

[c] At this point, the following paragraphs from the original manuscript were omitted from the published article:

> One can ponder the question: Why does man (when the social community suppresses almost every representation of rowdyism) not lose the capabilities and motivations in peacetime that enable him to commit mass murder in war? The reasons seem to me as follows: when I look into the mind of the well-intentioned average citizen, I see a moderately illuminated cozy room. In one corner stands a well-tended shrine,—the pride of the man of the house—and every visitor is loudly alerted to the presence of this shrine, upon which "patriotism" is written in huge letters. It is usually a taboo to open this shrine. Moreover, the master of the house barely knows, or does not know at all, that this shrine holds the moral requisites of bestial hatred and mass murder that he dutifully takes out in case of war to use them. You will not find this type of shrine in my little room, dear reader, and I would be happy if you would adopt the attitude that a piano or a bookshelf would be a more fitting piece of furniture in that same corner of your little room than the one you find tolerable only because you have been used to it since your early youth.
>
> I have no intention of making a secret of my international sentiments. How close I feel to a human being or a human organization depends only upon how I judge their intentions and capabilities. The state to which I belong as a citizen does not play any role in my feelings [*Gemütsleben*], because I see it more like a matter of business, such as the relations with a life insurance company. (From what I have said above, there should be no doubt that I must strive to be a citizen of a country that presumably does not force me to take part in a war.)

6. The Nightmare (1917)

Berliner Tageblatt (25 December 1917), p. 1. [*Collected Papers* Vol. 6, Doc. 49].

> This article was published in a special section of the *Berliner Tageblatt* discussing the possible abolition of the final secondary school exam ("Reifeprüfung" or "Maturitätsprüfung") in Germany.

I consider the final secondary school examination that succeeds the normal school education *not only unnecessary, but even harmful*. I consider it unnecessary because the school's teachers can undoubtedly judge the maturity of a young man who has attended that school for several years. The teachers' impression of a student derived during the school years together with the usual numerous papers from assignments (which every student must complete) are a succinctly comprehensive and superior basis on which to judge the student than any carefully executed examination.

I consider the final secondary school exam harmful for two reasons. The fear of exams as well as *the large number of topical subjects that must be assimilated through memorization harm the health of many young men to a considerable degree*. This fact is too well known to require detailed verification. But I will nevertheless mention the well-known fact that many men in the most varied professions have been plagued into their later years by nightmares whose origins trace back to the final secondary school exam. And these are men who have stood up to their responsibilities in life and by no means can be counted among the "neurasthenics."

The final secondary school exam is furthermore harmful because it *lowers the level of teaching in the last school years*. Instead of an exclusively substance-oriented occupation with individual subjects, one too often finds a lapse into shallow drilling of the students for the exam. Instead of in-depth teaching, one gets what amounts to an *exercise for show* designed to give a certain luster to the class for the benefit of the examiners.

Therefore, away with the final secondary school exam!

7. Motives for Research (1918)

Zu Max Plancks sechzigstem Geburtstag. Karlsruhe: C. F. Müllersche Hofbuchhandlung (1918), pp. 29–32.[a] [*Collected Papers* Vol. 7, Doc. 7].

> This address was delivered at a celebration of Max Planck's sixtieth birthday (1918) before the Physical Society in Berlin. Max Planck (1858–1947) was a professor of theoretical physics at the University of Berlin, who contributed greatly to physics with his quantum theory, advanced in 1900, which provided the foundation for developing modern atomic physics.

In the temple of science, there are many mansions, and those that dwell therein—and the motives that led them thither—are various indeed. Many take to science out of a joyful sense of superior intellectual power; science is their own special sport from which they seek vivid experience and the satisfaction of ambition. Many others found in the temple are those who have offered the products of their brains on this altar for purely utilitarian purposes. Were an angel of the Lord to come and drive all the people belonging to these two categories out of the temple, the assemblage would be seriously depleted, but there would still be some men, of both present and past times, left inside. Our Planck is one of them, and that is why we love him.

I am quite aware that we have just now lightheartedly expelled, in imagination, many excellent men who are largely, perhaps chiefly responsible for the building of the temple of science; and in numerous instances, our angel would find it a pretty ticklish job to decide. But of one thing, I feel sure: if the types we have just expelled were the only types there, the temple would never have come to be, no more than a forest consisting of nothing but creepers could grow. For these people, if it comes to a point, any sphere of human activity will do; whether they become engineers, officers, tradesmen, or scientists depends on circumstances.

Now, let us have another look at those who have found favor with the angel! Most of them are somewhat odd, uncommunicative, and solitary fellows, really less like each other, despite these common characteristics, than the hosts of the rejected. What has brought them to the temple? That is a difficult question, and no single answer will cover it. To begin with, I believe, with Schopenhauer, that one of the strongest motives that leads men to art and science is an escape from everyday life, with its painful crudity and hopeless dreariness, from the fetters of one's own ever-shifting desires. A finely tempered nature longs to escape from personal life into the world of objective perception. This desire may be compared to the townsman's irresistible longing to escape from his noisy, cramped surroundings into the silence of high mountains, where the eye ranges freely through the still, pure air and fondly traces out the restful contours apparently built for eternity.

With this negative motive goes a positive one. Man tries to make for himself, in the fashion that suits him best, a simplified and intelligible picture of the world; he then tries, to some extent, to substitute this cosmos of his for the world of experience and thus to overcome it. This is what the painter, the poet, the speculative philosopher, and the natural scientist do, each in his own fashion. Each makes this cosmos and its construction the pivot of his emotional life, to find the peace and security he cannot find in the narrow whirlpool of personal experience.

What place does the theoretical physicist's picture of the world occupy among all these possible pictures? It demands the highest possible standard of rigorous precision in the description of relations, such that only the use of mathematical language can give. On the other hand, regarding his subject matter, the physicist has to limit himself very severely: he must content himself with describing the simplest events that can be brought within the domain of our experience; all events of a more complex order are beyond the power of the human intellect to reconstruct with the subtle accuracy and logical perfection the theoretical physicist demands—supreme purity, clarity, and certainty at the cost of completeness. But what can be the attraction of getting to know such a tiny section of nature thoroughly, while leaving everything subtler and more complex shyly and timidly alone? Does the product of such a modest effort deserve to be called by the proud name of a theory of the universe?

In my belief, the name is justified, for the general laws on which the structure of theoretical physics is based claim to be valid for any natural phenomenon whatsoever. With them, it ought to be possible to arrive at the description—that is to say, the theory—of every natural process, including life, using pure deduction, were that process of deduction not far beyond the capacity of the human intellect. The physicist's renunciation of completeness for his cosmos is therefore not a matter of fundamental principle.

The supreme task of the physicist is to arrive, by pure deduction, at those universal elementary laws from which the cosmos can be built up. There is no logical path to these laws; only intuition resting on a sympathetic understanding of experience can reach them. In this methodological uncertainty, one might suppose there are any number of possible systems of theoretical physics all equally well justified, and this opinion is no doubt correct, theoretically. But the development of physics has shown that, at any given moment, out of all conceivable constructions, a single one has always proved itself decidedly superior to all the rest. Nobody who has really gone deeply into the matter will deny that the world of phenomena uniquely determines the theoretical system in practice, despite no logical bridge between phenomena and their theoretical principles; this is what Leibniz described so happily as a "pre-established harmony." Physicists often accuse epistemologists of not paying sufficient attention to this fact. Here, it seems to me, lie the roots of the controversy carried on some years ago between Mach and Planck.

The longing to behold this pre-established harmony is the source of the inexhaustible patience and perseverance with which Planck has devoted himself, as we see, to the most general problems of our science, refusing to let himself be diverted to more grateful and more easily attained ends. I have often heard colleagues try to attribute this attitude of his to extraordinary willpower and discipline—wrongly, in my opinion. The state of mind enabling a man to do work of this kind is akin to that of the religious worshiper or the lover; the daily effort comes from no deliberate intention nor program, but straight from the heart.

There he sits, our beloved Planck, and smiles inside himself at my childish playing-about with the lantern of Diogenes. Our affection for him needs no threadbare explanation. May the love of science continue to illuminate his path in the future and lead him to the solution of the most important problem in present-day physics he has himself posed and done so much to solve. May he succeed in uniting quantum theory with electrodynamics and mechanics into a single logical system.

Editorial Note

[a] This address was published in English in Einstein, *Ideas and Opinions*, trans. Sonja Bargmann (New York: Crown, 1982), pp. 224–227, where it was titled "Principles of Research."

8. Dialogue about Objections to the Theory of Relativity (1921)

Die Naturwissenschaften 6 (1918), pp. 697–702. [*Collected Papers* Vol. 7, Doc. 13].

Einstein wrote this article, in the form of a dialogue modeled on Galileo's *Dialogue Concerning the Two Chief World Systems* (1632), in reply to criticism of his theory of relativity by two outspoken German anti-relativists, Ernst Gehrcke and Philipp Lenard, who argued that relativity owed its success largely to propaganda, that Einstein had plagiarized Paul Gerber's theory of gravitation, and that relativity was counterintuitive and violated healthy common sense.

Kritikus: People like me have regularly expressed their various doubts about the theory of relativity in journals, but rarely has one of you relativists[1] responded. We do not want to go into the reasons for this neglect, whether it was arrogance, a feeling of weakness, or laziness. Perhaps it was an especially powerful mixture of all these mental forces, or maybe the criticism also revealed, often with clear evidence, that the critic really had too little knowledge of the subject matter. These issues, as I have said, shall not be discussed, but one thing, I want to tell you right away: today, I have come to you personally to make it impossible for you to shirk it as has happened before. Because, I assure you, I will not yield until you have answered all my questions.

---

[1] "Relativist" here means an adherent of the physical theory of relativity, not one of philosophical relativism.

8. Dialogue about Relativity Theory

But in order not to give you too much of a shock—since perhaps you even find a certain pleasure in the task (which you cannot escape anyway)—I will also tell you something consoling. I am not, like some of my colleagues, so steeped in the dignity of my guild as to come on like a superior being with otherworldly insight and self-assuredness (like a science writer or, worse, a theater critic). Rather, I will talk like a mortal human, particularly since I know that the father of criticism is quite often a lack in one's own ideas. I also do not intend—as one of my colleagues recently did—to attack you like prosecutors do, charging you with theft of intellectual property or other dishonorable acts. My intrusion was only motivated by a desire to clarify several points on which opinions are still too divergent. But I must also ask that you allow the publication of our dialogue, not the least because the shortage of paper is not the only deficiency that deprives my friend, the editor Berolinensis, of his sleep.

I can see your willingness, so let us come right to the substance. Since the special theory of relativity has been formulated, its result of the delaying influence of motion upon the rate of clocks has elicited protest and, as it looks to me, with good reason. This result seems necessarily to lead to a contradiction with the very foundations of the theory. To make things perfectly clear between us, let this result of the theory be phrased first and precisely enough.

Let K be a Galilean coordinate system in the sense of the special theory of relativity, i.e., a body of reference relative to isolated material points moving uniformly along straight lines. Furthermore, let U^1 and U^2 be two identical clocks free of any outside influence. They operate at the same rate when placed immediately side by side or at an arbitrary distance from each other, and both are at rest relative to K. But if one of the clocks—e.g., U^2—is in a state of uniform translatory motion relative to K, then it shall, according to the special theory of relativity, when judged from the coordinate system K, go at a slower rate than clock U^1, which is still placed at rest relative to K. This result in itself already strikes me as strange. Grave scruples arise if one next considers the following well-known thought experiment.

Let A and B be two mutually distant points in system K. To fix the conditions, let us assume A to be the origin of K, and B a point on the positive x-axis. At the beginning, both clocks shall rest at point A. They operate at the same rate, and their hands shall indicate the same time.

Now we shall give clock U^2 a constant velocity along the positive x-axis such that it moves toward B. In B, we imagine its velocity inverted such that U^2 again moves back to A. Upon arrival in A, the clock is braked and brought to rest relative to U^1. Since (judged from K) the change in the position of the hands of U^2 (which might occur during the velocity inversion of U^2) will certainly not exceed a certain amount, and since U^2 runs at a slower rate than U^1 during its uniform motion along distance AB (again as judged from K), U^2 must be late relative to U^1 after its return, provided distance AB is of sufficient length.—Do you agree with this conclusion?

Relativist: I agree, absolutely. It saddened me to see that some authors who otherwise stand on the ground of the theory of relativity wanted to avoid this inescapable result.

Krit.: Now comes the snag. According to the principle of relativity, the entire process must occur in exactly the same way when represented in reference to the coordinate system K', which partakes in the movement of clock U^2. Relative to K', it is then clock U^1 that moves to and fro while clock U^2 is at rest all the time. At the end of the movement, U^1 must be late against U^2, in contradiction to the result above. Even the devoutest adherents of the theory cannot claim that, of two clocks resting side by side, each one is late relative to the other.

Rel.: Your last assertion is, of course, incontestable. However, the entire line of reasoning is not legitimate because, according to the special theory of relativity, coordinate systems K and K' are not at all equivalent systems. In fact, this theory claims only the equivalence of all Galilean (nonaccelerated) systems, i.e., coordinate systems relative to which sufficiently isolated material points move uniformly in straight lines. Coordinate system K is indeed such a system, but not the intermittently accelerated system K'. Consequently, no contradictions in the foundations of the theory can be construed from the fact that U^2 is late against U^1 after the to and fro movement.

Krit.: I see, you have defused my objection. But, I must tell you, your argument leaves me more convicted than really convinced. Besides, my objection is immediately resurrected from the dead if one accepts the general theory of relativity. Because coordinate systems of arbitrary states of motion are equivalent according to that theory, and I can describe the previous process with respect to K' (permanently linked to U^2) as well as I can with respect to K.

Rel.: It is certainly correct that, in the general theory of relativity, we can use coordinate system K' as well as coordinate system K. But it is easily seen that systems K and K' are by no means equivalent when it comes to the process under consideration. Although the process is to be interpreted in the manner outlined above when seen from system K, the process is something entirely different when viewed from K', as shall be demonstrated in the juxtaposition and figure shown below.

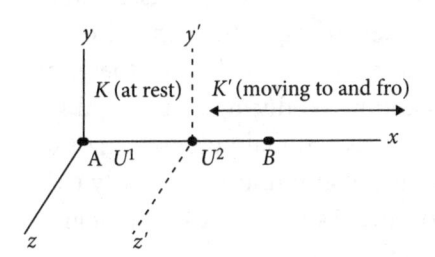

System of reference is K *System of reference is K'*

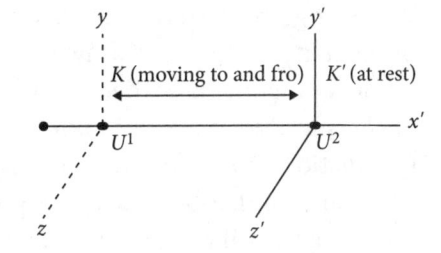

1. Clock U^2 gets accelerated by an external force along the positive x-axis until it attains velocity v. U^1 remains at rest.
2. U^2 moves with constant velocity v until it reaches point B on the positive x-axis. U^1 remains at rest.
3. Clock U^2 gets accelerated by an external force acting along the negative x-axis until it attains the velocity v in the negative x-direction. U^1 remains at rest.
4. U^2 moves back with constant velocity v in the direction of the negative x-axis, close to U^1. U^1 remains at rest.
5. An external force brings U^2 to rest.

1. There arises a gravitational field in the direction of the negative x-axis, in which clock U^1 falls accelerated until it attains velocity v. An external force along the positive x-axis acts upon clock U^2 and prevents it from moving. The gravitational field vanishes again as soon as clock U^1 has attained velocity v.
2. U^1 moves with the constant velocity v until it reaches point B' on the negative x-axis. U^2 remains at rest.
3. There arises a homogeneous gravitational field in the direction of the positive x-axis, in which clock U^1 gets accelerated in the direction of the positive x-axis until it has attained velocity v, whereupon the gravitational field again vanishes. An external force along the negative x-axis prevents clock U^2 from making any movement in this gravitational field.
4. U^1 moves with the constant velocity v in the direction of the positive x-axis, close to U^2. U^2 remains at rest.
5. There arises a gravitational field in the direction of the negative x-axis that brings clock U^1 to rest, whereupon the gravitational field vanishes again. During this time U^2 is kept at rest by an external force.

One must carefully keep in focus that both the left and right columns describe exactly the same process, but the description on the left refers to coordinate system K, the one on the right to coordinate system K'. According to both descriptions, it is clock U^2 that lags by a certain amount behind clock U^1 at the end of the process considered. With respect to the coordinate system K', the phenomenon is explained in the following manner: During procedural steps 2 and 4, U^1 clock, moving at velocity v, has indeed a slower rate than clock U^2, which is at rest. But the time lag gets overcompensated by the faster rate of U^1 during procedural step 3 because, according to the general theory of relativity, a clock has a more accelerated rate the higher the gravitational potential is at the clock's location; and during procedural step 3, U^2 is indeed at a location of higher gravitational potential than U^1. Calculation shows that this running-ahead amounts to precisely twice as much as the lag-behind during procedural steps 2 and 4. This analysis completely clarifies the paradox you referred to.

Krit.: I see. Indeed, you extricated yourself very skillfully, but I would be lying if I declared myself completely satisfied. The cause of contention is not removed, only pushed to another place. Because your analysis of the difficulty shows only its connection with another one, which also has been raised repeatedly. You solved the paradox by considering the influence on clocks of the gravitational field relative to K'. But isn't this gravitational field only fictitious? Its existence is, I should say, only simulated by the choice of coordinates. After all, real gravitational fields are always generated by masses and cannot be made to vanish by a suitable choice of coordinates. How should one believe that a merely fictitious field could influence the rate of clocks?

Rel.: First, I must point out that the distinction between real and non-real is not very productive. With respect to K', the gravitational field "exists" just as any other physical object that can only be defined in relation to a system of coordinates does, even though it may not be there relative to the system K. There is nothing particularly strange in this, as can easily be seen from the following example from classical mechanics. Nobody doubts the "reality" of kinetic energy; otherwise, one would have to deny the reality of energy per se. But it is clear the kinetic energy of a body depends upon the state of motion of the coordinate system; with a suitable choice of the latter, one obviously can cause the kinetic energy of a body in translatory movement to take on, at a certain moment, any given positive value or, say, the value zero. In the special case where all masses have velocities equal in size and direction, a suitable choice of the coordinate system can bring the entire kinetic energy to zero. I think the analogy is complete.

Instead of "real" and "non-real," we should rather distinguish more clearly between quantities inherent in a physical system per se (independent of the choice of coordinate systems) and other quantities that do depend upon the coordinate system. The obvious demand would be that physics should use only quantities of the first kind in its laws. History has shown that this cannot be realized in practice, as the development of classical mechanics has already clearly demonstrated. One could, for example, think of—and, actually, it has already been tried—to introduce only the distances between material points instead of the coordinates into the laws of classical mechanics; *a priori*, one might expect this to be the simplest way to achieve the goals of the theory of relativity. Scientific development, however, did not confirm this expectation. Relativity theory cannot dispense with coordinates and thus must use coordinates as quantities not resulting from definable measurements. According to the general theory of relativity, the four coordinates of the space-time continuum can even be chosen completely arbitrarily, as parameters devoid of any independent physical meaning. Part of this arbitrariness remains even in those quantities (field components) with whose aid we describe physical reality. Only certain ones, usually rather complicated expressions composed of field components and coordinates, are measurable (i.e., real) quantities independent of the system of coordinates. The components of the gravitational field in a space-time point, for example, have no equivalent quantity independent of the choice of coordinates; the gravitational field at a *certain location* represents nothing "physically real," but the gravitational field together with other data does. Therefore, one can neither say the gravitational field at a location is "real," nor that it is "only fictitious."

The main difficulty in studying the theory of relativity seems to lie in the fact that the connection in it between *quantities in equations* and *measurable quantities* is far more indirect than in the customary theories of old. Your last objection, too, is because you did not focus on this distinction.

You proclaimed the fields used in the clock experiment also as merely fictitious because lines of force in *real* gravitational fields must necessarily be generated by masses when there were no field-generating masses present in the example we discussed. There is a twofold response to this. *First*: It is not an *a priori* necessary requirement that the Newtonian concept of every gravitational field generated by masses should be kept in the general theory of relativity. This question, again, is linked to the circumstances mentioned above, that the meaning of field components is much less directly defined than it is in the Newtonian theory. *Second*: One cannot say there are no masses to which the generation of the field could be attributed. It is true, however, that accelerated coordinate systems cannot be employed as the real causes of the field, though a humorous critic once thought he could attribute that idea to me. But all stars in the universe can be thought of as taking part in the generation of the gravitational field because they are, during the acceleration phases of the coordinate system K', accelerated relative to the latter and thus can induce a gravitational field just as well as acceleratedly moving electric charges induce an electric field. Approximate integration of the gravitational equations has indeed shown that such inductive effects actually must occur with acceleratedly moving masses. From these considerations, it is clear that a complete clarification of the questions you raised can only be obtained if one forms a picture of the geometric-mechanical constitution of the universe as a whole that is compatible with the theory. This, I tried last year and have found, as it seems to me, a completely satisfying model, but going into it would lead too far.

Krit.: After your last explanations, it seems indeed that the clock paradox is not suited to deduce internal contradictions in the theory of relativity. In fact, it now does not seem improbable to me that the theory may have no internal contradictions, but this is not sufficient to take the theory into serious consideration. *I just cannot see why it should be necessary to take upon oneself, for the sake of one's intellectual predilection,—I mean the idea of relativity—such horrible complications and mathematical difficulties.* In your last answer, you yourself showed clearly enough that they are not minor. Will it, for example, occur to someone to make use of the possibility offered by the theory of relativity to relate the movement of the celestial bodies of our solar system to a geocentric system of coordinates, a system that, of all things, takes part in the rotational movement of the earth? Could such a coordinate system really be considered "at rest" and equivalent to others when, relative to it, the fixed stars race around the Earth at horrendous velocities? Would this idea not run counter to all common sense and the postulate of the economy of thinking? I cannot refrain from repeating some strong words Lenard recently uttered on this subject. After he discussed special relativity, where he characterized the "moving" coordinate system by a railroad train in motion, he said: "Now, let the imagined railroad train make a clearly non-uniform motion. When everything in the train then goes to pieces because of inertial effects while outside nothing is damaged, then—I believe—*no common sense* can escape the conclusion that it was the very train and not the environment that changed its motion with a jolt. Now, the generalized principle of relativity demands, in its simple elementary sense, to admit even in this case that it might have been the environment that suffered a change in velocity, and the whole disaster in the train was only a consequence of the jolt of the outside world, transmitted through the "gravitational effect" of the outside world upon the interior of the train. For the obvious question of why the church steeple next to the train did not collapse when its environment suffered the jolt, why such consequences of the jolt show themselves so *one-sidedly* only in the train, while *still* no one-sided, unequivocal conclusion about the placement of the change in motion should be possible. For this question, the principle apparently has no answer that could satisfy the simple mind."

Rel.: For several reasons, we willingly have to accept the complications to which the theory leads us. For one, it is a great satisfaction for a consistently thinking man to understand that the concept of absolute movement—which kinematically makes no sense anyway—need not be introduced into physics. It cannot be denied that the foundation of physics benefits by the avoidance of this concept in its logical structure. Furthermore, the fact of the equivalence of inertia and gravity of bodies irresistibly demands clarification. Aside from this, physics needs a method of getting to a local field theory for gravitation. Theoreticians could not attack this problem without an effective limiting principle because one could postulate *many theories* that all agree with the rather limited experience in this field. *Embarras de richesse* is one of the most malicious foes, making the life of a theoretician difficult. The postulate of relativity limited these possibilities to such an extent that the path the theory *had to take* was predetermined. Finally, the secular movement of the perihelion of the planet Mercury had to be explained. Its existence had been established with certainty by the astronomers, but the Newtonian theory could not find a satisfying explanation for it.—The postulate of the equivalence of coordinate systems as a *matter of principle* does not claim that every coordinate system is equally *convenient* for the investigation of a specific physical system; this is already the case in classical mechanics. Strictly speaking, one should, for example, not say the Earth moves in an ellipse around the Sun, as this statement assumes a coordinate system in which the Sun is at rest, whereas classical mechanics also allows for systems relative to which the Sun *moves* uniformly in a straight line. But nobody would seriously want to use the latter coordinate system for an investigation of the motion of the Earth. Similarly, nobody would conclude from this example that coordinate systems in which the center of gravity of the system under consideration remains at rest at the origin are in principle distinguished from other coordinate systems. The same applies to the example you mentioned. Nobody would investigate our solar system in a coordinate system in which the Earth is at rest, because this would be impractical. But *in principle*, such a coordinate system would still be equivalent to any other one in the general theory of relativity. The phenomenon that, in such coordinate systems, fixed stars would race around at tremendous velocities is no argument against its *admissibility*, but merely against the usefulness of this choice of coordinates. The same is true of the complicated structure of the gravitational field relative to such a coordinate system that, for example, would also have components corresponding to the centrifugal forces. A similar situation prevails in Herr Lenard's example. In the theory of relativity, one cannot interpret the case in the sense that "*possibly it was* the environment (of the train) that suffered the change in velocity." We do not have two mutually exclusive hypotheses about the location of movement, but rather two equivalent ways in principle to describe the same factual phenomenon.[2]

---

[2] That the steeple does not topple over is caused—in the second way of description—by the fact that, together with the ground and the whole Earth, it is in free fall in a gravitational field (present during the jolt) while, on the other hand, the train is prevented from a free fall by outer forces (break forces). A free-falling body acts with respect to its inner processes like a freely floating body, removed from all outer influences.

Only utilitarian reasons can decide which representation has to be chosen, but not arguments about principles. The following counterexample will show how inadvisable it is to appeal to so-called common sense as an arbiter in such things. Lenard himself says: So far, no pertinent objections have been found to the validity of the *special* principle of relativity (i.e., the principle of relativity between uniformly translatory motions of coordinate systems). The uniformly moving train could as well be seen "at rest" and the tracks, including the landscape, as "uniformly moving." Will the "common sense" of the locomotive engineer allow this? He would object that he does not go on to heat and grease the landscape, but rather the locomotive, and that it consequently must be the latter whose movement shows the effect of his labor.

Krit.: I have to admit, after this discussion, that it is not as simple to disprove your perspective as I previously thought. Still, I have several objections in mind, but I would rather not bother you with them until I have thoroughly thought about today's exchange. But before we part, I have one more question—not an objection, just pure curiosity: What is the present situation of the sick man of theoretical physics, the ether, whom some of you have declared as finally dead?

Rel.: The sick man has had a fluctuating fate; I don't at all think one could say he is dead yet. Before Lorentz, he was there as an all-penetrating liquid, a gas-like fluid, and in various other forms of existence, changing from author to author. Under Lorentz, he became rigid, embodying coordinate systems "at rest," or a preferred state of motion of the universe. With the special theory of relativity, there was no longer a preferred state of motion; this meant denying the ether in the sense of previous theories, because if there were an ether, it would have to have a distinct state of motion at every point in space-time since it plays an important role in optics. But, as the special theory of relativity taught, a preferred state of motion does not exist, and that is why there is no ether in the old sense of the word. The general theory of relativity does not know of a preferred state of motion in a distinct point either, which possibly could be interpreted as the velocity of some ether. But while, in the special theory of relativity, a portion of space without matter and devoid of an electromagnetic field is truly empty (i.e., not characterized by any physical quantities), it is quite different in general relativity. There, empty space, in the previous sense, has physical qualities, mathematically characterized by the components of the potential of gravitation, that determine the metrical behavior of that portion of space, as well as its gravitational field. This situation can very well be interpreted by speaking of an ether whose state varies from point to point. However, one has to be careful not to attribute any matterlike properties (e.g., a distinct velocity at each point) to this "ether."

9. Time, Space, and Gravitation (1919)

The Times, London (28 November 1919), pp. 13–14. [*Collected Papers* Vol. 7, Doc. 26].

> After the final results of the British eclipse expedition led by Arthur S. Eddington were announced at the joint meeting of the Royal Society and the Royal Astronomical Society in Burlington House on 6 November 1919, and as a result of "the very wide scientific and popular interest in the difficult subject," Einstein agreed to write a brief account of his theory and its implications to the Berlin correspondent of *The Times*. For his English audience, Einstein felt compelled to soothe the uproar in British scientific circles and the popular press, which reported on his alleged destruction of Newton's theory.

I respond with pleasure to your Correspondent's request that I should write something for *The Times* on the Theory of Relativity.

After the lamentable breach in the former international relations existing among men of science, it is with joy and gratefulness that I accept this opportunity to communicate with English astronomers and physicists. It was by the high and proud tradition of English science that English scientific men should have given their time and labor, and English institutions should have provided the material means, to test a theory that had been completed and published in the country of their enemies in the midst of war. Although the investigation of the influence of the solar gravitational field on rays of light is a purely objective matter, I am nonetheless very glad to express my personal thanks to my English colleagues in this branch of science; for without it, I could hardly have lived to see the most important implication of my theory tested.

There are several kinds of theories in physics. Most of them are constructive. These attempts build up a picture of complex phenomena out of some relatively simple proposition. The kinetic theory of gases, for instance, attempts to refer to molecular movement and the mechanical, thermal, and diffusional properties of gases. When we say that we understand a group of phenomena, we mean that we have found a constructive theory that embraces them.

Theories of principle

But in addition to this most weighty group of theories, there is another group consisting of what I call theories of principle. These employ the analytic, not the synthetic method. Their starting point and foundation are not hypothetical constituents, but empirically observed general properties of natural phenomena, principles from which mathematical formulae are deduced of such a kind that they apply to every case that presents itself. Thermodynamics, for instance, starting from the fact that perpetual motion never occurs in ordinary experience, attempts to deduce from this, by analytic processes, a theory that will apply in every case. The merit of constructive theories is their comprehensiveness, adaptability, and clarity; that of the theories of principle, their logical perfection, and the security of the foundation.

The theory of relativity is a theory of principle. To understand it, the principles on which it rests must be grasped. But before stating these, it is necessary to point out that the theory of relativity is like a house with two separate stories, the special theory and the general theory of relativity.

Since the time of the ancient Greeks, it has been well known that, in describing the motion of a body, we must refer to another body. The movement of a railway train is described with reference to the ground; of a planet, with reference to the total assemblage of visible fixed stars. In physics, the bodies to which motions are spatially referred are termed systems of coordinates. The laws of the mechanics of Galileo and Newton can be formulated only by using a system of coordinates.

System of coordinates

The state of motion of a system of coordinates cannot be chosen arbitrarily if the laws of mechanics are to hold good (it must be free from twisting and from acceleration). The system of coordinates employed in mechanics is called an inertia-system. The state of motion of an inertia-system, so far as mechanics are concerned, is not restricted by nature to one condition. The condition in the following proposition suffices: a system of coordinates moving in the same direction and at the same rate as a system of inertia is itself a system of inertia. The special relativity theory is therefore the application of the following proposition to any natural process:—Every law of nature which holds good with respect to a coordinate system K must also hold good for any other system K', provided that K and K' are in uniform movement of translation.

The second principle on which the special relativity theory rests is that of the constancy of the velocity of light in a vacuum. Light in a vacuum has a definite and constant velocity, independent of the velocity of its source. Physicists owe their confidence in this proposition to the Maxwell-Lorentz theory of electrodynamics.

The two principles which I have mentioned have received strong experimental confirmation, but do not seem to be logically compatible. The special relativity theory achieved their logical reconciliation by making a change in kinematics, that is to say, in the doctrine of the physical laws of space and time. It became evident that a statement of the coincidence of two events could have a meaning only in connection with a system of coordinates, that the mass of bodies and the rate of movement of clocks must depend on their state of motion with regard to the coordinates.

The older physics

But the older physics, including the laws of motion of Galileo and Newton, clashed with the relativistic kinematics that I have indicated. The latter gave origin to certain generalized mathematical conditions with which the laws of nature would have to conform if the two fundamental principles were compatible. Physics had to be modified. The most notable change was a new law of motion for (very rapidly) moving mass points, and this soon came to be verified in the case of electrically laden particles. The most important result of the special relativity system concerned the inert mass of a material system. It became evident that the inertia of such a system must depend on its energy content, and thus we were driven to the conception that inert mass was nothing else than latent energy. The doctrine of the conservation of mass lost its independence and became merged with the doctrine of the conservation of energy.

The special relativity theory, which was simply a systematic extension of the electrodynamics of Maxwell and Lorentz, had consequences that reached beyond itself. Must the independence of physical laws with regard to a system of coordinates be limited to systems of coordinates in uniform movement of translation with regard to one another? What has nature to do with the coordinate systems that we propose and with their motions? Although it may be necessary for our descriptions of nature to employ systems of coordinates that we have selected arbitrarily, the choice should not be limited in any way so far as their state of motion is concerned. (General theory of relativity.) The application of this general theory of relativity was found to conflict with a well-known experiment, according to which it appeared that the weight and the inertia of a body depended on the same constants (identity of inert and heavy masses). Consider the case of a system of coordinates which is conceived as being in stable rotation relative to a system of inertia in the Newtonian sense. The forces which, relative to this system, are centrifugal must, in the Newtonian sense, be attributed to inertia. But these centrifugal forces are, like gravitation, proportional to the mass of the bodies. Is it not, then, possible to regard the system of coordinates as at rest and the centrifugal forces as gravitational? The interpretation seemed obvious, but classical mechanics forbade it.

This slight sketch indicates how a generalized theory of relativity must include the laws of gravitation, and actual pursuit of the conception has justified the hope. But the way was harder than was expected, because it contradicted Euclidean geometry. In other words, the laws according to which material bodies are arranged in space do not exactly agree with the laws of space prescribed by the Euclidean geometry of solids. This is what is meant by the phrase "a warp in space." The fundamental concepts "straight," "plane," & *e.*, accordingly, lose their exact meaning in physics.

In the generalized theory of relativity, the doctrine of space and time, kinematics, is no longer one of the absolute foundations of general physics. The geometrical states of bodies and the rates of clocks depend in the first place on their gravitational fields, which again are produced by the material systems concerned.

In terms of principles, the new theory of gravitation diverges considerably from Newton's theory. But in practical application the two agree so closely that it has been difficult to find cases in which the actual differences could be subjected to observation. As yet, only the following have been discovered:—

1. The distortion of the oval orbits of the planets round the Sun (confirmed in the case of the planet Mercury).
2. The deviation of light-rays in a gravitational field (confirmed by the photographs taken during the English solar eclipse expedition).
3. The shifting of spectral lines toward the red end of the spectrum in the case of light coming to us from stars of appreciable mass (not yet confirmed).

The great attraction of the theory is its logical consistency. If any deduction from it should prove untenable, it must be given up. A modification of it seems impossible without destruction of the whole.

No one must think that Newton's great creation can be overthrown in any real sense by this or by any other theory. His clear and wide ideas will forever retain their significance as the foundation on which our modern conceptions of physics have been built.

A final comment. The description of me and my circumstances in *The Times*[a] shows an amusing feat of imagination on the part of the writer. By an application of the theory of relativity to the taste of readers, today in Germany I am called a German man of science, and in England I am represented as a Swiss Jew. If I come to be regarded as a *bête noire*,[b] the descriptions will be reversed, and I shall become a Swiss Jew for the Germans and a German man of science for the English!

Editorial Notes

[a] This reference is to a short note in *The Times* (London), 8 November 1919, p. 12, titled "Dr. Albert Einstein," wherein he was called "a Swiss Jew." His academic appointments were listed as: "for some time professor in mathematical physics at the Polytechnic at Zurich, and then professor at Prague. Afterwards he was nominated a member of the Kaiser Wilhelm Academy for Research in Berlin," and he was described politically as "an ardent Zionist."

[b] Someone who is notably detested.

10. Induction and Deduction in Physics (1919)

Berliner Tageblatt (25 December 1919), p. 1. [*Collected Papers* Vol. 7, Doc. 28].

> Published together with contributions by Fritz Haber, Walther Nernst, and Max Planck, intended as statements on the achievements of German science for inclusion in the newspaper's special Christmas issue. The views expressed here by Einstein on theory formation, the role of intuition and deduction, and the unknowability of scientific truth remained essentially unchanged in later decades.

The simplest idea one can form about the origin of an empirical science is that it is based on the inductive method. Individual facts are selected and grouped together so that the laws governing their connections show clearly. By grouping these laws, one can establish more general laws, until a uniform system for the existing individual facts has been created, a system that would allow the reflective mind to reverse the process and arrive back at the individual facts from the ultimate generalities gained in this way through pure thought.

But even a casual examination of actual developments teaches us that the important achievements in scientific knowledge only rarely came about in this way; for if a researcher were to approach things without any preconceived notion, how could he pick those facts simple enough to reveal the natural laws behind their connections from the tremendous number of complicated experiences? Galileo would have never found the law of free fall without his preconceived notion that the conditions as we find them are complicated by the effects of air resistance, and one therefore must focus on those cases in which this effect is minimized.

The truly great advances in our understanding of nature have come in ways almost diametrically opposed to induction. The scientist is led to propose one or several hypothetical basic laws through an intuitive understanding of the essential features of a large complex of facts. From the basic law (system of axioms), he derives the consequences, as completely as possible, in a purely logical-deductive manner. These conclusions, which often must be derived from the basic law through lengthy derivations, can then be compared to experience and thus provide a criterion for the justification of the assumed basic law. Basic law (axioms) and conclusions together form what is called a "theory." Every expert knows that the greatest advances in natural science—e.g., Newton's theory of gravitation, thermodynamics, the kinetic theory of gases, modern electrodynamics, etc.—all originated this way and that their basis has a principally hypothetical character. The researcher always starts with facts, and the aim of his efforts is their connection. However, he does not find his system of thought in a methodical, inductive way, but adapts it to the facts by an intuitive selection from among all conceivable axiom-based theories.

Thus, a theory can very well be found incorrect if there is a logical error in its deductions or if an empirical fact is not in accordance with one of its conclusions. The *truth* of a theory, however, can never be proven, for it is impossible to know whether an experience that lies in the future will not contradict its conclusion. There are always other conceivable systems of thought that can connect the very same facts. If two theories that both agree with the same given facts are available, there are no other criteria but the intuitive eye of the researcher to prefer one over the other. This explains why astute researchers who master the theories and facts can still be passionate supporters of opposing theories.

In these tumultuous times, I offer the reader this brief, objective, and dispassionate reflection because I believe quiet devotion to the eternal goals shared by all civilized men can serve today's political recovery better than political considerations and confessions.

11. Immigration from the East (1919)

Berliner Tageblatt (30 December 1919), p. 2. [*Collected Papers* Vol. 7, Doc. 29].

> This first public statement by Einstein on Jewish affairs was written at the request of the Workers' Welfare Bureau of Jewish Organizations, an umbrella agency established in January 1918, to which the major Jewish interest groups in Germany, including the Zionists, belonged.
>
> Einstein's statement was solicited in response to a steady drumbeat of attacks by right-wing newspapers, in particular the *Deutsche Tageszeitung*, organ of the German National People's Party.[a] In late September, for instance, it had called for a halt to Jewish immigration into Germany "daily by the hundreds and the thousands," claiming that "today the Jews feel themselves to be rulers of Germany."

Among German public voices, one increasingly hears a demand for legal measures against Eastern European Jews. It is claimed that, in Berlin alone, there are 70,000 Russian—i.e., Eastern European—Jews and these Eastern European Jews (*Ostjuden*) are alleged to be profiteers, black marketers, Bolsheviks,[b] or elements averse to work.

Measures that devastate so many individuals must not be triggered by slogan-like assertions, even less so when an objective reexamination has shown that we have a case here of agitation by demagogues that does not reflect the actual situation and is not a suitable means for counteracting existing wrongs. Agitation against Eastern European Jews especially raises suspicion that calm judgment is being dimmed by strong antisemitic instincts and, simultaneously, that a *specific* method is chosen that, by influencing the mood of the people, diverts from the true problems and the real causes of the general calamity.

As far as is known, an official inquiry by the authorities that would undoubtedly reveal the baselessness of the accusations has not been conducted. It may very well be true that 70,000 Russians live in Berlin, but according to competent observers, only a small fraction of them are Jews, while the overwhelming majority are of *German descent*. According to authoritative estimates, no more than 15,000 Jews have emigrated from the East since the signing of the peace treaty. Almost without exception, they were forced to flee by the horrible conditions in Poland and seek refuge here *until allowed to emigrate elsewhere*. Hopefully, many of them will find a true homeland as free sons of the Jewish people in the newly established Jewish Palestine.

It is quite likely that there are Bolshevist agents in Germany, but they undoubtedly hold foreign passports, have at their disposal ample funds, and cannot be seized by any administrative measures. The big profiteers among the Eastern European Jews have certainly taken precautions long ago to elude arrest by officials. The only ones affected would be *those poor and unfortunate ones* who, in recent months, made their way to Germany under inhumane privations to seek employment here. Only these elements, certainly harmless to the German national economy, would fill the concentration camps and perish there, both physically and spiritually. Then one will complain about the self-made "parasitic existences" who no longer know how to take their place in a normally functioning economy. The misguided policy of suddenly laying off thousands of Eastern European Jewish laborers—who were coerced into coming to Germany during the war—and thus depriving them of their means of livelihood, leaving them with nothing to eat, and systematically denying them job opportunities—has indeed forced people into the black market to keep themselves and their families from starving. The German economy, too, is certainly best served if the public supports the efforts of those who try to channel Eastern European Jewish immigrants into productive work (as does, for example, the often mentioned "Jewish Labor Office"). Any "order of expulsion"—now so vigorously demanded—would only have the effect that the worst and most harmful elements remain in the country, while those willing to work would be driven into bitter misery and despair.

The public conscience is so dulled toward appeals to humanity that it no longer even senses the horrible injustice being contemplated. I'll refrain from going into details. But it is disturbing when even leading politicians do not consider how much their proposed treatment of Eastern European Jews will damage Germany's *political and economic position*. Has it already been forgotten how much the deportation of Belgian laborers undermined the moral credibility of Germany? And today, Germany's situation is incomparably more critical. Despite all efforts, it is extremely difficult to reestablish the disrupted international relations; in all nations, only a few intellectuals among the peoples of the world are initiating some first attempts; the hope for new economic connections (e.g., the material help of America) is still feeble today. The expulsion of the Eastern European Jews—resulting in unspeakable misery—would only appear to the whole world as new evidence of "German barbarism" and provide it with a pretext, in the name of humanity, to hamper Germany's reconstruction.

The recuperation of Germany truly cannot be accomplished by force against a small and defenseless fraction of the population.

Editorial Notes

[a] The German National People's Party (Deutschnationale Volkspartei) was a far-right nationalist political party active between the years 1918 and 1933. In 1929, it formed a political alliance with the Nazi Party that allowed Nazi ascension to power.

[b] This refers to the antisemitic canard of Jewish Bolshevism, which asserts that the Russian Revolution was incited by Russian Jews and that they were the primary power-holders within the Bolsheviks.

12. Uproar in the Lecture Hall/An Explanation (1920)

8-Uhr-Abendblatt 73, no. 38 (13 February 1920), pp. 2–3. [*Collected Papers* Vol. 7, Doc. 33].

Although Einstein had no obligation to teach as a member of the Prussian Academy of Science, he regularly offered lectures at the University of Berlin. University regulations allowed only registered students, auditors, and faculty of the university to attend such lectures, but Einstein himself announced an open admission policy.

My popular lectures on the theory of relativity were attended not only by students, but also by many individuals who were, in fact, not authorized to attend. For this reason, the student council declared it would no longer tolerate this. I pointed out that the large hall has enough room for all who want to listen and that this should hardly cause any problems. The student council was not satisfied with this and turned this question to the *rector of the university*. The rector sent me a *letter* wherein he pointed out that, according to existing regulations, these individuals had no right to enter the hall. This is *formally correct*. I, however, took the perspective that I find it reprehensible to deny these individuals the right to listen without good reason. Therefore, yesterday, instead of lecturing, I began a discussion with my audience that, however, did not lead to a definite result. Consequently, I found it necessary to cancel further lectures and declare to the student body that they could request a refund of their tuition fees. However, I have no intention of canceling my lectures in general; I plan to resume them in a different form. The lecture hall has yet to be determined. Should an incident such as yesterday's occur again, I will cease my lectures altogether. What happened yesterday cannot be called a scandal, even though some remarks that were made demonstrated a certain animosity toward me. *Antisemitic* remarks per se did not occur, but the *undertone* could be interpreted that way.

13. A Confession (1920)

Israelitisches Wochenblatt, no. 39 (24 September 1920), p. 10. [*Collected Papers* Vol. 7, Doc. 37].

An invitation from the Central-Verein deutscher Staatsbürger jüdischen Glaubens (CV) was extended in late March 1920 to several prominent Jewish professors in Berlin to attend a meeting of its working committee. On the agenda was the question of how to enlighten an antisemitic German professoriate, which was "completely ignorant of Jewish ways and of the Jewish nature" (Central-Verein to Einstein, 29 March 1920).

13. A Confession

Today, I received your invitation to attend a meeting on the 14th of this month[a] devoted to the fight against antisemitism in academic circles. I would gladly attend, were I to believe in the possible success of such an endeavor. But first, we should fight antisemitism and a servile mentality among us Jews through enlightenment. More dignity and more independence in our ranks! Only when we have the courage to see ourselves as a nation, only when we respect ourselves, can we earn the respect of others, or, rather, they will come of themselves. Antisemitism as a psychological phenomenon will persist as long as Jews come into contact with non-Jews—and what can it hurt? Maybe we owe it to antisemitism that we continue to maintain ourselves as a race; at least, I believe so.

I cannot suppress a pained smile whenever I read "German citizens of the Jewish faith."[1] What is at the bottom of this fine expression? What is the Jewish faith, after all? Is there a form of "non-faith" by which one stops being a Jew? No. But behind this expression lies a double confession by beautiful souls, namely: first, I would rather not have anything to do with my poor Jewish (Eastern European Jewish)[b] brothers; and second, I don't want to be seen as a child of my people, but solely as a member of a religious community. Is this sincere? Can the Aryan have respect for such pussyfooters? I am not a German citizen ... but I am a Jew, and I am glad to belong to the Jewish people, even though I do not consider it a chosen one. Let the Aryan be happy in his antisemitism, and let us keep the love for people like us.

Don't make angry faces because of this confession! It is not meant to be either malicious or unfriendly!

Very respectfully yours!

A. Einstein

Editorial Notes

[a] "14th of the month" refers to April, not September.

[b] In both typescripts, "Jewish" is omitted and "Eastern European Jewish" is not in parentheses.

---

[1] The invitation had been extended by the Central-Verein deutscher Staatsbürger Jüdischen Glaubens—Central Association of German Citizens of Jewish Faith.

14. Ether and the Theory of Relativity (1920)

Aether und Relativitaetstheorie. Berlin: Springer Berlin Heidelberg (1920), pp. 3–15. [*Collected Papers* Vol. 7, Doc. 38].

This document presents the text of Einstein's inaugural lecture at the University of Leyden, where he had been appointed Extraordinary Professor (*Bijzonder Hoogleeraar*). During Einstein's visit to Leyden in October 1919, Lorentz urged him to present his current views on the ether to the public. Einstein promised to do so as soon as an opportunity arose. In an article in the *Nieuwe Rotterdamsche Courant*, Lorentz expressed his conviction that a search for "mutations and movements that might take place in it [the ether] … abandoned at present, will once more be followed with good results…. Einstein's theory need not keep us from so doing; only the ideas about the ether must accord with it."

When Lorentz wrote to Einstein concerning the Leyden appointment the following month, he mentioned that an inaugural lecture, though not absolutely required, would be much appreciated. Einstein wrote back that he would hold an inaugural lecture on the topic of the ether. As he was writing the lecture, Einstein characterized it as a "more or less colored retrospective look at the development of our opinions about the physical properties of space." The lecture, originally scheduled for 5 May 1920, had to be postponed because of delays both in the approval of the appointment and in issuing visa documents, and because of Einstein's other commitments. The position was officially created only on 24 June 1920, and Einstein was appointed shortly thereafter. Einstein finally delivered the lecture on 27 October 1920.

How does it come about that the physicists set alongside the idea of ponderable matter, which is derived by abstraction from everyday life, the idea of the existence of another kind of matter, the ether? The explanation is probably to be sought in those phenomena that have given rise to the theory of action at a distance, as well as the properties of light that have led to the undulatory theory. Let us devote a little while to the consideration of these two subjects.

14. Ether and Relativity

Outside of physics, we know nothing of action at a distance. When we try to connect cause and effect in the experiences natural objects afford us, it seems at first as though there are no mutual actions other than those of immediate contact—e.g., the communication of motion by impact, push and pull, heating or inducing combustion through a flame, etc. It is true that weight—which is, in a sense, action at a distance—plays a critical part even in everyday experience. But since the weight of bodies in daily experience meets us as something constant, something not linked to any cause variable in time or place, we do not speculate as to the cause of gravity in everyday life and therefore do not become conscious of its character as action at a distance. It was Newton's theory of gravitation that first assigned a cause to gravity by interpreting it as action proceeding from masses at a distance. Newton's theory is probably the greatest stride ever made in the effort toward the causal nexus of natural phenomena. And yet, this theory evoked a lively sense of discomfort among Newton's contemporaries, as it seemed to be in conflict with the principle, springing from the rest of experience, that there can be reciprocal action only through contact and not through immediate action at a distance.

It is only with reluctance that man's desire for knowledge endures a dualism of this kind. How was unity to be preserved in his comprehension of the forces of nature? Either by trying to look upon contact forces as being themselves distant forces admittedly observable only at a very small distance—and this was the road Newton's followers, entirely under the spell of his doctrine, mostly preferred to take—or by assuming that the Newtonian action at a distance is only *apparently* immediate action at a distance and, in truth, is conveyed by a medium permeating space, whether by movements or by elastic deformation of this medium. Thus, the endeavor toward a unified view of the nature of forces leads to the hypothesis of an ether. This hypothesis, to be sure, did not at first bring with it any advance in the theory of gravitation or physics generally, so it became customary to treat Newton's law of force as an axiom that was not further reducible. But the ether hypothesis was always bound to play some part in physical science, even if only a latent part, at first.

When the far-reaching similarity subsisting between the properties of light and those of elastic waves in ponderable bodies was revealed in the first half of the nineteenth century, the ether hypothesis found fresh support. It appeared beyond question that light must be interpreted as a vibratory process in an elastic, inert medium that filled up universal space. It also seemed that this medium of ether, as a necessary consequence of the fact that light is capable of polarization, must be of the nature of a solid body, as transverse waves are not possible in a fluid, only in a solid. Thus, the physicists were bound to arrive at the theory of the "quasi-rigid" luminiferous ether, the parts of which can carry out no movements relative to one another except the small movements of deformation corresponding to light-waves.

This theory—also called the theory of the stationary luminiferous ether—moreover found strong support in an experiment also of fundamental importance in the special theory of relativity: the experiment of Fizeau, from which one was obliged to infer that the luminiferous ether did not take part in the movements of bodies. The phenomenon of aberration also favored the theory of the quasi-rigid ether.

The development of the theory of electricity along the path opened up by Maxwell and Lorentz gave quite a peculiar and unexpected turn to the development of our ideas concerning the ether. For Maxwell himself, the ether indeed still had purely mechanical properties, though of a much more complicated kind than the mechanical properties of tangible solid bodies. But neither Maxwell nor his followers succeeded in elaborating a mechanical model for the ether that might furnish a satisfactory mechanical interpretation of Maxwell's laws of the electromagnetic field. The laws were clear and simple, the mechanical interpretations clumsy and contradictory. Almost imperceptibly, the theoretical physicists adapted themselves to a situation that, from the standpoint of their mechanical program, was very depressing. They were particularly influenced by the electrodynamical investigations of Heinrich Hertz. For, whereas they previously had required a conclusive theory to content itself with the fundamental concepts belonging exclusively to mechanics (e.g., densities, velocities, deformations, stresses), they gradually accustomed themselves to admitting electric and magnetic force as fundamental concepts side-by-side with those of mechanics without requiring a mechanical interpretation for them. Thus, the purely mechanical view of nature was gradually abandoned. But this change led to a fundamental dualism that was insupportable in the long run. A way of escape was now sought in the reverse direction, by reducing the principles of mechanics to those of electricity—and this, especially—as confidence in the strict validity of the equations of Newton's mechanics was shaken by the experiments with β-rays and rapid cathode rays.

This dualism still confronts us in unextenuated form in the theory of Hertz, where matter appears not only as the bearer of velocities, kinetic energy, and mechanical pressures, but also as the bearer of electromagnetic fields. Since such fields also occur *in vacuo*,—i.e., in free ether—the ether also appears as the bearer of electromagnetic fields. The ether appears indistinguishable in its functions from ordinary matter. Within matter, it takes part in the motion of matter; and in empty space, it has a velocity everywhere so that the ether has a definitely assigned velocity throughout the whole of space. There is no fundamental difference between Hertz's ether and ponderable matter (which in part subsists in the ether).

The Hertz theory suffered not only from the defect of ascribing to matter and ether,—mechanical states on the one hand, and electrical states that do not stand in any conceivable relation to each other on the other—it was also at variance with the result of Fizeau's important experiment on the velocity of the propagation of light in moving fluids, as well as other established experimental results. [a]

[b] Such was the state of things when H. A. Lorentz entered upon the scene. He brought theory into harmony with experience through a wonderful simplification of theoretical principles. He achieved this, the most important advance in the theory of electricity since Maxwell, by taking its mechanical properties from ether, and its electromagnetic qualities from matter. As in empty space, so, too, is ether—not matter—the exclusive seat of electromagnetic fields in the interior of material bodies when viewed atomistically. According to Lorentz, the elementary particles of matter alone can carry out movements; their electromagnetic activity is entirely confined to carrying electric charges. Thus, Lorentz succeeded in reducing all electromagnetic happenings to Maxwell's equations for free space.

As to the mechanical nature of the Lorentzian ether, it may be said, in a somewhat playful spirit, that immobility is the only mechanical property of which it has not been deprived by H. A. Lorentz. It may be added that the whole change in the conception of the ether brought about by the special theory of relativity consisted in taking away the ether's last mechanical quality, namely, its immobility. How this is to be understood will forthwith be expounded.

The space-time theory and the kinematics of the special theory of relativity were modeled on the Maxwell-Lorentz theory of the electromagnetic field. This theory therefore satisfies the conditions of the special theory of relativity. But when viewed from the latter, it acquires a novel aspect. For if K be a coordinate system relative to which the Lorentzian ether is at rest, the Maxwell-Lorentz equations are valid primarily with reference to K. But by the special theory of relativity, the same equations also hold in relation to any new coordinate system K' moving in uniform translation relative to K without any change of meaning.[c] Now comes the anxious question:—Why must I distinguish the K system above all K' systems, which are physically equivalent to it in all respects, in the theory by assuming that the ether is at rest relative to the K system? For the theoretician, such an asymmetry in the theoretical structure with no corresponding asymmetry in the system of experience is intolerable. If we assume the ether to be at rest relative to K, but in motion relative to K', the physical equivalence of K and K' seems to me, from the logical standpoint, not indeed downright incorrect, but nevertheless unacceptable.

The next possible position to take up in the face of this state of things appeared to be the following: the ether does not exist at all. The electromagnetic fields are neither states of a medium, nor bound down to any bearer, but are independent realities irreducible to anything else, exactly like the atoms of ponderable matter. This conception suggests itself more readily because, according to Lorentz's theory, electromagnetic radiation brings impulse and energy with it much like ponderable matter does. And since both matter and radiation are but special forms of distributed energy according to the special theory of relativity, ponderable mass loses its isolation and appears as only a special form of energy.

More careful reflection teaches us, however, that the special theory of relativity does not compel us to deny the ether. We may assume the existence of an ether; we must only give up ascribing a definite state of motion to it, i.e., we must take from it, by abstraction, the last mechanical characteristic Lorentz had still left it. We shall see later that this point of view—the conceivability of which I shall at once endeavor to make more intelligible by a somewhat halting comparison—is justified by the results of the general theory of relativity.

Think of waves on the surface of water. Here, we can describe two entirely different things. Either we may observe how the undulatory surface that forms the boundary between water and air alters over time, or we instead—with the help of small floats, for instance—can observe how the position of the separate particles of water alters over time. If the existence of such floats for tracking the motion of the particles of a fluid were a fundamental impossibility in physics—if, in fact, nothing else whatsoever were observable, save for the shape of the space occupied by the water as it varies in time—we should have no ground for the assumption that water consists of movable particles. But all the same, we could characterize it as a medium.

We have something like this in the electromagnetic field. For we may picture the field to ourselves as consisting of lines of force. If we wish to interpret these lines of force to ourselves as something material in the ordinary sense, we are tempted to interpret the dynamic processes as motions of these lines of force so that each separate line of force is tracked across time. It is well known, however, that this way of regarding the electromagnetic field leads to contradictions.

Generalizing, we must say this: It can be supposed that there exist extended physical objects to which the idea of motion cannot be applied. They may not be thought of as consisting of particles that allow themselves to be separately tracked through time. In Minkowski's idiom, this is expressed as follows: Not every extended conformation in the four-dimensional world can be regarded as composed of worldlines. The special theory of relativity forbids us to assume the ether consists of particles observable through time, but the hypothesis of ether in itself does not conflict with the special theory of relativity. We only must be on our guard against ascribing a state of motion to the ether.

Certainly, from the standpoint of the special theory of relativity, the ether hypothesis appears at first to be an empty hypothesis. In the equations of the electromagnetic field, *only* the intensities of the field occur in addition to the densities of the electric charge. The career of electromagnetic processes *in vacuo* appears to be completely determined by these equations, uninfluenced by other physical quantities. The electromagnetic fields appear as ultimate, irreducible realities, and it seems at first superfluous to postulate a homogeneous, isotropic ether medium and envisage electromagnetic fields as states of this medium.

But on the other hand, there is a weighty argument to be adduced in favor of the ether hypothesis.[d] To deny the ether is ultimately to assume that empty space has no physical qualities whatsoever. The fundamental facts of mechanics do not harmonize with this view, for the mechanical behavior of a corporeal system hovering freely in empty space depends not only on relative positions (distances) and relative velocities, but also on its state of rotation, which physically may be taken as a characteristic not appertaining to the system in itself. To be able to look upon the rotation of the system as something real, at least formally, Newton objectivizes space. Since he classes his absolute space together with real things, rotation relative to an absolute space is also something real for him. Newton might no less well have called his absolute space "Ether"; what is essential is merely that besides observable objects, another imperceptible thing must be looked upon as real to enable acceleration or rotation to be looked upon as something real.

Indeed, Mach tried to avoid accepting something unobservable as real by endeavoring to substitute, in mechanics, a mean acceleration with reference to the totality of the masses in the universe in place of an acceleration with reference to absolute space. But inertial resistance opposed to relative acceleration of distant masses presupposes action at a distance. And as the modern physicist does not believe that he may accept this action at a distance, he comes back once more, if he follows Mach, to the ether that must serve as a medium for the effects of inertia. But this conception of the ether, to which we are led by Mach's way of thinking, differs essentially from ether as conceived by Newton, Fresnel, and Lorentz. Mach's ether not only *conditions* the behavior of inert masses, but *is also conditioned* in its state by them.

Mach's idea finds its full development in the ether of the general theory of relativity. According to this theory, the metrical qualities of the continuum of space-time differ in the environment of different points of space-time and are partly conditioned by the matter existing outside the territory under consideration. This space-time variability of the reciprocal relations of the standards of space and time—or, perhaps, the recognition of the fact that "empty space," in its physical relation, is neither homogeneous nor isotropic, compelling us to describe its state by ten functions (the gravitation potentials $g_{\mu\nu}$)—has, I think, finally disposed of the view that space is physically empty. But therewith, the conception of the ether has again acquired an intelligible content, though this content differs widely from that of the ether of the mechanical undulatory theory of light. The ether of the general theory of relativity is a medium devoid of *all* mechanical and kinematic qualities itself, but helps to determine mechanical (and electromagnetic) events.[e]

What is fundamentally new in the ether of the general theory of relativity, as opposed to the ether of Lorentz, consists in this: The state of the former is determined at every place by law-like connections with the matter and the state of the ether in neighboring places in the form of differential equations; whereas the state of the Lorentzian ether, in the absence of electromagnetic fields, is conditioned by nothing outside itself and is the same everywhere. The ether of the general theory of relativity is transmuted conceptually into Lorentz's ether if we substitute constants for the functions of space that describe the former, disregarding the causes conditioning its state. Thus, I think we may also say that the ether of the general theory of relativity is the outcome of the Lorentzian ether through relativization.

As to the part the new ether will play in the physics of the future, we are not yet clear. We know that it determines the metrical relations in the space-time continuum—e.g., the configurative possibilities of solid bodies, as well as gravitational fields—but we do not know whether it has an essential share in the structure of the electrical elementary particles constituting matter, nor do we know whether it is only in the proximity of ponderable masses that its structure differs essentially from that of the Lorentzian ether, whether the geometry of spaces of cosmic extent is approximately Euclidean. But we can assert, because of the relativistic equations of gravitation, that there must be a departure from Euclidean relations with spaces of a cosmic order of magnitude if there exists a positive mean density of the matter in the universe, no matter how small. In this case, the universe must necessarily be spatially unbounded and of finite magnitude, its magnitude determined by the value of that mean density.

If we consider the gravitational field and electromagnetic field from the standpoint of the ether hypothesis, we find a remarkable difference between the two. There can be no space, nor any part of space, without gravitational potentials; for these confer upon space its metrical qualities, without which it cannot be imagined at all. The existence of the gravitational field is inseparably bound to the existence of space. On the other hand, a part of space may very well be imagined without an electromagnetic field; thus, in contrast with the gravitational field, the electromagnetic field seems only secondarily linked to the ether, the formal nature of the electromagnetic field being, as yet, in no way determined by that of gravitational ether. From the present state of theory, it looks as though the electromagnetic field, as opposed to the gravitational field, rests upon an entirely new formal *motif*, as though nature might just as well have endowed the gravitational ether with fields of quite another type, for example, with fields of a scalar potential instead of fields of the electromagnetic type.[f]

Since the elementary particles of matter are also essentially nothing other than condensations of the electromagnetic field according to our present conceptions, our present view of the universe presents two realities completely separated from each other conceptually, though connected causally, namely, gravitational ether and electromagnetic field—or, as they might also be called, space and matter.

Of course, it would be a great advance if we could succeed in comprehending the gravitational field and electromagnetic field together as one unified conformation. Then, for the first time, the epoch of theoretical physics founded by Faraday and Maxwell would reach a satisfactory conclusion. The contrast between ether and matter would fade away, and, through the general theory of relativity, the whole of physics would become a complete system of thought like geometry, kinematics, and the theory of gravitation. An exceedingly ingenious attempt in this direction has been made by the mathematician H. Weyl, but I do not believe that his theory will hold its ground in relation to reality. Further, in contemplating the immediate future of theoretical physics, we ought not unconditionally reject the possibility that the facts comprised in the quantum theory may set bounds to the field theory beyond which it cannot pass.

Recapitulating, we may say that, according to the general theory of relativity, space is endowed with physical qualities; in this sense, therefore, there exists an ether. According to the general theory of relativity, space without ether is unthinkable; for in such space, there not only would be no propagation of light, but also no possibility of existence for standards of space and time (measuring-rods and clocks), nor therefore any space-time intervals in the physical sense. But this ether may not be thought of as endowed with the quality characteristic of ponderable media, as consisting of parts that may be tracked through time. The idea of motion may not be applied to it.[g]

Editorial Notes

[a] In Hertz's theory, the Earth drags along the ether in its vicinity. The theory is therefore incompatible with the result of Fizeau's experiment.

[b] The next two paragraphs were written on a sheet interleaved with the manuscript. The passage replaces an earlier draft, which was deleted in the manuscript:

> That was how things stood when H. A. Lorentz intervened. He boldly did away with the aforementioned dualism. For him, only the omnipresent ether, not ponderable matter, carried electromagnetic fields. On the other hand, the latter, which Lorentz thought to be atomistically constituted, was incapable of processes other than mechanical ones (movements) and only electromagnetically effective because its elementary particles carried electric charges. For its part, the ether was completely immobile; the only states it could assume were electromatic fields. One can facetiously say this ether's immobility is its only mechanical property. Lorentz, so to speak, vanished the ether by removing its mechanical properties; insofar as it carried electromagnetic fields, ether was space, not matter. The change in perspective can be summarized as follows: For Hertz, matter (and ether) carries the states of both motion and electric fields; according to Lorentz, only ether (or space) carries electromagnetic fields, whereas ponderable matter's sole electrical property is that it carries electric charges.

[c] The published paper has K^1 instead of K'.
[d] The passage "But on the other hand, ..." was written on a sheet interleaved with the manuscript. It replaces a deleted portion of the manuscript:

> On the other hand, however, one can already present an irrefutable important argument for the ether hypothesis from the standpoint of special relativity or classical mechanics, the fundamental significance of which Newton had already clearly recognized when establishing classical mechanics. Mach particularly emphasized this, although he did not relate it to the ether hypothesis. To deny the ether ultimately means to assume that empty space has no physical properties. According to this assumption, the behavior of a body freely floating in space (or inside an evacuated hollow sphere) must be independent of its motion, especially its rotational motion. In truth, however, this is not the case; the occurrence of centrifugal effects on the body shows that the behavior of the body is not yet completely causally determined by the conditions that can be observed solely on it. Newton's 'absolute space' is actually nothing other than what we today call 'ether'; it is based on the understanding that only facts compel us to ascribe physical qualities to space, even if not a specific state of motion, as Newton seems to have believed. The circumstances observable on the body itself do not yet fully satisfy the requirements of causality.
>
> A satisfactory insight into the role that 'empty space' plays in the causal nexus of physical processes has only been provided by general relativity.

[e] In the manuscript, this paragraph had an additional sentence, which was subsequently deleted:

> This concept of ether is so far removed from the notions we instinctively form through everyday experience that we can only grasp it clearly with the help of mathematical language.

[f] In the manuscript, there are two more sentences between "electromagnetic type" and "Since ..." By drawing a line from "electromagnetic type" and "Since ..." Einstein indicates that these two sentences should be replaced with the two paragraphs that follow:

> In this sense, the theory in its current form is still not entirely satisfactory; it lacks a coherent foundational unity. The theory would only be satisfactory if the field quantities used for description formed a formal unity—similar to how the electric and magnetic fields do in special relativity.

[g] In response to Paul Ehrenfest's suggestion in a letter on 10 March 1920, Einstein prepared the following closing remarks, which are preserved in a two-page autograph manuscript [1 006] and are not included in the printed text:

> Gentlemen Curators of the University! Gentlemen Curators of the Chair!
> I feel the need to sincerely thank you. I have long felt at home in this old and venerable institution of science, which I have often had the privilege to visit as a guest. However, through my appointment to your university, you have brought me into an even closer relationship with it, to my great joy. I will strive to justify the trust you have placed in me.
> Dear Colleagues!
> For a long time, several of you have become such close friends that it is hardly possible for me to address you on this occasion as a newcomer. This is especially true for you, Herr Lorentz. Your life's work has been so decisive for my own scientific endeavors that I cannot imagine how my development would have unfolded without your research.
> The scientific and personal relationships with you, my Leyden colleagues, for which I have traveled to Leyden so many times, have meant more to me for years than I can express; and I am sure that this will remain the case.
> Ladies and Gentlemen Students!
> When I was asked whether I would accept an invitation to this university, my first question was: What do I have to offer where physics is already taught so excellently? But I was told that my main task would be to participate in the scientific work here and, of course, to offer my advice to the students engaged in specialized studies. In this regard, I truly hope to be of help to you in your work from time to time. Furthermore, numerous interactions will arise naturally in the colloquia and discussions. However, you will find me at one place together with my younger colleagues, sitting beside you: namely, in Mr. Lorentz's lecture course.
> "Dixi."

15. To the General Association for Popular Technical Education (1920)

Neue Freie Presse (24 July 1920), p. 8. [*Collected Papers* Vol. 7, Doc. 42].

> Wilhelm Exner (1840–1931), one of the founders of the Technical Museum for Industry and Commerce (Technologisches Gewerbemuseum) in Vienna and President of the Austrian Technical Research Institute (Technische Versuchsanstalt) had asked Einstein to join the association and its executive committee in early May 1920 and again in early July. He enclosed an appeal signed by prominent individuals and requested a statement from Einstein that might help in overcoming resistance from humanist circles, which were unwilling to acknowledge the equal value of a technical education.

I am convinced that your Association[a] can be highly beneficial to people's lives. I would rather not judge its direct practical importance. As a teacher, however, I can see benefits in more than one respect.

The educational system is always in danger of detaching itself from the world of sensual experience. Every education creates its own world of concepts. At their inception, these are closely linked to those realities for whose comprehension they were created. However, the linguistically fixed concept tends to generalization, which broadens its domain of application, but weakens its connection with sensual experience.

Thus, especially in times of an aging culture, terms become empty and formal and lose their relationship with sensual experience. Who could deny that our *Gymnasia*,[b] with their emphasis on language, are especially vulnerable to this danger? But even the practice of mathematics, detached from real applications, faces the same danger; and so, over the centuries, geometricians were able to forget that their science ultimately deals with solid bodies and rays of light. A geometrician who fundamentally denies this degrades his science to a mere play of words, devoid of any content. Science can only then stay healthy and progressive when its connection to the world of sensual experience is maintained, no matter how indirect this connection may be.

A preoccupation with technology is thus highly suited to counteract the degeneration of science in the above sense.

On the other hand, technology should be turned into a true element of civilization by raising public awareness of its rich intellectual and aesthetic content. What does a fine person associate with the word "technology?" Greed, exploitation, social divisiveness, class hatred, soulless mechanization, racial degeneration, and meaningless, hectic activity. Is it a surprise that our educated friend of mankind detests technology as an ill-bred child of our era that threatens to destroy the more refined pleasures of life? We must not allow this robust child to grow up wildly if society is to benefit from it. One must attempt to understand it to gain influence over it. It has the power to improve life. In this, I see the second great task of your association.

Editorial Notes

[a] The General Association for Popular Technical Education.

[b] A term used for secondary schools preparing students for university studies.

16. On New Sources of Energy (1920)

Berliner Tageblatt (25 July 1920), p. 4. [*Collected Papers* Vol. 7, Doc. 43].

> The *Berliner Tageblatt* requested expert opinions of the most eminent physicists and chemists of the University of Berlin, among whom was Einstein. According to the introductory part of the long article, the reason for the inquiry was a search for alternative sources of energy to replace coal, which the French were demanding as reparations at a recent conference in Spa, Belgium. A final agreement required that Germany deliver two million tons within the next six months.

It is a fact that Rutherford[a] has discovered that nitrogen can be smashed by alpha-particles; in air irradiated by alpha-rays, small amounts of secondary rays with a range different from that of alpha-rays are generated. Rutherford interpreted some of these particles to have masses even smaller than the particles of alpha-rays, which consist of helium. It turned out that their mass was equal to that of hydrogen atoms. From this, one could conclude that nitrogen atoms were smashed while hydrogen atoms, among others, were generated. Recently, Rutherford demonstrated that atoms of atomic weight 3 are also generated. Consequently, we have here an entirely new kind of atom, a *new element*. As far as is known now, the energy to achieve the decay is provided by the alpha-particles. As to the energy set free, opinion is still divided. One can by no means exclude the possibility that *quite significant amounts of energy are freed by this process*. It seems possible and not at all unlikely that *new sources of energy* with tremendous power can be found here, but based upon presently known facts, there is no direct support yet for this idea. It is extremely difficult to make prophecies, but the possibility seems to be there. If it is at all possible to free the inner-atomic energy, it would likely have *immense significance for the energy resources available to human labor*. Meanwhile, however, these processes can only be observed by the most intricate means. This must be emphasized because, otherwise, people always immediately get very excited. However, if this trace leads further, especially if the rays freed by the alpha-particles can cause the same effects, then one cannot know whether such a *development cannot move ahead rapidly*.

Editorial Note

[a] Ernest Rutherford (1871–1937), Professor of Physics at the University of Cambridge and Director of its Cavendish Laboratory, first reported on his findings on artificial disintegration to the Royal Institution of Great Britain on 6 June 1919 (Rutherford 1919a, Rutherford 1919b).

17. My Response: On the Anti-Relativity Company (1920)

Berliner Tageblatt (27 August 1920), pp. 1–2. [*Collected Papers* Vol. 7, Doc. 45].

> Einstein's theory of relativity became the target of attacks by Paul Weyland, a novelist and German nationalist who inserted himself into German scientific discourse with the aim "to cleanse German science of the Jews."[a] For this purpose, he formed a front organization called the Association of German Natural Scientists for the Preservation of Pure Science (*Arbeitsgemeinschaft deutscher Naturforscher zur Erhaltung reiner Wissenschaft e.V.*). In August 1920, Weyland delivered a speech at the Berlin Philharmonic in a meeting hosted by his Association of German Natural Scientists—which Einstein attended out of curiosity—denouncing Einstein's theory of relativity as "Jewish arrogance" and "scientific Dadaism."[b] Weyland's attack on relativity was immediately followed by one from experimental physicist Ernst Gehrcke,[c] who opposed the theory of relativity because it would discredit the theory of luminiferous ether. In direct response to both men, Einstein published this article in which he refers to the Association of German Natural Scientists as "the Anti-Relativity Company" (*die antirelativistische G.m.b.H.*).

Under the pretentious name "*Arbeitsgemeinschaft deutscher Naturforscher*," a colorful society has assembled for the provisional purpose of denigrating the theory of relativity and me, as its originator, in the eyes of non-physicists. Mr. Weyland and Mr. Gehrke[d] recently delivered a first lecture at the Philharmonic in this sense, which I also attended. I am very well aware both speakers are unworthy of an answer from my pen, as I have good reason to believe motives other than the search for truth underlie this undertaking. (Were I a German nationalist, with or without a swastika, instead of a Jew of liberal, international persuasion, then ...). I am only answering because well-meaning circles have repeatedly urged me to make my perspective known.

First, I will remark that, as far as I know, there is hardly a scientist today who has made substantial contributions to theoretical physics who would not admit that the theory of relativity is logically constructed in its entirety and in agreement with the reliably experimental facts so far. The most important theoretical physicists—namely, H. A. Lorentz, M. Planck, Sommerfeld, Laue, Born, Larmor, Eddington, Debye, Langevin, and Levi-Civita—support the theory, and most have made valuable contributions to it. Among physicists of international reputation, Lenard is the only vocal opponent of the theory of relativity I could name.[e]

I admire Lenard as a master of experimental physics, but he has not yet produced anything in theoretical physics, and his objections against the general theory of relativity are of such superficiality that, until now, I did not think it necessary to answer them in detail. I intend to make up for this.

I have been reproached for running a tasteless advertising campaign for the theory of relativity. I can only say that I have always been a friend of well-chosen, rational words and concise presentation. Pompous phrases and words give me goosebumps, whether they deal with the theory of relativity or with anything else. I have often made fun of effusions that are now attributed to me. Besides, I am happy to let the gentlemen of the Company have their fun.

Now, to the lectures. Mr. Weyland, who does not seem to be an expert at all (physician, engineer, politician? I could not find out), has brought up nothing of substance whatsoever. He broke out in coarse crudeness and base accusations. The second speaker, Mr. Gehrke, has produced incorrect statements on the one hand, and made a one-sided selection of material on the other, trying to create a false impression among uninformed laymen. The following examples may demonstrate this:

Mr. Gehrke claims the theory of relativity leads to solipsism, an assertion that will make every expert laugh. He bases this on the well-known example of the two clocks (or twins), in which one of them makes a round trip with respect to the inertial system; the other does not. He claims the theory would lead to the truly senseless result that, of two clocks resting side by side, each one would run slower relative to the other, even though he has been repeatedly informed by the best experts in the theory, verbally and in writing, that they refute this statement. I can see this only as a deliberate attempt to mislead the lay public.

Furthermore, Mr. Gehrke alludes to the objections made by Mr. Lenard, many of which relate to examples from the mechanics of everyday life. These are already invalid due to my general proof that the statements of the general theory of relativity correspond, in a first approximation, to those of classical mechanics.

However, what Mr. Gehrke had to say about the experimental confirmation of the theory is, to me, the most convincing proof that his goal was not to unveil the true state of affairs.

Mr. Gehrke wants us to believe that the motion of the perihelion of Mercury could also be explained without the theory of relativity. There are two possibilities. The first would be to invent special interplanetary masses so large and distributed in such a way that they would result in a motion of the perihelion of the observed magnitude. This is, of course, a highly unsatisfactory way out when compared with that of the theory of relativity, which explains the motion of the perihelion of Mercury without any special assumptions. The other would be to refer to a paper by Gerber, who gave the correct formula for the movement of Mercury's perihelion before I did. Yet experts not only agree that Gerber's derivation is faulty from beginning to end, but also that the formula cannot be obtained as a consequence of the assumptions from which Gerber started out.[f] Mr. Gerber's paper is therefore completely worthless, a failed and irreparable theoretical attempt. I state that the general theory of relativity provided the first real explanation of the motion of the perihelion of Mercury. Originally, I did not mention Gerber's paper because I was not aware of it when I wrote my paper on the motion of the perihelion of Mercury. But even had I known about it, I would have had no reason to mention it. The personal attack that Mr. Gehrke and Mr. Lenard directed against me on these grounds has generally been judged unfair by the true experts; until now, I considered it beneath my dignity to waste even a word on it.

In his lecture, Mr. Gehrke inappropriately presented the reliability of the masterfully executed English measurements of the deflection of light rays by the Sun when he mentioned only *one* of *three* independent observation groups, the one that produced inaccurate results because of distortions in the heliostat mirror. He did not mention that the English astronomers themselves interpreted their results as a brilliant confirmation of the general theory of relativity in their official report.

Regarding the red shift of spectral lines, Mr. Gehrke did not reveal that current measurements still contradict each other, nor that a final decision is still outstanding. He quoted only witnesses *against* the existence of the line shift predicted by the theory of relativity, but he concealed that, according to the most recent investigations by Grebe and Bachem[g], as well as those by Perot, the previous results are no longer convincing.

Finally, I would like to mention that a discussion about the theory of relativity will take place at the *Naturforscherversammlung*,[h] upon my suggestion, in Nauheim. Anybody who dares appear in front of a scientific forum can present his objections there.

It will cause a strange impression in foreign countries—in particular, with my Dutch and British colleagues, H. A. Lorentz and Eddington, who both studied the theory of relativity in detail and have lectured on it repeatedly—that the theory, as well as its originator, are disparaged in such a way in Germany itself.

Editorial Notes

[a] See Kleinert 1993 and Goenner 1993 for thorough accounts of the campaign against Einstein and his theory of relativity.
[b] Weyland 1920.
[c] Gehrcke 1920.
[d] Einstein spells Ernst Gehrcke's surname as "Gehrke" throughout his response.
[e] See his anti-relativistic pamphlet Lenard 1918 and the critical remarks he added in Lenard 1920.
[f] Paul Gerber (1854–before 1917) was a high school teacher in Stargard, Pomerania. His formula for the precession of planetary perihelia was first published in Gerber 1898; Gehrcke republished Gerber's lengthier study, Gerber 1902, in *Annalen der Physik*. In Gehrcke 1916, he compared Gerber's formula with Einstein's, showing that they were identical.
[g] Grebe and Bachem 1920, p. 51.
[h] The Convention of Natural Scientists, hosted by the Society of Natural Scientists and Physicians (*Gesellschaft Deutscher Naturforscher und Ärzte*) in Bad Nauheim on 23 September 1920.

18. On the Contribution of Intellectuals to International Reconciliation (1920)

Thoughts on Reconciliation. New York: Deutscher Gesellig-Wissenschaftlicher Verein von New York (1920), pp. 10–11. [*Collected Papers* Vol. 7, Doc. 47].[a]

This statement was solicited by the German Social and Scientific Society of New York for a memorial volume on the occasion of the fiftieth anniversary of its founding. Respondents were to answer the following

> question: "What and how can intellectuals of all countries contribute practically to the true reconciliation of former belligerent nations and the permanent fraternity of mankind?" Proceeds from the sale were to flow to the Schiller Foundation in Weimar for the support of needy intellectuals in Germany and Austria.

The most valuable contribution to a reconciliation of the nations and a permanent fraternity of mankind is in my opinion contained in their scientific and artistic creations, because they raise the human mind above personal and national aims of a selfish character. Concentration of energy upon those problems and aims common to all people of intellect produces quite naturally a feeling of comradeship, which is bound always to re-unite the true scholars and artists of all countries, though it is inevitable that the less great-hearted and less independent among them will always be temporarily estranged by political and other passions. The intellectuals should never weary of emphasizing the internationality of mankind's most beautiful treasures, and their corporations should never stoop to foster political passions by public declarations or other demonstrations.

Finally, I would remark that, in my opinion, it would be of value for the reconciliation of the nations if young students and artists were to study to a greater extent than heretofore in what were enemy countries before. Direct experience counteracts more effectively those fatal ideologies which under the influence of the World War have settled in the brains.

Editorial Note

[a] Translation was provided alongside the original German in the publication.

19. The Common Element in Artistic and Scientific Experience (1921)

Menschen. Zeitschrift neuer Kunst 4 (1921), p. 19. [*Collected Papers* Vol. 7, Doc. 51].

> Einstein's draft poses the title as a question. The co-editor of *Menschen* (Dresden), Walter Hasenclever (1890–1940), a playwright and poet, solicited the comment from Einstein in the belief that modern painting and poetry served as "an intuitive premonition of your discoveries." He hoped that Einstein might help convince the literary public of the kinship between art and science.

When the world ceases to be a stage for personal hopes, aspirations, and desires and we stand before it as free creatures full of admiration, questions, and contemplation, we enter the realm of art and science. If we describe what we see and experience in the language of logic, we do science; if we convey connections through forms inaccessible to the rational mind, but intuitively recognizable as making sense, we do art. Common to both is the loving devotion to what is beyond the personal, beyond our will.

20. Geometry and Experience (1921)

Geometrie und Erfahrung. Berlin: J. Springer (1921), pp. 3–20.[a] [*Collected Papers* Vol. 7, Doc. 52].

> An expanded version of a lecture delivered on 27 January 1921 at a public session of the Prussian Academy of Sciences. The last part appeared first in a reprint by Springer, Berlin, 1921. In his analysis of foundational issues and principles, Einstein here analyzes the relationship between geometry and physical space in the light of general relativity. Although he agrees with some aspects of Henri Poincaré's conventionalism, he

> argues that one is not at liberty to choose any physical theory in order to fit experience to a fixed geometry. A new fundamental theory of matter is required.

One reason mathematics enjoys special esteem above all other sciences is that its laws are absolutely certain and indisputable, while those of all other sciences are, to some extent, debatable and in constant danger of overthrow by newly discovered facts. Despite this, the investigator in another department of science would not need to envy the mathematician if the laws of mathematics referred to objects of our mere imagination rather than objects of reality. For it cannot cause surprise that different individuals should arrive at the same logical conclusions when they have already agreed upon the fundamental laws (axioms) as well as the methods by which other laws are to be deduced therefrom. But there is another reason for the high repute of mathematics, as it is mathematics that affords the exact natural sciences a certain measure of security which, without mathematics, they could not attain.

At this point, an enigma presents itself that has agitated inquiring minds in all ages. How can it be that mathematics—being a product of human thought independent of experience, after all—is so admirably appropriate to the objects of reality? Can human reason, without experience, ascertain the properties of real things through mere thinking?

In my opinion, the brief answer to this question is: Insofar as the laws of mathematics refer to reality, they are not certain, and insofar as they are certain, they do not refer to reality. It seems to me that complete clarity as to this state of things first became a common property through that new departure in mathematics known by the name of mathematical logic, or "Axiomatics."[b] The progress achieved by axiomatics consists in its neat separation of the logical-formal from its objective or intuitive content; according to axiomatics, the logical-formal alone forms the subject matter of mathematics, which is not concerned with the intuitive or other content associated with the logical-formal.

Let us consider for a moment, from this perspective, any axiom of geometry—for instance, the following: Through two points in space there always passes one, and only one, straight line. How is this axiom to be interpreted, in the older sense and the more modern sense?

The older interpretation: Everyone knows what a straight line and point are. Whether this knowledge springs from an ability of the human mind, experience, some collaboration of the two, or some other source is not for the mathematician to decide. He leaves the question to the philosopher. Being based upon this knowledge, which precedes all mathematics, the axiom stated above is, like all other axioms, self-evident; that is, it is the expression of a part of this *a priori* knowledge.

The more modern interpretation: Geometry treats of entities which are denoted by the words straight line, point, etc. These entities do not take any knowledge or intuition whatsoever for granted, but presuppose only the validity of the axioms such as the one stated above, which are to be taken in a purely formal sense, i.e., as void of all intuitive or experiential content. These axioms are free creations of the human mind. All other propositions of geometry are logical inferences from the axioms (which are to be taken in the nominalist sense only). The matter of which geometry treats is first defined by the axioms. Schlick, in his book on epistemology, has therefore characterized axioms very aptly as "implicit definitions."[c]

This view of axioms, advocated by modern axiomatics, purges mathematics of all extraneous elements and thus dispels the mystic obscurity that formerly surrounded the principles of mathematics. But a presentation of its principles thus clarified also makes it evident that mathematics, as such, cannot predicate anything about perceptual or real objects. In axiomatic geometry, the words "point," "straight line," etc., stand only for empty conceptual schemata. What gives them substance is not relevant to mathematics.

Yet, on the other hand, it is certain that mathematics generally—and, particularly, geometry—owes its existence to the felt need of learning something about the relations of real things to one another. The very word "geometry"—which, of course, means earth-measuring—proves this; for earth-measuring has to do with the possibilities of the disposition of certain natural objects with respect to one another, namely, with parts of the earth, measuring-lines, measuring wands, etc. The system of concepts of axiomatic geometry alone cannot make any assertions as to the relations of real objects of this kind, which we will call practically rigid bodies. To be able to make such assertions, the geometry must be stripped of its merely logical-formal character by the coordination of real objects of experience with the empty conceptual framework of axiomatic geometry. To accomplish this, we need only add the proposition: Solid bodies are related with respect to their possible dispositions, as are bodies in Euclidean geometry of three dimensions. Then the propositions of Euclid contain affirmations as to the relations of practically rigid bodies.

The geometry supplemented in this way is evidently a natural science; we may, in fact, regard it as the most ancient branch of physics. Its affirmations essentially rest on induction from experience, not solely on logical inferences. We will call this supplemented geometry "practical geometry" and shall distinguish it from "purely axiomatic geometry" in what follows. The question whether the practical geometry of the universe is Euclidean or not has a clear meaning, and its answer can only be furnished by experience. All linear measurement in physics is practical geometry, in this sense; so, too, is geodetic and astronomical linear measurement, should we call to our help the law of experience that light is propagated in a straight line, and indeed in a straight line in the sense of practical geometry.

I attach special importance to the view of geometry I have just set forth because, without it, I would have been unable to formulate the theory of relativity. Without it, the following reflection would have been impossible: In a system of reference rotating relatively to an inert system, the laws of disposition of rigid bodies do not correspond to the rules of Euclidean geometry due to the Lorentz contraction; thus, if we admit non-inert systems, we must abandon Euclidean geometry. The decisive step in the transition to general covariant equations would certainly not have occurred if the above interpretation had not served as a stepping-stone. If we deny the relation between the body of axiomatic Euclidean geometry and the practically rigid body of reality, we readily arrive at the following view entertained by that acute and profound thinker, H. Poincaré: Euclidean geometry is, above all, distinguished from other imaginable axiomatic geometries by its simplicity. Now, since axiomatic geometry by itself contains no assertions as to the reality that can be experienced, but can do so only in combination with physical laws, it should be possible and reasonable—whatever the nature of reality may be—to retain Euclidean geometry. For if contradictions between theory and experience manifest themselves, we should decide to change physical laws rather than axiomatic Euclidean geometry. If we deny the relation between the practically rigid body and geometry, we shall indeed not easily free ourselves from the convention that Euclidean geometry is to be retained as the simplest.

Why is the equivalence of the practically rigid body and the body of geometry, which suggests itself so readily, denied by Poincaré and other investigators? Simply because, under closer inspection, the real solid bodies in nature are not rigid, as their geometrical behavior—that is, their possibilities of relative disposition—depend upon temperature, external forces, etc. Thus, the original, immediate relation between geometry and physical reality appears destroyed, and we feel impelled toward the following, more general view characterizing Poincaré's standpoint. Geometry (G) predicates nothing about the relations of real things; only geometry together with the purport (P) of physical laws can do so. Using symbols, we may say that only the sum of (G) + (P) is subject to the control of experience. Thus, (G) may be chosen arbitrarily, as well as parts of (P); all these laws are conventions. All that is necessary to avoid contradictions is to choose the remainder of (P) so that (G) and the whole of (P) are together in accord with experience. Envisaged in this way, axiomatic geometry and the part of natural law given a conventional status appear epistemologically equivalent.

Sub specie aeterni Poincaré, in my opinion, is right. The ideas in the theory of relativity of the measuring rod and the clock coordinated with it do not find their exact correspondence in the real world. It is also clear that the solid body and the clock do not play the part of irreducible elements in the conceptual edifice of physics, but that of composite structures, which may not play any independent part in theoretical physics. But it is my conviction that, in the present stage of development of theoretical physics, these ideas must still be employed as independent ideas; for we are still far from possessing such certain knowledge of theoretical principles as to be able to give exact theoretical constructions of solid bodies and clocks.

Further, as to the objection that there are no really rigid bodies in nature and that the properties predicated on rigid bodies therefore do not apply to physical reality, this objection is by no means as radical as it might appear from a hasty examination, for it is not a difficult task to determine the physical state of a measuring rod so accurately that its behavior relative to other measuring-bodies shall be sufficiently free from ambiguity as to allow it to be substituted for the "rigid" body. It is to measuring-bodies of this kind that statements as to rigid bodies must be referred.

All practical geometry is based upon a principle accessible to experience, which we will now try to realize. We will call that which is enclosed between two boundaries marked upon a practically rigid body a distance. We imagine two practically rigid bodies, each with a distance marked out on it. These two distances are said to be "equal to one another" if the boundaries of the one distance can be brought to coincide permanently with the boundaries of the other. We now assume that:

If two distances are found to be equal once and anywhere, they are equal always and everywhere.[d]

Not only does the practical geometry of Euclid rest upon this assumption,[e] but also its nearest generalization, the practical geometry of Riemann, and, therewith, the general theory of relativity.

Of the experimental reasons warranting this assumption, I will mention only one. The phenomenon of the propagation of light in empty space assigns a distance—namely, the appropriate path of light—to each interval of local time and vice versa. Thence it follows that the above assumption for distances must also hold good for intervals of clock-time in the theory of relativity. Consequently, it may be formulated as follows: If two ideal clocks are going at the same rate at any time and at any place (being, then, in immediate proximity to each other), they will always go at the same rate, regardless of where and when they are again compared with each other at one place.—If this law were not valid for real clocks, the proper frequencies for the separate atoms of the same chemical element would not be in such exact agreement as experience demonstrates. The existence of sharp spectral lines is a convincing experimental proof of the above-mentioned principle of practical geometry. This is the ultimate foundation, in fact, that enables us to speak with meaning of the mensuration, in Riemann's sense of the word, of the four-dimensional continuum of space-time.

Properly speaking, the question whether the structure of this continuum is Euclidean by Riemann's general scheme or otherwise is, according to the view advocated here, a physical question that must be answered by experience, not a question of a mere convention to be selected on practical grounds. Riemann's geometry will be the right thing if the laws of disposition of practically rigid bodies are transformable into those of the bodies of Euclid's geometry with an exactitude that increases in proportion as the dimensions of the part of space-time under consideration diminish.

This proposed physical interpretation of geometry indeed breaks down when applied immediately to spaces of sub-molecular order of magnitude. Nevertheless, even in questions as to the constitution of elementary particles, it retains part of its importance. For even when it is a question of describing the electrical elementary particles constituting matter, the attempt may still be made to ascribe physical importance to those ideas of fields that have been physically defined for the purpose of describing the geometrical behavior of bodies that are large as compared with the molecule. Success alone can decide as to the justification of such an attempt, which postulates physical reality for the fundamental principles of Riemann's geometry outside the domain of their physical definitions. It might turn out that this extrapolation has no better warrant than the extrapolation of the idea of temperature to parts of a body of molecular order of magnitude.

It appears less problematic to extend the ideas of practical geometry to spaces of cosmic order of magnitude. It might, of course, be objected that a construction composed of solid rods departs more and more from ideal rigidity in proportion as its spatial extent becomes greater.

But it will hardly be possible, I think, to assign fundamental significance to this objection. Therefore, the question whether the universe is spatially finite or not seems, to me, a decidedly pregnant question in the sense of practical geometry. I do not even consider it impossible that this question will be answered before long by astronomy. Let us call to mind what the general theory of relativity teaches in this respect. It offers two possibilities:—

1. The universe is spatially infinite. This can be so only if the average spatial density of the matter in universal space concentrated in the stars vanishes, i.e., if the ratio of the total mass of the stars to the magnitude of the space through which they are scattered approximates indefinitely to the value zero when the spaces considered are constantly greater and greater.
2. The universe is spatially finite. This must be so if there is a mean density of the ponderable matter in universal space differing from zero. The smaller the mean density, the greater the volume of universal space.

I must not fail to mention that a theoretical argument can be adduced in favor of the hypothesis of a finite universe. The general theory of relativity teaches that the inertia of a given body is greater as there are more ponderable masses in proximity to it; thus, it seems very natural to reduce the total effect of the inertia of a body to action and reaction between it and the other bodies in the universe, as, indeed, ever since Newton's time, gravity has been completely reduced to action and reaction between bodies. From the equations of the general theory of relativity, it can be deduced that this total reduction of inertia to reciprocal action between masses—as required by E. Mach, for example—is possible only if the universe is spatially finite.

On many physicists and astronomers, this argument makes no impression. Experience alone can ultimately decide which of the two possibilities is realized in nature. How can experience furnish an answer? At first, it might seem possible to determine the mean density of matter by observing that part of the universe that is accessible to our perception. This hope is illusory. The distribution of the visible stars is extremely irregular so that we never may venture to set down the mean density of star-matter in the universe as equal, let us say, to the mean density in the Milky Way. In any case, however great the space examined may be, we could not feel convinced that there were no more stars beyond that space. Therefore, it seems impossible to estimate that mean density.

But there is another road that seems, to me, more practicable, though it also presents great difficulties. For if we inquire into the deviations shown by the consequences of the general theory of relativity accessible to experience when compared with the consequences of the Newtonian theory, we, first of all, find a deviation that shows itself close to gravitating mass and has been confirmed in the case of the planet Mercury. But if the universe is spatially finite, there is a second deviation from the Newtonian theory that, in the language of the Newtonian theory, may be expressed thus: The gravitational field is, in its nature, such as though it were produced not only by the ponderable masses, but also by a mass-density of negative sign distributed uniformly throughout space. Since this fictitious mass-density would have to be incredibly small, it could make its presence felt only in gravitating systems of very large dimensions.

If we know, let us say, the statistical distribution of the stars in the Milky Way as well as their masses, then we can calculate according to

Newton's law the gravitational field as well as the stars' mean velocities so that the Milky Way should not collapse under the mutual attraction of its stars, but should maintain its actual expanse. Now, if the actual velocities of the stars—which can, of course, be measured—were smaller than the calculated velocities, we should have proof that the actual attractions at great distances are smaller than by Newton's law. From such a divergence, it could be proven indirectly that the universe is finite. It would even be possible to estimate its spatial dimensions.[f]

Can we picture, to ourselves, a three-dimensional universe that is finite, yet unbounded?

The usual answer to this question is "No," but that is not the right answer. The purpose of the following remarks is to show that the answer should be "Yes." I want to show that, without any extraordinary difficulty, we can illustrate the theory of a finite universe using a mental image to which, with some practice, we shall soon grow accustomed.

First, an observation of epistemological nature. A geometrical-physical theory is as such incapable of being directly pictured, being merely a system of concepts. But these concepts serve the purpose of bringing a multiplicity of real or imaginary sensory experiences into connection in the mind. To "visualize" a theory, or bring it home to one's mind, therefore means to give a representation to that abundance of experiences for which the theory supplies the schematic arrangement. In the present case, we have to ask ourselves how we can represent that relation of solid bodies with respect to their reciprocal disposition (contact) corresponding to the theory of a finite universe. There is really nothing new in what I have to say about this, but innumerable questions addressed to me prove that the requirements of those who thirst for knowledge of these matters have not yet been completely satisfied. So, will the initiated pardon me, please, if part of what I shall bring forward has long been known?

What do we wish to express when we say that our space is infinite? Nothing more than that we might lay any number whatever of bodies of equal sizes side by side without ever filling space. Suppose we are provided with a great many wooden cubes, all the same size. In accordance with Euclidean geometry, we can then place above, besides, and behind one another to fill a part of space of any dimensions; but this construction would never be finished; we could go on adding more and more cubes without ever finding that there was no more room. That is what we wish to express when we say that space is infinite. It would be better to say that space is infinite in relation to practically rigid bodies, assuming that the laws of disposition for these bodies are given by Euclidean geometry.

Another example of an infinite continuum is the plane. On a plane surface, we may lay squares of cardboard so that each side of any square has the side of another square adjacent to it. The construction is never finished; we can always go on laying squares if their laws of disposition correspond to those of plane figures of Euclidean geometry. The plane is therefore infinite in relation to the cardboard squares. Accordingly, we say that the plane is an infinite continuum of two dimensions and space is an infinite continuum of three dimensions. What is meant here by the number of dimensions, I think, I may assume to be known.

Now, we take an example of a two-dimensional continuum that is finite, but unbounded. We imagine the surface of a large globe and a number of small paper discs, all the same size. We place one of the discs anywhere on the surface of the globe. If we move the disc about, anywhere we like, on the surface of the globe, we do not come upon a limit or boundary anywhere on the journey. We therefore say that the spherical surface of the globe is an unbounded continuum. Moreover, the spherical surface is a finite continuum. For if we stick the paper discs on the globe so that no disc overlaps another, the surface of the globe will finally become so full that there is no room for another disc. This simply means that the spherical surface of the globe is finite in relation to the paper discs. Further, the spherical surface is a non-Euclidean continuum of two dimensions. The laws of disposition for the rigid figures lying in it do not agree with those of the Euclidean plane. This can be shown in the following way: Place a paper disc on the spherical surface, and place six more discs, each of which is to be surrounded in turn by six discs, around that disc in a circle, and so on. If this construction is made on a plane surface, we have an uninterrupted disposition in which six discs are touching every disc except those that lie on the outside.

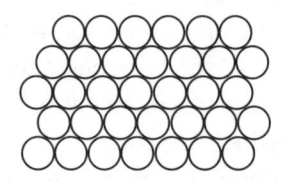

Fig. 1

On the spherical surface, the construction also seems to promise success at the outset, and the smaller the radius of the disc proportionally to that of the sphere, the more promising it seems. But as the construction progresses, it becomes more and more patent that the disposition of the discs, in the manner indicated, without interruption, is not possible, as it should be by Euclidean geometry of the plane surface. In this way, creatures that cannot leave the spherical surface and cannot even peep out from the spherical surface into three-dimensional space might discover, merely by experimenting with discs, that their two-dimensional "space" is not Euclidean, but spherical space.

From the latest results of the theory of relativity, our three-dimensional space is probably also approximately spherical—that is, the laws of disposition of rigid bodies in it are not given by Euclidean geometry, but approximately by spherical geometry—only if we consider parts of space which are sufficiently large. Now, this is the place where the reader's imagination boggles. "Nobody can imagine this thing," he cries indignantly. "It can be said, but cannot be thought. I can represent to myself a spherical surface well enough, but nothing analogous to it in three dimensions."

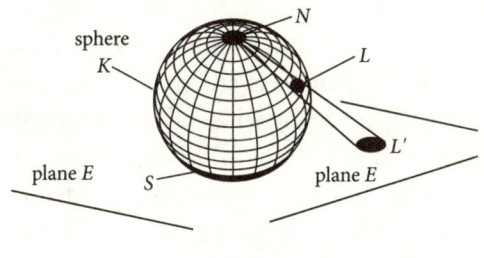

Fig. 2

We must try to surmount this barrier in the mind, and the patient reader will see that it is by no means a particularly difficult task. For this purpose, we will first give our attention once more to the geometry of two-dimensional spherical surfaces. In the above figure, let K be the spherical surface touched at S by a plane E, which, for the facility of presentation, is shown in the drawing as a bounded surface. Let L be a disc on the spherical surface. Now, let us imagine that, at the point N of the spherical surface, diametrically opposite to S, there is a luminous point throwing a shadow L' of the disc L upon the plane E.

Every point on the sphere has its shadow on the plane. If the disc on sphere K is moved, its shadow L' on plane E also moves. When disc

L is at S, it almost exactly coincides with its shadow. If it moves on the spherical surface away from S upwards, the disc shadow L' on the plane also moves away from S on the plane outward, growing bigger and bigger. As disc L approaches the luminous point N, the shadow moves off to infinity and becomes infinitely great.

Now we ask: What are the laws of disposition of the disc-shadows L' on the plane E? Evidently, they are exactly the same as the laws of disposition of the discs L on the spherical surface, as, for each original figure on K, there is a corresponding shadow figure on E. If two discs on K are touching, their shadows on E also touch. The shadow-geometry on the plane agrees with the disc-geometry on the sphere. If we call the disc-shadows rigid figures, then spherical geometry holds good on the plane E with respect to these rigid figures. Moreover, the plane is finite with respect to the disc-shadows, since only a finite number of shadows can find room on the plane.

At this point, somebody will say, "This is nonsense. The disc-shadows are *not* rigid figures. We have only to move a ruler about on the plane E to convince ourselves that the shadows constantly increase in size as they move away from S on the plane toward infinity." But what if the ruler were to behave on the plane E in the same way as the disc-shadows L'? It would then be impossible to show that the shadows increase in size as they move away from S; such an assertion would then no longer have any meaning whatsoever. In fact, the only objective assertion that could be made about the disc-shadows would be just this: that they are related in the same way as the rigid discs on the spherical surface in the sense of Euclidean geometry.

We must carefully bear in mind that our statement on the growth of the disc-shadows as they move away from S toward infinity has no objective meaning if we are unable to employ Euclidean rigid bodies that can be moved about on the plane E to compare the size of the disc-shadows. In respect to the laws of disposition of the shadows L', the point S has no special privileges on the plane, any more than on the spherical surface.

The representation given above of spherical geometry on the plane is important for us because it readily allows itself to be transferred to the three-dimensional case.

Let us imagine a point S of our space and a considerable number of small spheres L', which can all be brought to coincide with one another.

But these spheres are not to be rigid in the sense of Euclidean geometry; their radius is to increase (in the sense of Euclidean geometry) when they are moved away from S toward infinity, and this increase is to take place in exact accordance with the same law as applies to the increase of the radii of the disc-shadows L' on the plane.

After having gained a vivid mental image of the geometrical behavior of our L' spheres, let us assume there are no rigid bodies in our space at all, in the sense of Euclidean geometry, but only bodies having the behavior of our L' spheres. Then, we shall have a vivid representation of three-dimensional spherical space—or rather, of three-dimensional spherical geometry. Here, our spheres must be called "rigid" spheres. Their increase in size as they depart from S is not to be detected by measuring with measuring-rods, any more than in the case of the disc-shadows on E behaving in the same way as the spheres. Space is homogeneous—that is to say, the same spherical configurations are possible in the environment of all points.[1] Our space is finite because, in consequence of the "growth" of the spheres, only a finite number of them can find room in space.

In this way, by using as stepping-stones the practice in thinking and visualization that Euclidean geometry gives us, we have acquired a mental picture of spherical geometry. We may impart more depth and vigor to these ideas without difficulty by carrying out special imaginary constructions. It would also not be difficult to analogously represent the case of what is called elliptical geometry. My only aim today has been to show that the human faculty of visualization is by no means bound to capitulate to non-Euclidean geometry.

Editorial Notes

[a] The first part of this document was originally published on 3 February 1921 in *Preußische Akademie der Wissenschaften* (Berlin). *Sitzungsberichte* (1921), pp. 123–130. An alternative English translation can be found in Einstein, *Ideas and Opinions*, trans. Sonja Bargmann (New York: Crown, 1982).

[b] The modern axiomatic method was introduced in Hilbert 1899.

[c] Moritz Schlick refers to Hilbert's axiomatic method in Schlick 1918, pp. 30–37.

[d] In the manuscript, this paragraph was originally marked with "a)" and followed by the deleted sentence: "b) If two lines are equal to one and the same third line they are equal among themselves."

---

[1] This is intelligible without calculation—but only for the two-dimensional case—if we revert once more to the case of the disc on the surface of the sphere.

[e] In the manuscript, "assumptions" ("Voraussetzungen") is followed by the deleted passage: "for which there are important reasons from physical experience. We will cite only one of these, the strongest."

[f] The next paragraph begins the part that was added to the version in the Sitzungsberichte. The manuscript for this part is not continuous with the first.

21. How I Became a Zionist (1921)

Jüdische Rundschau (21 June 1921), pp. 351–352. [*Collected Papers* Vol. 7, Doc. 57].

> In the subsequent issue—*Jüdische Rundschau* (28 June 1921)—the editors point out that they neglected to mention that the article "was not written by him [Einstein] personally, that it was an interview, which he granted to the representative of the Jewish Correspondence Bureau in New York. The text communicated to us by the JCB in this form was submitted to Prof. Einstein by our editors and approved by him with several emendations." An English-language translation, with some emendations but presumably based on the German-language original that is not preserved, appeared on 17 June 1921 in *The Jewish Chronicle* with the title "Jewish Nationalism and Antisemitism. Their Relativity."

Until a generation ago, the Jews in Germany did not consider themselves members of the Jewish people. They considered themselves merely members of a religious community, and many still hold on to this perspective today. They are indeed far better assimilated than Russian Jews. They have attended mixed schools and have adapted to the German people and their cultural life. Yet, despite the official equal rights they enjoy, there is a strong social antisemitism. And it is especially the educated circles that have become the carriers of the antisemitic movement. They have even established a "science" of antisemitism,[a] while the intellectuals of Russia—at least, before the war—were usually philosemitic and frequently made honest attempts to fight the antisemitic movement. This has to do with many causes. To some extent, this phenomenon is caused by the fact that Jews exert an influence on the spiritual life of the German people that far exceeds their number. While, in my opinion, the economic position of the German Jews is highly overestimated, their influence on the press, literature, and science in Germany is indeed very strong and impresses itself even on the superficial observer. There are very many who are not really antisemites and are honest in their opinions. They consider Jews to be of a nationality that differs from the German and therefore feel their national identity[b] threatened by the growing Jewish influence. While the percentage of Jews, for example, in England is perhaps not much less significant than in Germany, English Jews certainly do not have the same effect on English society and culture, even though they have access to the highest official positions there and a Jew can be nominated—something that is unthinkable in Germany—to be Lord Chief Justice or Viceroy of India.

Antisemitism is frequently a matter of political calculation. Whether somebody subscribes to antisemitism often depends on the political party he belongs to. A socialist, even if he is a convinced antisemite, will not confess or act on his conviction, as it does not fit into the program of the party. For conservatives, however, antisemitism often arises only from the desire to exploit instincts[c] that already exist in the general population. In a country like England, where Jewish influence is smaller and the reaction of non-Jews is therefore much lesser,[d] the existence of old, deep-rooted liberal traditions hinders the rapid growth of antisemitism.

I say this without claiming personal knowledge of the country. I was never in England.[e] Nevertheless, the attitude English science and media adopted toward my theory was already quite indicative. While the assessment of my theory in Germany generally depended on the political party orientation of the papers, the attitude of the English scientists has proven that their sense of objectivity cannot be muddied by political viewpoints. The English have—as I would like to add here—actually influenced the development of our science to a high degree and have approved the testing of the theory of relativity with great vigor and remarkable success. While in America, antisemitism appears only on a social level, in Germany, the political, rather than social, antisemitism[f] is much more evident. As I see it, the reality of the Jews' racial peculiarity necessarily influences their social relations with the non-Jews.[g] The conclusion Jews should draw is, in my opinion, to accept the peculiarities in their social way of life and acknowledge their cultural contributions. First of all, they should show a certain noble reserve and not strive so hard to intermingle socially, which the others desire little, if at all. On the other hand, antisemitism in Germany also has effects that should be welcomed from a Jewish perspective. I believe German Jewry owes its continued existence to antisemitism. Religious forms, which have prevented Jews in the past from intermingling and from fully integrating into their environment, increasingly vanish with growing prosperity and education. So nothing else remains to cause social separation beyond this contrast to their environment called antisemitism. Without this contrast, the integration of the Jews in Germany would happen fast and unhindered.

I experienced this myself. Until seven years ago,[h] I lived in Switzerland; and as long as I lived there, I did not become aware of my Jewishness, and there was nothing in my life that would have stirred my Jewish sentiment or stimulated it. This changed as soon as I relocated to Berlin. There, I saw the predicament of many young Jews. I saw how the antisemitic environment led them to struggle for a secure basis for their livelihood or prevented them from pursuing orderly studies. This is especially true for Eastern Jews, who are continuously subjected to harassment. I do not believe they are numerous in Germany, except in Berlin, which may have a larger number. Nevertheless, their presence has become a question that increasingly preoccupies the German general public. In meetings, at conferences, and in newspapers, their quick removal or internment is asked for. The housing shortage and the economic depression are used as arguments to justify these harsh demands. Those facts are deliberately overstated to prejudice public opinion against Eastern Jewish immigrants. Eastern Jews are being made into scapegoats for certain defects in present-day German economic life that, in reality, are after-pains of the war. The confrontational attitude against these unfortunate refugees, who have escaped today's hell of Eastern Europe, has become an efficient political weapon successfully wielded by demagogues. When the government contemplated measures against Eastern Jews, I stood up for them in the *Berliner Tageblatt*,[i] wherein I pointed out the inhumanity and irrationality of these measures.

Together with a few colleagues—Jews and non-Jews—I held university courses for Eastern Jews, and I would like to add that our endeavor was met with the official recognition of and full support of the Ministry of Education.

These and similar experiences have awakened in me the Jewish-national feeling. I am not a Jew in the sense that I would demand the preservation of the Jewish or any other nationality as an end in itself. Rather, I see Jewish nationality as a fact, and I believe every Jew must draw the consequences from this fact. I consider the raising of the Jewish self-consciousness a necessity also in the interest of a natural coexistence with non-Jews. This was the main motive for my joining the Zionist movement. Zionism, to me, is not just a colonizing movement directed toward Palestine. The Jewish nation is a living fact in Palestine, as well as in the Diaspora, and the Jewish national feeling must be kept alive everywhere Jews live.[j] Members of a tribe or a people must, in today's living conditions, have a lively tribal awareness in order not to lose their dignity and moral rectitude. It was only the unbroken vitality of the masses of American Jewry that made it clear to me how ailing German Jewry is.

We live in an age of exaggerated nationalism, and, being a small nation, we must take this into account. But my Zionism does not exclude cosmopolitan conceptions. I assume the reality of a Jewish nationality and believe that every Jew has responsibilities toward his fellow Jews. The significance of Zionism is, of course, manifold. It opens up the possibility of a dignified human existence to many Jews who presently suffer in the hell of Ukraine or are economically ruined in Poland. By repatriating Jews to Palestine and providing healthy and normal economic circumstances, Zionism is a productive activity that enriches human society. But the main point is that Zionism strengthens the dignity and self-confidence of Jews that is necessary for their existence in the Diaspora and that the Jewish center in Palestine creates a strong bonding agent providing stability for the Jews. The undignified quest for adaptation among many members of my circle[k] has always been very repulsive to me.

The founding of a free Jewish community in Palestine will again provide the opportunity for the Jewish people to develop their creative abilities fully and unhindered. The establishment of the Hebrew University and similar institutes will not only lead the Jewish people to their own national renaissance but will also allow them to contribute, on a freer basis, to the spiritual life of the world.[l]

Editorial Notes

[a] "Movement" ("Bewegung") at this point in the text is translated as "disease" in *The Jewish Chronicle*; where "'science' of antisemitism" ("'Wissenschaft' des Antisemitismus") appears, it has been translated as "'culture' of antisemitism" in *The Jewish Chronicle*.

[b] Where "national identity" ("nationalen Eigenart") appears in the text, it is translated as "national entity" in *The Jewish Chronicle*.

[c] "Instincts" ("Instinkten") in the text is translated as "ill-feeling" in *The Jewish Chronicle*.

[d] The phrase "the reaction of non-Jews is therefore much lesser" ("die Reaktion der Nichtjuden darum viel geringer ist") does not appear in the English-language version printed by *The Jewish Chronicle*.

[e] An editor's note here indicates that Einstein gave the interview on which this article is based in the period between 2 April, when he disembarked in New York, and 30 May, when he departed for England.

[f] Where "the political antisemitism" ("der politische Antisemitismus") occurs in the text, it is translated as "the communal antisemitism" in *The Jewish Chronicle*.

[g] The phrase "with the non-Jews" ("zu den Nichtjuden") does not occur in *The Jewish Chronicle* text. In place of the following four sentences, the English text is: "I believe that German Judaism is thus being influenced to a great extent by antisemitism."

[h] "Until seven years ago" ("bis vor sieben Jahren") has incorrectly been reduced to "two" in *The Jewish Chronicle*.

[i] This would be referring to Einstein 1919c (Doc. 11 of this volume).

[j] In place of the following two sentences, *The Jewish Chronicle* writes: "To deny the Jew's nationality in the Diaspora is, indeed, deplorable. If one adopts the point of view of confining Jewish ethnic nationalism to Palestine, then one, to all intents and purposes, denies the existence of a Jewish people. In that case, one should have the courage to carry through, most quickly and completely, entire assimilation."

[k] "My peers" ("Standesgenossen") is translated as "friends" in *The Jewish Chronicle*.

[l] After the document is faithfully translated until "not only lead the Jewish people to their own national renaissance" ("das jüdische Volk nicht bloß seine eigene nationale Renaissance herbeiführen"), the last portion of the sentence in *The Jewish Chronicle*'s translation reads: "but will enrich their moral culture and knowledge; and, as centuries ago, be directed to new and better ways than those which present world-conditions necessarily entail for them."

22. The Development and Present Position of the Theory of Relativity (1921)

King's College Lecture, 13 June 1921. [*Collected Papers* Vol. 7, Doc. 58].

> Einstein was invited to speak at King's College by Principal Ernest Barker (1874–1960), who presided at a gala dinner afterward. Lord Haldane introduced Einstein to an enthusiastic crowd, proclaiming that "what Newton was to the 18th century, Einstein was to the 20th," and calling his conception of the universe "more revolutionary than that of Galileo, Copernicus, or Newton." The audience included many distinguished scientists, who listened to Einstein speak extemporaneously for more than an hour. Einstein's lecture and the dinner marked the high point of his triumphal ten-day visit to England, as reported not only in England but also in Germany.

It is a special joy for me to be able to speak in the capital of the country from where the most important basic ideas of theoretical physics were brought into the world. I think of the theories of the motion of masses and gravitation that Newton gave us, and of the concept of the electromagnetic field by Faraday and Maxwell, which provided physics with a new foundation. One may well say the theory of relativity brought a kind of conclusion to Maxwell's and Lorentz's grand framework of ideas by trying to extend the physics of fields to all of its phenomena, gravitation included.

Turning to the subject of the theory of relativity, I want to emphasize that this theory has no speculative origin; rather, it owes its discovery only to the desire to adapt theoretical physics to observable facts as closely as possible. This is by no means a revolutionary process, merely the natural development of a trail that can be traced through the centuries. The abandonment of certain concepts of space, time, and motion—treated until now as fundamental—must not be perceived as voluntary, but only as enforced by observed facts.

The law of the constancy of the speed of light—which was corroborated through the development of electrodynamics and optics, along with Michelson's famous experiment, which decisively demonstrated the equality of all inertial systems (principle of special relativity)—relativized the concept of time, where every inertial system had to be given its own special time. During the development of this idea, it became obvious that the previous connection between the immediate experience on the one hand and the combination of coordinates and time on the other had not been analyzed with sufficient rigor.—It is, by the way, an essential trait of the theory of relativity to more rigorously clarify the relation between the general concepts and experienced facts. The valid principle here is to base the justification of physical concepts exclusively upon their clear and unique relation to experienced facts.

According to the theory of special relativity, spatial coordinates and time have an absolute character only insofar as they can be directly measured using rigid bodies and clocks.—But they are relative insofar as they depend upon the state of motion of the selected inertial system. According to the theory of special relativity, the four-dimensional continuum formed by the union of space and time retains that kind of absolute character that, in previous theories, space and time held separately (Minkowski). From an interpretation of the coordinates and time through measurements, one finds what influence motion (relative to the coordinate system) has upon the shape of bodies and the rate of clocks; also the equivalence of energy and inertial mass.

The theory of general relativity owes its origin primarily to the experimental fact of the numerical equality of inertial and gravitational mass of a body, a fundamental fact for which classical mechanics has given no interpretation. Such interpretation is obtained by extending the principle of relativity to inertial systems accelerated relative to each other. The introduction of coordinate systems that are accelerated relative to inertial systems causes the appearance of gravitational fields relative to these coordinate systems. This is the reason the theory of general relativity, based upon the equality of inertia and gravity, also provides a theory of the gravitational field.

The introduction of coordinate systems accelerated relative to each other as equally admissible coordinate systems—as it is implied by the identity of inertia and gravity—leads in combination with the results of the theory of special relativity to the conclusion that the laws of positioning rigid bodies in the presence of gravitational fields do not obey the laws of Euclidean geometry. An analogous result follows with respect to the rate of clocks. From this arises the need for a second generalization of the theory of space and time, as the immediate interpretation of space and time coordinates as measurements with rods and clocks breaks down.

This generalization of metric—already worked out in the field of pure mathematics by the research of Gauss and Riemann—is essentially based upon the fact that the metric of the theory of special relativity is still valid for the general case, provided it is confined to small domains.

The course of development we described here deprives the space-time coordinate system of all independent reality. Now, the metric reality is only defined when the space-time coordinates are combined with the mathematical quantities that describe the gravitational field.

There is a second root to the basic ideas of the theory of general relativity. As Ernst Mach has already emphasized, there is the following unsatisfying point in Newton's theory: when motion is viewed from a purely descriptive view rather than a causal one, there is a relative motion of things against each other. However, the acceleration in Newton's laws of motion cannot be understood in the concept of relative motion. This forced Newton to imagine a hypothetical physical space relative to which an acceleration should exist. This concept of absolute space, introduced *ad hoc*, is admittedly logically correct, but it is not satisfying. Therefore, Mach looked for a modification of the equations of mechanics such that the inertia of bodies is not derived from their motion against absolute space, but rather against the totality of all the other gravitating bodies. Given the state of the art as it existed then, his attempts had to fail.

But it seems very reasonable to pose the problem at all. This train of thought presses itself with even stronger intensity upon the theory of general relativity, as according to this theory, the physical properties of space are influenced by ponderable matter. This speaker is convinced that the theory of general relativity can satisfactorily solve this problem only by viewing the world as spatially closed. The mathematical results of the theory lead to this view if one assumes the mean density of ponderable matter in the universe has a finite value, no matter how small this value may be.

23. On a Jewish Palestine (1921)

Jüdische Rundschau (1 July 1921), p. 371. [*Collected Papers* Vol. 7, Doc. 60].

The address was delivered on the evening of 27 June 1921 in the Blüthner-Saal in Berlin to a capacity crowd.

Ladies and Gentlemen!

Rebuilding Palestine is not a mere charity issue or a colonial affair for us Jews, but a problem of cardinal importance for the Jewish people. Palestine is not primarily a refuge for Eastern Jews, but the manifestation of a reawakened national sense of community of all Jews.

But is it opportune and necessary *to experience and strengthen this sense of community*? I believe I can answer this question with *an unconditional "yes,"* not only based on my immediate sentiment, but also on sound reasons.

Let us briefly look at the *development of the German Jews* during the past hundred years. Just a century ago, most of our ancestors lived, with few exceptions, in the ghetto. They were poor, politically disenfranchised,[a] and separated from non-Jews by a wall of religious traditions, secular forms of living, and legal restraints; in their spiritual development, they were limited to their own literature and only marginally influenced by the enormous cultural renewal experienced by European spiritual life during the Renaissance. However, these little recognized, modestly living people were ahead of us in one respect: *every one of them belonged, with every fiber of his soul, to a community* in which he was completely immersed, wherein he felt fully accepted as a member of whom nothing that would have been against his natural way of thinking was demanded. Our ancestors back then were physically and intellectually rather stunted, but in social respects, were in an enviable emotional equilibrium.

Then came emancipation. It offered unanticipated opportunities for development. Individual people soon gained positions[b] in the upper economic and social strata of society. Greedily, they absorbed the magnificent achievements that art and science had created in the

Occident. With glowing passion, they participated in this development by creating lasting values themselves. In the process, they adapted to the external forms of life of the non-Jewish world; increasingly turned away from their religious and social traditions; and adopted non-Jewish customs, forms, and ways of thinking. It seemed as though they completely dissolved into the far more populous, politically and culturally better-organized host nations so that nothing recognizable would remain of them after a few generations. A complete dissolution of Jewish national traditions ("*Volkstum*") in Central and Western Europe appeared unavoidable.

But it turned out differently. Apparently, a variety of races have instincts that counteract intermixing. The adaptation of the Jews to the European nations in terms of language, customs, and, in part, even religious forms *could not extinguish that feeling of otherness* that existed between the Jews and their European host nations.[c] In short, this spontaneous feeling of strangeness originates, in its last instance, in a reduction of energy and therefore *cannot be eliminated from the world through well-intended treatises.*[d] Nationalities do not want to be mixed; they want to go their own way. A state of peace can only be achieved by mutual tolerance and respect.

A prerequisite for that is, most of all, that we Jews *regain awareness of our nationality and the self-respect* we so desperately need for our existence. We must relearn to acknowledge our ancestors and our history, and *as a people, we have to again take upon ourselves cultural tasks* that are apt to strengthen our sense of community. It is not enough to take part in the development of mankind as individuals. We also have to face those tasks capable of furthering the national whole. This is the only way for Jewry to socially recuperate.

It is from this perspective that I ask you to look at the Zionist movement. *History today has ascribed us the task* of actively participating in the economic and cultural construction of Palestine.[e] Enthusiastic and highly gifted men have prepared the work. Many excellent tribal comrades are ready to completely dedicate themselves to it. May every one of them fully appreciate the importance of this task and contribute to its success to the best of his abilities![f]

Now, I would like to say a few things about the experiences I had on a trip to America, which I undertook in the past months, in the service of the Zionist organization especially for the advancement of the University in Jerusalem.

My greatest experience was *seeing a Jewish people for the first time in my life*. Ladies and Gentlemen! I've seen a great many Jews (laughter), but I have not seen a Jewish people in Berlin or elsewhere in Germany. This Jewish people that I saw in America had immigrated to America from Russia, Poland, and Eastern Europe in general. These people still carry a healthy national feeling within them that has not yet been destroyed by the customization and division of the individual. I have witnessed how exceptionally self-sacrificing and creative these people are. In very little time, for example, they were able to secure the *planned university*, at least as far as the medical faculty is concerned. Furthermore, I have witnessed that, *for the most part, it is the middle class* and common folk in general, not those with special social standing or exceptional natural gifts, who still have the strongest healthy sense of community and self-sacrifice. It was there that I had the impression that, if we could succeed in establishing a center for the Jewish people in Palestine, we would have again a spiritual center, and the feeling of isolation would leave us even though most of us are scattered in all countries. This is the great liberating effect I expect from the rebuilding of Palestine.

Editorial Notes

[a] The phrase "politically disenfranchised" ("politisch entrechtet") was inserted into the autograph draft, but was not present in the printed speech in *Jüdische Rundschau*.

[b] The *Jüdische Rundschau* publication has "contacts" ("Fühlung") where Einstein wrote "positions" ("Stellungen") in the draft.

[c] In place of the following sentence, Einstein wrote "Ultimately, antisemitism spontaneously arises from this sense of otherness" ("Auf dieses spontane Gefühl der Fremdheit ist in letzter Instanz der Antisemitismus zurückzuführen") in the draft. The typesetter apparently bowdlerized the sentence, first, by allowing the sense of "otherness" ("Fremdheit") to arise from the "diminution of energy" ("Energieverminderung"), and second, by reversing the cause-and-effect relationship of antisemitism and otherness in the draft.

[d] Where Einstein wrote "treatises" ("Traktate") in the draft, *Jüdische Rundschau* printed "pressures" ("Druckmittel").

[e] In place of "Palestine," Einstein wrote "our homeland" ("unseres Stammlandes") in the draft.

[f] The corresponding portion of the autograph text ends at this point. The remainder of the speech was, according to Einstein, delivered extemporaneously (*Berliner Tageblatt*, 28 June 1921, Evening Edition, p. [2]).

24. The Impact of Science on the Development of Pacifism (1921)

K. Lenz and W. Fabian, eds. *Die Friedensbewegung: Ein Handbuch der Weltfriedensströmungen der Gegenwart.* Berlin: Schwetschke (1922), pp. 78–79.[a] [*Collected Papers* Vol. 7, Doc. 69].

> Initially solicited by a publishing house, which requested a finished manuscript by 15 October, the document is dated by the reference to its completion in Einstein to Henri Barbusse, 9 December 1921. The preface to the published volume is dated March 1922. The list of other contributors includes seven of Einstein's wartime colleagues in the *Bund "Neues Vaterland,"* as well as Henri Barbusse.

Wars present the most severe obstacle for the development of all endeavors—in particular, all cultural goals that are essentially based on the cooperation of people from all nations. War deprives the intellectual worker of the external and internal conditions upon which his work must be based. If he is still of sufficient youth and vigor, war will make him a slave of an organization aimed at annihilation, or surround him with an atmosphere of excitement and hatred. In addition, war creates an oppressive economic dependency lasting for years due to the impoverishment associated with it. Consequently, a human being whose highest values are spiritual ones must be a pacifist. This is also verified by history, wherein men of the past are not just counted, but rather weighed.

What about the effect of science on the development of pacifism? In this respect, the influence of the humanities has evidently been very slight. One easily observes that most representatives of the science that comes to mind first—namely, history—have by no means advanced the ideals of pacifism. Many representatives of this science— even though not quite the best ones—have publicly come forth with astounding, strongly chauvinistic, and militaristic pronouncements, especially during the recent great war.

The situation is quite different in the natural sciences. Due to the universal character of their subject matter and the need for internationally organized cooperation, they are inclined toward an international understanding and thus favor pacifist goals. A similar attitude prevails among national economists, who necessarily must view war as a disturbance of the economic processes caused by lack of organization.

But the deepest effect of the natural sciences upon the historical process that interests us here is not of a spiritual, but of a material kind. Technical inventions arising from the natural sciences have chained the economic dynamics together internationally and thus determined— also by military technology—that all war must become a matter of international significance. When this development has entered the consciousness of human beings after sufficient calamitous experiences, then men will also find the energy and goodwill to create organizations that have the power to prevent wars.

Editorial Note

[a] A slightly modified French version appeared 18 January 1922 as "La science et le pacifisme," *Clarté* 1 (1921–1922), p. 118.

25. The Plight of German Science: A Danger for the Nation (1921)

Neue Freie Presse (25 December 1921), p. 1. [*Collected Papers* Vol. 7, Doc. 70].

> Requested to write an article that might further the "scientific community of interest," Einstein directed his comments to a general public in contrast to earlier, more limited appeals for an exchange of scientific literature and greater tolerance. His broader approach reflected in part the degenerating state of scientific research in Germany.

German-speaking countries face a danger that needs to be emphatically pointed out by those familiar with it. The economic distress related to the heavy political blows does not affect everybody equally. Especially hard hit are institutions and individuals whose material existence depends directly on the state; among them are the scientific institutes and researchers whose work relies in great part not only on the flourishing of the economy, but also on Germany's and Austria's cultural standing.

To appreciate the full gravity of the situation, one has to take the following into account: In times of suffering, people tend to take care only of their immediate needs.

One pays for work that *directly* creates material value. Science, however, must not focus on immediate practical results, lest it risk wasting away. *The insights and methods created by science mostly have only indirect practical value, and often only to future generations*; neglecting science will later cause a lack of intellectual workers who, by virtue of their independent views and judgments, could discover novel economic ways and adapt to new situations.

If scientific research becomes stunted, this nation's intellectual life will be choked up, thereby limiting numerous possibilities for future development. This must be prevented. With the weakening of the state due to developments in foreign policy, it is the economically stronger citizens whom it behooves to assist and prevent scientific life from withering away.

Insightful men have clearly understood the prevailing conditions and created institutions to support all research in Germany and Austria. Please help to make these efforts a beautiful success. In my teaching activity, I see with admiration that economic worries have not been capable of suffocating the will and love for scientific research. Quite the contrary! *It appears that the heavy blows have even strengthened the love of ideals.* Everywhere, people work with glowing enthusiasm under very difficult circumstances. Please ensure that the willpower and talent of today's youth do not waste away to the terrible detriment of everyone.

26. Preface to Bertrand Russell's *Political Ideals* (1922)

B. Russell. *Politische Ideale*. Berlin: Deutsche Verlagsgesellschaft für Politik und Geschichte m.b.H. (1922), p. 5. [*Collected Papers* Vol. 13, Doc. 2].

Bertrand Russell (1872–1970) was a British philosopher, mathematician, socialist, pacifist, and freelance writer and lecturer.

The availability of this great English mathematician's lucid discourse to the German public is very welcome.[a] It is not some wavering professor vacillating between "on the one hand and the other" speaking to us here, but one of those resolute, straightforward individuals who exist independent of the period into which they happened, almost by chance, to have been born. Unbending consistency and warm human sensitivity prescribe his path. He follows down this path unperturbed by the consequences his stance may bring. He did not play the martyr when he let himself be robbed of his professorship and found himself in prison for antimilitaristic propaganda.

He wants military force to be completely eliminated and recommends consistent training of the population in organized passive resistance as the means of counteracting aggressive military force from abroad. This solution will not appear utopian to those who experienced the Kapp putsch[b] in Germany.

Russell furthermore addresses the problem of social policy. Driven by an ardent interest in the progress of human organization, he traveled through Bolshevist Russia to learn its lessons. His ideal is the development of the free creative powers of the individual within a social order that banishes fear about one's livelihood without lapsing into a hypertrophic bureaucratism, the worst enemy of socialistic endeavors.

We may or may not agree with individual details of Russell's opinions. What a delight it is to become acquainted with the thoughts of a sharp-minded and truly noble man of our time on issues that touch all serious people today. Let each person make up his own mind about this great Englishman.

A. Einstein

Editorial Notes

[a] Russell's *Political Ideals* was published in English in 1917.

[b] A failed right-wing putsch against the German government on 13 March 1920, led by Wolfgang Kapp (1858–1922).

27. Review of Wolfgang Pauli, *The Theory of Relativity* (1922)

Die Naturwissenschaften 10 (1922), pp. 184–185. [*Collected Papers* Vol. 13, Doc. 62].

Wolfgang Pauli (1900–1958) published the 240-page exposition of the theory of relativity while still a graduate student in Munich. He would become one of the most important physicists of the twentieth century.

Pauli, W., Jr., Relativitätstheorie. Offprint from *Enzyklopädie der mathematischen Wissenschaften*. Leipzig, B. G. Teubner, 1921. IV, pp. 539 to 775. 17 × 25 cm. Price bound 40.– marks; bound 50.– marks.

Whoever studies this mature and broadly conceived work would not believe the author is a man of twenty-one years of age.[a] One does not know what to admire most: the psychological grasp of the development of ideas, the certainty of mathematical deduction, the profound physical vision, the capacity for clear systematic presentation, the knowledge of the literature, the factual completeness, or the certainty of the critique.

This exhaustive exposition of roughly 230 pages is arranged as follows:

I. Development of the special theory of relativity, with careful account taken of the definitive observational data for its foundation.

II. Mathematical aids for the special and general theories of relativity. The paragraphs on affine tensors and infinitesimal transformations are especially recommended to knowledgeable readers.
III. Further development of the special theory of relativity, exhaustive from both the formal and physical points of view.
IV. General theory of relativity (75 pages). Model account of the development of the ideas. Complete presentation of the mathematical methods for solving specific problems. The discussions on the energy equation and the criticism of Weyl's theory are particularly valuable.

Anyone working creatively in the area of relativity should consult Pauli's edition, likewise anyone who wants to familiarize himself accurately with the principal issues.[b]

A. Einstein, Berlin.

Editorial Notes

[a] Pauli completed his article, Pauli 1921, in December 1920, while still a graduate student in Munich. He added final corrections at the proof stage, taking into consideration criticism expressed by Felix Klein, shortly before presenting himself for the doctoral examination on 25 July 1921 (Hermann et al. 1979, pp. 27–32). The published text of the article in the *Encyclopädie der mathematischen Wissenschaften* was issued on 15 September 1921, and then independently as a special offprint, with a preface by Arnold Sommerfeld.

[b] Almost forty years later, Pauli published an English translation of this review with critical comments (Pauli 1958), which was subsequently reissued as a Dover reprint.

28. In Memoriam Walther Rathenau (1922)

Neue Rundschau 33 (1922), pp. 815–816. [*Collected Papers* Vol. 13, Doc. 317].

Walther Rathenau (1867–1922), an engineer by training, became a leading industrialist as chairman of the board of the AEG company. After World War I he was one of the founders of the German Democratic Party.

> He was appointed minister of reconstruction and charged with participating in the Versailles war reparations negotiations. He was appointed foreign minister in 1922. He was assassinated on 24 June 1922, two months after having signed the Treaty of Rapallo, by two members of the radical right-wing *Deutschvölkischer Schutz- und Trutzbund* and of the Organisation Consul. Rathenau had been the central target of vitriolic attacks directed against Jews and against the government's policy of appeasement toward the Allies. Einstein was deeply affected by the brutal killing of his friend, with whom he last visited a month earlier.

My feelings for Rathenau were, and still are, those of delighted admiration and gratitude for his giving me hope and consolation in the current dismal European situation. As a clear-sighted and warm person, he bestowed on me the gift of unforgettable shared hours. His synoptic view of the major economic correlations, his psychological appreciation of the peculiarities of nations, for all social spheres, and his knowledge of the individual were admirable. And he loved them all, even though he knew them, as one who had the strength to say yes to this life. When he chatted at the table among friends, a delightful mixture of earnestness and genuine Berlin humor made his conversation a unique pleasure. It is easy enough to be an idealist when one is living in cloud-cuckooland; he, however, was an idealist even though he was living on Earth and knew its odor as rarely another does.

I regretted that he became a minister. Given the attitude taken by a large part of the educated class in Germany against Jews, it is my conviction that proud reserve by Jews in public life would be the natural thing. Still, I would not have thought hate, delusion, and ingratitude would go so far. To those who have led the ethical education of the German people over the last fifty years, however, I would call out: By their fruits ye shall know them.

29. On the Present Crisis of Theoretical Physics (1922)

Kaizo 4, no. 12 (December 1922), pp. 1–8.[a] [*Collected Papers* Vol. 13, Doc. 318].

> In 1921, the progressive monthly journal *Kaizo* invited Einstein to visit Japan, where he arrived on the steamship *Kitano Maru* on 17 November 1922. The tour included a scientific lecture series in Tokyo and in several other cities. The trip resulted in the first publication of Einstein's collected papers, published in Japanese translation. In this paper, and in related works and correspondence at this time, Einstein made it clear that quantum theory, though problematic in its conceptual foundations, is absolutely necessary given its empirical successes. In more detail, he expressed the inevitability of the Planck-Bohr-Rutherford quantum theory and ended with a discussion of how electrodynamics might have to be modified to account for the quantum.

The goal of theoretical physics is to create a logical system of concepts, based on the fewest possible mutually independent hypotheses, allowing for a causal understanding of the entire complex of physical processes. In response to the question of how this scientific system developed and grew in the time before Maxwell, one would have been able to give the following answer.

One first grounds the immovable basis of all exact science on uncontested facts of human observation or thought, namely, on geometry and analysis. The Greeks already did this, so those coming afterward had little left to do of principal novelty beyond developing infinitesimal calculus. Then came the establishment of the actual fundamental laws of physics, i.e., the fundamental laws of mechanics by Galileo, Newton, and their contemporaries. Until around the end of the 19th century, physicists were convinced these fundamental laws of mechanics would have to be the basis of absolutely all theoretical physics, i.e., that every physical theory would ultimately have to be based on mechanics.

So the basis of physics seemed to have been finally erected, and a theoretical physicist's work seemed to be to conform the theory to the ever-growing abundance of investigated phenomena through specialization and differentiation. No one thought about the possibility that the foundations of physics would have to be changed. Then, because of Faraday's and Maxwell's research, it gradually became obvious that the foundations of mechanics failed for electromagnetic phenomena. This change took place in several stages. Initially, both of the above-named pioneers realized electromagnetic processes could not be represented with the theory of unmediated forces by action-at-a-distance.

According to Newton, all forces that can produce accelerations of material points are attributable to an instantaneous influence that the rest of the material points individually exert on the observed material point. As an alternative to this theory of unmediated action-at-a-distance, Maxwell and Faraday posited the theory of electromagnetic fields. According to this theory, all dissemination of electric force is based not on instantaneous action-at-a-distance, but on a propagation-conducive state of space or ether: the electromagnetic field. The energy-endowed field, describable by continuous space functions, assumed physical reality alongside the movable mass point that, according to Newton's theory, had figured as the sole carrier of energy. It is known that Hertz helped bring about general recognition of this conception through his experiments on the propagation of electric force.[b]

At first, the extent of the revolutionary character of field theory was not fully realized. Maxwell himself was still convinced electrodynamic processes were conceivable as motions of the ether; he even used mechanics to forge ahead to the field equations. In the interim, however, it became more and more evident that it was impossible to derive the electromagnetic equations from mechanical ones. In search of a unified basis for physics, one was eventually compelled to try the opposite and derive the mechanical equations from electromagnetic ones. This attempt suggested itself even more after J. J. Thomson realized an electromagnetic inertia existed for electrically charged bodies and after M. Abraham showed that the inertia of electrons was interpretable purely electromagnetically.[c] With the attribution of inertia to electromagnetic processes, a complete revolution in the foundations of physics had come to pass, at least in principle. In place of the mass point as the ultimate of reality, the electromagnetic field entered as a fundamental elementary building block of theoretical physics. A theoretical construction of matter on a purely electromagnetic basis is generally known to have succeeded to a certain degree. Specifically, we know today that the cohesive forces are purely electromagnetic.

These are not the only fruits of the Faraday-Maxwellian field theory. Recognition of the covariance of Maxwell's electromagnetic equations with Lorentz transformations led to the special theory of relativity and hence to an awareness of the equivalence between inertia and energy; the extension of field theory to gravitation, considering the identity of inertial and ponderous mass, led to the general theory of relativity. With the general theory of relativity, one pillar of Newton's theory that had hitherto always been believed to be one of the necessary foundations of all science—namely, Euclidean geometry—sank away. It had emerged in early ages, from primitive experiments on solids, and was silently assumed by physicists to be an exactly pertinent law for the orientation of solid bodies of even temperature not subjected to external influences; but now, because of important considerations based indirectly on experiment, a doctrine already proposed by Gauss and Riemann had to replace it. With the general theory of relativity, the developmental phase of theoretical physics founded by Faraday and Maxwell seems to have come to a close.

In the last two decades, it has been recognized that even a basis for physics characterized by the Faraday-Maxwellian field theory cannot hold up to experience any better than the mechanics grounded on it. It is rather to be expected that scientific progress will demand a fundamental change no less profound than the one we have summarized under the name "field theory." As we are still far away from a logically clear foundation, however, we must content ourselves here with showing how the foundation has thus far proven inadequate and how well account has been taken of important groups of physical phenomena through successful yet still groping attempts subsumed under the term "quantum theory."

Quantum theory found its origins in the theory of thermal radiation, for which a unification of mechanics and electromagnetic field theory delivers a law that is irreconcilable with experience and even intrinsically unreasonable. The fundamental problem of heat radiation can be formulated as follows: Thermodynamics teaches that there is radiation in the interior of a cavity surrounded by opaque bodies at temperature T whose composition is entirely independent of the nature of the bodies forming the cavity's walls. If ρ is the monochromatic radiation density,—i.e., ρdv is the radiation energy in the cavity per unit volume whose frequency lies between v and $v + dv$—then ρ is a very specific function of v and T. It is not determinable by purely thermodynamic considerations; rather, its derivation presupposes an insight into the nature of the process of generation and absorption of the radiation:

Classical mechanics connected with Maxwellian electrodynamics yields an expression of the form:

$$\rho = \alpha v^2 T, \qquad (1)$$

which cannot be right, as it produces an infinitely large value for the total density of the cavity radiation $\int_{v=0}^{v=\infty} \rho dv$. Planck found the expression corroborating all foregoing observations, in conformance with experience,

$$\rho = \frac{8\pi h v^3}{c^3} \cdot \frac{1}{e^{\frac{hv}{kT}} - 1}, \qquad (2)$$

[d]where k means a constant connected to the absolute size of atoms, and h a natural constant hitherto unknown in physics, which one may well describe as the fundamental constant of quantum theory. In 1900, Planck offered a theory for this formula that implicitly contains a hypothesis irreconcilable with physics up to that time, which we can retrospectively interpret based on the experimental and theoretical research of the past two decades. Wherever a sine-type oscillating process of frequency v exists in nature, its energy always carries an integral multiple of hv; intermediate energy values do not occur in nature for sine-type oscillating processes.

Based on this hypothesis, it was possible to correctly derive not only Planck's formula (2) of heat radiation, but also the law of the specific heats of crystalline solids.[e] But these derivations are all intrinsically contradictory; while making use of this new hypothesis, they always rely on the foundations of classical physics incompatible with them.

Considering all the major advances that Maxwell's electrodynamics and Newton's mechanics have made in physics, as well as their present indispensability, one is obliged to doubt the basic hypothesis of quantum theory as much as possible. But phenomena directly confirming quantum theory do exist, even though their incompatibility with the foundations of classical physics is immediately clear.

The energy density of radiation emitted from a radiating source diminishes, according to Maxwell's theory, with the reciprocal square of the distance. The energy available at one place for processes of absorption per unit of time would thus have to diminish infinitely with distance. As, for example, the chemical decomposition of a molecule or the release of an electron from an atom require a definite amount of energy, radiation that has been sufficiently weakened by propagation away from the light source similarly should no longer be capable of generating such a chemical process. Experience shows, on the contrary,[f] that the chemical and photoelectric effectiveness of the radiation is completely independent of its density; the total chemical action of radiation permeated by matter is only dependent on its total energy, not at all on its spatial energy density. Experiments by E. Warburg have furthermore demonstrated that the energy absorbed per chemical elementary process is always equal to $h\nu$, independent of the spatial energy of the radiation. This result also follows from experiments on the photoelectric effect and on the generation of cathode rays from X-rays.

We know today that this energy really does originate from the radiation and is not, for example, gradually accumulated. The absorption of light constitutes indivisible elementary processes, during each of which the energy $h\nu$ is completely transformed. We know nothing about the details of such an elementary process. If just the energetic properties of radiation were known, we would then see ourselves obliged to postulate a kind of molecular theory of radiation of the kind of Newton's emission theory of light. However, finding explanations for diffraction and interference processes poses insurmountable obstacles. Moreover, it should probably be kept in mind that the field theory of radiation is no more incorrect than the theory of elastic waves in solids, which establishes their thermal content, for both theories collide to the same degree with the quantum relation and must be combined with the latter in the same way to arrive at an appropriate interpretation of the results of experience.

The grand development of our knowledge of the composition of atoms, which we essentially owe to the great masters Rutherford and Bohr, has led to a highly significant generalization of the quantum rule, which we now want to consider. Even before the Rutherford-Bohr theory, the assumption was that absorption or emission of a spectral line would have to correspond to a transition of an atom or molecule from one preferred state into another. As the states of the elementary forms were certainly not interpretable as sine-type oscillations, the problem arose of how to extend quantum theory to mechanical systems of more general character, which Bohr's, Sommerfeld's, Epstein's, and Schwarzschild's investigations have succeeded in doing, step by step. The results obtained by these researchers are being elevated to secure findings through precise confirmation in the field of spectroscopy.

If a mechanical system is describable by coordinates q_ν which experience cyclic changes over time, and if, for each degree of freedom ν, the momentum $p_\nu \left(= \frac{\partial L(q \cdot p)}{\partial q_\nu} \right)$ belonging to q_ν is describable as a function of exclusively the one coordinate q_ν, then for $\int p_\nu q_\nu$ any ν, the integral over one cycle is an integral multiple of Planck's constant h. Thus, the "permissible" states according to quantum theory are defined for so-called "quasi-periodic" mechanical systems. This general rule has proven its value in such very subtle and varied special cases (e.g., Epstein's theory of the Stark effect)[8] that its general correctness has become quite probable.

From the general theoretical point of view, the following is especially puzzling. On one hand—as already pointed out—mechanics does not appear to be generally valid, as statistical mechanics, which is based on it, leads to results that contradict experience (e.g., specific heats of solids). On the other hand, the mechanical laws, most astonishingly, stand the test within the scope of validity of the above rule. Are there only supposed to be quasi-periodic elementary processes in nature or, more generally, only such mechanical systems as have as many complete integrals as degrees of freedom? Such a thought seems absurd, given the kinetic theory of gases. The problem of how the validity of classical mechanics (and electrodynamics) is restricted by the demands of quantization is shrouded in as deep darkness today as it was fifteen years ago.

It has often been noted that, in the present state of our knowledge, the possibility of representing natural laws by differential equations appears doubtful. Indeed, we must consider a complete cycle of the system according to the quantum rule just indicated, in order to judge whether a particular state of the system is quantum-theoretically permissible or not. To really do justice to the quantum relations, a new mathematical language seems necessary; in any case, it seems preposterous to express the laws through a combination of differential laws and integral conditions as we do today. Once again, the foundations of theoretical physics are shaken, and experience is calling for the expression of a higher level of lawfulness. When will the saving idea be granted to us? Happy are those who may live to see it.

(August 1922)

Editorial Notes

[a] The original title of the manuscript was "On the Present Crisis of the Theory of Light" ("Über die Krise der Licht-Theorie.").
[b] See, e.g., Hertz 1889 and Hertz 1892.
[c] See Thomson 1881, Abraham 1902.
[d] See Planck 1900.
[e] See Einstein 1907 (*Scientific Writings*, Doc. 6).
[f] See Einstein 1905a (*Scientific Writings*, Doc. 2).
[g] See Epstein 1916.

30. Musings on My Impressions in Japan (1923)

Kaizo 5 (1923), pp. 338–343.[a] [*Collected Papers* Vol. 13, Doc. 391].

> Einstein visited Japan on a five-week lecture tour in late 1922 at the invitation of the progressive Japanese journal *Kaizo*, in whose pages this article appeared. He visited temples in Nikko and Nara and attended an imperial chrysanthemum viewing in his honor in the garden of the Akasaka Palace. In Tokyo, Einstein attended performances of Noh and Kabuki theater and of ancient Japanese court music and dance.

I have been traveling much around the world in the last few years, really more than befits a scholar. The likes of me really ought to stay quietly put in their study and ponder. For my earlier trips, there was always an excuse that could easily pacify my not-very-susceptible conscience. But when Yamamoto's[b] invitation to Japan arrived, I immediately resolved to go on such a great voyage that must demand months, even though I am unable to offer *any* excuse other than that I would never have been able to forgive myself for letting a chance to see Japan with my own eyes go by unheeded.

Never in my life have I been more envied in Berlin, and genuinely so, than the moment it became known that I was invited to Japan. For in our country, this land is shrouded more than any other in a veil of mystery. Among us, we see many Japanese living a lonely existence, studying diligently, smiling in a friendly manner. No one can fathom the feelings concealed behind this guarded smile. And yet, it is known that, behind it, there lies a soul different from ours which reveals itself in the Japanese style, as is apparent in numerous small Japanese products and Japanese-influenced literature coming into fashion from time to time. Everything I knew about Japan could not give me a clear picture. My curiosity was in utmost suspense when, on board the *Kitanu Maru*, I passed through the Japanese channel and saw the countless delicate green islets glowing in the morning sun. But glowing most of all were the faces of all the Japanese passengers and the ship's entire crew. Many a tender young lady, who otherwise would never be seen before breakfast time, was roaming restlessly and blissfully about on deck at six o'clock in the morning, heedless of the raw morning wind, to catch the first possible glimpse of her native soil. I was moved to see how overcome they all were with deep emotion. A Japanese loves his country and his nation most of all; and despite his linguistic proficiency and great curiosity about everything foreign, away from home, he still feels more alien than anyone else. How is this explained?

I have been in Japan for two weeks now,[c] and yet so much is still as mysterious to me as on the first day. Some things I did learn to understand, though, most of all, the shyness that a Japanese feels in the company of Europeans and Americans. At home, our entire upbringing is tuned toward our ability to tackle life's struggles as single beings under the best possible circumstances. Particularly in cities, individualism in the extreme, cutthroat competition drawing on our utmost energy, feverish laboring to acquire as much luxury and pleasure as possible. Family bonds are loosened; the influence of artistic and moral traditions on daily life is relatively slight. The isolation of the individual is seen as a necessary consequence of the struggle for survival; it robs a person of that carefree happiness only integration into a community can offer. The predominantly rationalistic education—indispensable for practical living under our conditions—lends this attitude of the individual even more poignancy; it makes the isolation of the individual weigh even more keenly on the conscience.

It is quite the contrary in Japan. The individual is left far less to his own devices than in Europe or America. Family ties are very much closer than at home, even though they are actually only provided very weak legal protection. But the power of public opinion is much stronger here than at home, assuring that the family fabric is not undone. Published and unpublished rumors help enforce what is, for the most part, already sufficiently secured by a Japanese upbringing and an innate kindheartedness.

The cohesion of extended families in material respects—mutual support—is facilitated by an individual complaisance about bed and board. A European can generally accommodate one person in his apartment without a perceptible disruption to the household order. Thus, a European man can only care for his wife and children at most, all being well. Often, wives—even women of higher status—must help earn a living and leave the children's education to the servants. It is even rare for adult siblings to provide for one another, let alone more distant relatives.

But there is also a second reason that makes closer protective ties between individuals easier in this country than where we live. It is the characteristic Japanese tradition not to express one's feelings and emotions, but to stay calm and relaxed in all circumstances. This is the basis upon which many people, even those not in mental harmony with one another, can live under a single roof without embarrassing frictions and conflicts arising. Herein lies, it appears to me, the deeper sense of the Japanese smile so puzzling to a European.

Does this upbringing to suppress the expression of an individual's feelings lead to an inner loss, a suppression of the individual himself? I do not think so. The development of this tradition was surely facilitated by a refined sensitivity characteristic of this nation and by an intense sense of compassion, which seems more potent than for a European. A rough word does not injure a European any less than it does a Japanese. The former immediately counters by stepping forward to the offensive, amply repaying in kind. A Japanese withdraws, wounded, and weeps. How often is the Japanese inability to utter sharp words interpreted as falseness and dishonesty!

For a foreigner like me, it is not easy to delve deeply into the Japanese mind. Being received everywhere with the greatest attention in festive garb, I hear more carefully weighed words than meaningful ones that inadvertently slip out of the depths of the soul. But what escapes me in direct experience with people is completed by the impressions of art, which is so richly and diversely appreciated in Japan as in no other country. With "art," I mean all things of permanence that human hands create here by aesthetic intent or secondary motivation.

In this regard, I never cease to be astonished and amazed. Nature and people seem to have united to bring forth a uniformity in style as nowhere else. Everything truly originating from this country is delicate and joyful; not abstractly metaphysical, but always quite closely connected to what is given in nature. Delicate is the landscape with its small green islets or hills; delicate are the trees; delicate is the most carefully farmed land, with its precise little parcels; but most especially so, the little houses standing on it; and, finally, the people themselves in their speech, their movements, their dress, including all objects they make use of. I took a particular liking to the Japanese house, with its very segmented smooth walls and many little rooms laid out throughout with soft mats. Each little detail has its own sense and import there. To this come the dainty people with their picturesque smiling, bowing, sitting—all things only to be admired, not imitated. You, o foreigner, try to do so in vain!—Yet Japan's dainty dishes are indigestible to you; better content yourself with watching.—Compared to our people, the Japanese are cheerful and carefree in their mutual relations; they live not in the future, but in the present. This cheerfulness always expresses itself in a refined form, never boisterously. Japanese wit is directly comprehensible to us. They, too, have much sense for the droll, for humor. I was astonished to note that, regarding these surely deeply lying psychological things, there is no great difference between the Japanese and the European. The soft-heartedness of the Japanese reveals itself here in that this humor does not assume a sarcastic note.

[d]Of greatest interest to me was Japanese music, which had developed partly, if not entirely, independently of ours. Only upon listening to completely strange art does one come closer to the ideal of separating the conventional from the essential that is conditioned by human nature. The differences between Japanese music and ours are indeed fundamental. While chords and architectonic structure appear to be

universal and indispensable in our European music, they are absent in Japanese music. On the other hand, they have in common the same thirteen tonal steps into which the octaves are divided. Japanese music seems to me to be a kind of emotive painting of inconceivably direct impressions. The exact tonal pitch does not even seem so absolutely imperative for the artistic effect. It much rather suggests to me a stylized depiction of the passions expressed by the human voice, as well as by such natural sounds conjured up in the human mind—emotive impressions, such as birdsong or the beating surf of the sea. This impression is amplified by the important role played by the percussive instruments, which have no specific tonal pitch of their own, but are particularly suited for rhythmic characterization. To me, the main attraction of Japanese music lies in the extremely refined rhythms. I am fully aware of the circumstance that the most intimate subtleties of this kind of music still elude me. Added to the fact that long experience is always presupposed for distinguishing the purely conventional from a performer's personal expression, the relation to the spoken and sung word—which plays a considerable role in most Japanese pieces of music—also escapes me. As I see it, a characteristic of the Japanese soul's artistic approach is the unique appearance of the mild flute, not the much more strident wind instruments made of metal. Here, too, is manifested the particular characteristic preference for the charming and the dainty that is especially prominent in Japanese painting and design of products for everyday life. I was most greatly affected by the music when it served as an accompaniment to a theater piece or a mime (dance), particularly in a *Noh* play. What stands in the way of the development of Japanese music into a major form of high art is, in my opinion, a lack of formal order and architectonic structure.

To me, the area of most magnificence in Japanese art is painting and wood carving. Here, it is properly revealed that the Japanese is a visual person who delights in form, who untiringly refashions events in artistic form then converted into stylized lines. Copying nature, in the sense of our realism, is foreign to the Japanese, just as is religious repudiation of the sensual, which is intrinsically so foreign to the Japanese mind despite the influence of Buddhism from the Asian continent. For him, everything is experienced in form and color, true to Nature, yet foreign to it insofar as stylization broadly predominates. He loves clarity and the simple line above all else. A painting is strongly perceived as an integral whole.

But among the great impressions I had during these weeks, so few could I mention, saying nothing about political and social problems. On the exquisiteness of the Japanese woman—this flower-like creature—I was also silent; for here, the common mortal must cede the word to the poet. But there is one more thing I have at heart. The Japanese rightfully admires the intellectual achievements of the West and immerses himself in the sciences successfully and with great idealism. But let him not thereby forget to keep pure the great attributes he possesses over the West: the artful shaping of life, the modesty and unpretentiousness in his personal needs, and the purity and calm of the Japanese soul.

Editorial Notes

[a] Pages are numbered in reverse order.
[b] Sanehiko Yamamoto (1885–1952), founder of the Democratic Cooperative Party and publisher of *Kaizo*.
[c] Einstein arrived on 17 November 1922.
[d] In the manuscript, this entire paragraph on Japanese music is missing. Instead, Einstein wrote: "(On music dictated to Mrs. I. Missing here. E.)." "Frau I." is probably a reference to Toni Inagaki, the wife of Morikatsu Inagaki who was Einstein's interpreter in Japan and chief secretary of the newly established Japanese Association of the League of Nations.

31. My Impressions of Palestine (1923)

New Palestine 4 (1923), p. 341.[a] [*Collected Papers* Vol. 14, Doc. 20].

On his return voyage from Japan, Einstein visited British Mandate Palestine between 2 and 14 February 1923 at the invitation of the Zionist Organization's Palestine Bureau. He visited Jerusalem, where he delivered the Hebrew University's inaugural lecture at the future site of the university on Mount Scopus. He also visited Tel Aviv, the Dead Sea, Haifa, and Nazareth. In his diary, he wrote: "I am absolutely wanted in Jerusalem.... The heart says yes but reason says no."

I cannot begin these notes without expressing my heartfelt gratitude to those who have shown so much friendship toward me during my stay in Palestine. I do not think I shall ever forget the sincerity and warmth of my reception, for they were to me an indication of the harmony and healthiness that reign in the Jewish life of Palestine.

No one who has come into contact with the Jews of Palestine can fail to be inspired by their extraordinary will to work, nor by their determination, which no obstacle can withstand. *Given that strength and spirit, there can be no question of the success of the colonization work.*

The Jews of Palestine are divided between the urban workers and the village colonizers. Among the achievements of the former, the city of Tel Aviv made a singularly profound impression on me. The rapidity and energy that has marked the growth of this town has been so remarkable that the young S. Ginsberg, son of Ahad Ha'am, jokingly referred to it as "Our Chicago."[b]

A remarkable tribute to the real power of Palestine is the fact that those who arrived over the last decades in the country are distinctly above average, both in the matter of culture and in their display of energy.

And among the Jewish "sights" of Jerusalem, none struck me more pleasantly than did the school of arts and crafts—*Bezalel—and the Jewish workingmen's* cooperatives. It was amazing to see the work that had been accomplished by young workers who, when they entered the country, could have been classified accurately as "unskilled labor." I noted that, besides wood, other building material is being produced in the country. But my pleasure was tempered somewhat when I learned of the fact that the American Jews who lend money for building purposes exact a high rate of interest.

To me, there was something wonderful in the spirit of self-sacrifice displayed by *our workers on the land*. One who has actually seen these men at work must bow before their unbreakable will and the determination that they show in the face of their difficulties, from debts to malaria. In comparison with these two evils, the *Arab question* practically almost vanishes. And regarding the last, I must remark that I have myself encountered more than once friendly relations between Jewish and Arab workers. In fact I believe that most of the difficulty comes from the intellectuals—and, at that, not from the Arab intellectuals alone.

Much care must be devoted to the study of combating malaria. This is an evil that affects not only the rural but also the urban population. During my visit to Spain some time ago, I was told that the Spanish Jews, in collaboration with the Spanish government, would send, at their expense, a specialist on the subject of malaria to Palestine,[c] and that this specialist should carry on his work in the University of Jerusalem. The malaria evil is still so strong that one may say that it affects one third of the workforce.

But the debt question is particularly depressing. Take, for instance, the workers in Deganiah. These splendid people groan under the weight of their debts and must live in the direst need in order not to contract new ones. One man, even with moderate means, were he large-hearted enough, could relieve this group of its heartbreaking burden. The spirit that reigns among the land and building workers is admirable. They take boundless pride in their work and have a feeling of profound love for the country and for the little locality in which they work.

In the matter of architectural taste as displayed in the buildings, in the towns, and on the land, there has been not a little to regret. But in this regard, the architect Kauffmann[d] fights vigorously to bring good taste and a love of beauty to the buildings.

To the government, considerable credit must be accorded for its construction of roads, for its fight against malaria, and, in general, for its sanitary work as a whole. Here the government has no light task before it. One can hardly find another such small country that is so heterogeneous and so beset by the various interests of those not living in it.

The greatest need today is for *skilled* labor. It is hoped that the completion of the Technion[e] will contribute mightily toward meeting these needs.

I am convinced *the colonization work will succeed*, in the sense that we shall create a unified community in Palestine that shall be a moral and spiritual center for the Jewish people. Here, and not in its economic achievement, lies, in my opinion, the significance of this work. Naturally, we cannot neglect the question of our economic position in Palestine, but we must at no time forget that all this is but a means to an end. To me, it seems of secondary importance that Palestine shall become economically independent at the greatest possible speed. I believe that it is of infinitely greater importance that Palestine becomes a powerful moral and spiritual center for the whole of the Jewish people. In this direction, the rebirth of the Hebrew language must be regarded as a splendid achievement. Now must follow institutions for the development of art and science. From this point of view, we must regard as of primary importance the founding of the University, which can begin its work in Jerusalem thanks largely to the enthusiastic devotion of the Jewish physicians of America. The University already possesses a journal of science,[f] which is produced with the earnest collaboration of Jewish scientists in many fields and countries.

Palestine will not solve the Jewish question, but its development will mean the revival of the soul of the Jewish people and I am happy to have been present during this period of rebirth.

Editorial Notes

[a] The above translation differs substantially from its original English version. Originally published in Hebrew on 20 April 1923 and in German on 24 April in *Jüdische Rundschau* 33 (1923): 195–196.

[b] Shlomo Ginzberg, the son of the journalist and writer Ahad Ha'am (1856–1927), the founder of cultural Zionism, opposed to Theodor Herzl's political Zionism.

[c] The expert in question was likely the Sephardi physician Angel Pulido, whom Einstein met in Madrid in March 1923.

[d] Richard Kauffmann (1877–1958) was a German-Jewish architect and urban planner.

[e] Einstein visited the Technion on 10 and 11 February 1923.

[f] Einstein edited the first volume dedicated to mathematics and physics (see Einstein 1923c).

32. Antisemitism and Academic Youth (1923)

Document sent from Berlin, after 15 July 1923.[a] [*Collected Papers* Vol. 14, Doc. 80].

This document was to be included in an almanac on the problems faced by Eastern European Jewish students in Western Europe. The almanac was the brainchild of two medical students from the University of Königsberg, Wolf Ze'ev Rabinowitsch and Lipmann Halpern. Rabinowitsch stimulated the interest of mathematician Jakob Grommer, a collaborator of Einstein, and together they visited Einstein to ask him

> to write an article on "antisemitism at the German university" for the almanac. However, the almanac would never be printed, and Einstein's manuscript would not be published until 1934 in the collection *The World as I See It* (German title *Mein Weltbild*).

As long as we lived in the ghetto, our belonging to the Jewish people brought material difficulties and, sometimes, physical danger, but not social or emotional problems. This situation changed with emancipation, quite particularly for Jews devoting themselves to the intellectual professions.

A young Jew at school and university is subjected to the influence of a society, one he highly respects and admires, that is nationalistically tinged—a society from which he receives his intellectual nurturing, to which he feels he belongs, and by which, at the same time, he sees himself treated with a certain deprecation and aversion as an alien. Driven more by the suggestive influence[b] of this psychological superiority than by utilitarian considerations,[c] he turns his back on his people and its traditions and regards himself to completely belong among the others, trying in vain to hide from himself and those others that this relationship is not mutual. Thus takes shape the deplorable baptized Jewish privy councillor of yesterday and today. Most of the time, it is not a lack of character nor over-ambition that make him into what he is, but rather—as I said—the suggestive power of an environment superior in numbers and influence.[d] He certainly does know that many excellent sons of the Jewish people have contributed substantially toward the blossoming of Europe's culture. But have they not all, with few exceptions, done roughly the same as he has?

As with many[e] mental afflictions, here, too, the cure lies in a clear awareness of its essence and its causes. We must[f] become clearly conscious of our otherness and accept the consequences. There is no sense in trying to convince the others of our mental and intellectual equality through reasoning, for the root of their conduct is not lodged in the cerebrum. Rather, we must emancipate ourselves *socially* and mainly satisfy our social needs[g] ourselves. We should have our own student associations and exercise a courteous, but consistent reserve toward non-Jews. In doing so, let us live according to our own ways rather than copy the customs of drinking and dueling that are essentially foreign to us.

One can be the carrier of European civilization, a good citizen of a state, and simultaneously a loyal Jew, who loves his tribe and honors his fathers. If we are conscious of this and act accordingly, then the problem of antisemitism, insofar as it is of a social nature, is solved for us.

Editorial Notes

[a] An alternative translation can be found in A. Einstein. *The World as I See It*. Translated by A. Harris. New York: Friede Covici 1934, pp. 154–156.

[b] In the manuscript draft, "Influence" ("Einfluss") was originally "power" ("Macht").

[c] Einstein had originally written "perspectives" ("Gesichtspunkten") in place of "considerations" ("Rücksichten").

[d] "Influence" once again replaces "power."

[e] Einstein originally wrote "most" ("den meisten") rather than "many" ("vielen").

[f] "Must" ("müssen") replaces "can" ("können").

[g] The words "but by" ("aber aus") have been excised.

33. Review of Josef Winternitz's *Relativity Theory and Epistemology* (1923)

Deutsche Literaturzeitung 1 (1924), pp. 20–22. [*Collected Papers* Vol. 14, Doc. 149].

Josef Winternitz (1896–1952) was former editor-in-chief of the *Vorwärts* in Reichenbach and a functionary of the German Communist Party. In the preface to the book, he thanks Einstein for their conversations: "I am especially obliged to Professor Einstein for being kind enough to indicate to me his views on almost all of the problems treated here." He had submitted the manuscript to Teubner in spring 1922, and suggested that the publisher ask Einstein "whether he can recommend [the manuscript] wholeheartedly based on the discussions that [Einstein] had with the author." Teubner included a copy of the preface and the table of contents and mentioned that Philipp Frank had already praised the manuscript. In his response, Einstein declared that he agreed with Frank and that he warmly recommended the publication.

33. Relativity and Epistemology

A lack of creative power often manifests itself among philosophical authors in that they do not present their subject systematically on their own, but instead adopt their theses from others and only attempt to comment upon and criticize them. The powerful author, however, grapples with his subject matter himself, creates a systematic representation of it, and only compares the results of his own analysis against the theses of others once he has formed and implemented his own point of view through independent work.

The author of this book belongs, in this sense, among the independent, creative ones and thereby has thorough knowledge of the subject at his disposal in its physical and philosophical aspects. In his philosophical approach, which is related to Schlick and Reichenbach,[a] he is, as far as my knowledge extends, the only one who has done full justice to Kant[b] without losing his own independence to him. The following statements, which I surely may take out of context, give a notion of the author's stance toward Kant on the one hand, and toward empiricists on the other:

"For Nature is not, as must defiantly be repeated again and again to all empiricists for whom Kant did not come to life, just something that 'experience' gives, but a system that thought builds out of individual facts that are, according to its principles, themselves meaningless and unrelated."

"The characteristics of extension, of spatial arrangement and motion as we encounter them in visual perception, exactly like color qualities, are certainly not distinguished in any way from these; but when we construct a *regularity* out of mathematical concepts of magnitude not borrowed from this worldview—particularly variables that represent spatiotemporal order, on which what are to us directly givens (in Mach's meaning), conformally depend—that we arrive at a not merely registering, descriptive, and classificatory science of Nature."

"'A priori' means ... that one should recognize these principles by their indispensability in the principle of cognition."

"There follows the assertion that we can indicate at least one such principle—namely, the law of causality."[c]

Like Kant, Winternitz asserts that science is a mental construct based on *a priori* principles. That this structure of our science rests and must rest on principles that do not themselves originate from experience will surely be acknowledged without a doubt. For me, doubts only begin to arise when the dignity of those principles—that is, their indispensability—is questioned. Are those principles, at least in part, composed in such a way that science is incompatible with their mutability, or are they altogether mere conventions, such as the ordering principle for words in a lexicon? Winternitz is inclined to the former conception; I, to the latter. W.'s notions of space and motion seem to me very appropriate.

In praiseworthy succinctness of expression, the author succeeded in shedding light on the topic from all sides in 230 pages, as the following table of contents proves:

I. Introductory remarks on the tasks, methods, and limits of scientific knowledge.

II. The meaning of the relativity of space and time.

III. Absolute space in physics.

IV. The fundamental idea of Einstein's special theory.

V. The four-dimensional world.

VI. Temporal order and causal connections.

VII. Geometry and observation.

VIII. Geometry as a physical hypothesis.

IX. General relativity and gravitation.

X. Time, space, and causality in the general theory.

XI. Relativity theory in the controversy among schools of thought.

Berlin.

Editorial Notes

[a] Moritz Schlick (1882–1936) was Professor of the Philosophy of Inductive Sciences at the University of Vienna; Hans Reichenbach (1891–1953) was Privatdozent at the Technische Hochschule in Stuttgart. Both were critical of Winternitz's book.

[b] Immanuel Kant (1724–1804).

[c] The quotes are from pp. 11–14 of the first chapter. The last quotation is not a faithful rendition of the original.

34. Sound Recording for the Prussian State Library (1924)

Forschungen und Fortschritte 40 (1924), pp. 133–135. [*Collected Papers* Vol. 14, Doc. 208].

> The recording was carried out in Berlin on 6 February 1924 by Wilhelm Doegen, director of the Sound Division of the Prussian State Library and founder of its Sound Archive. On this occasion, Einstein, accompanied by his wife, was very interested in recording technology, especially the oscillograph. Although they did not talk about politics, when Doegen mentioned that in his collection the recording of the Kaiser's voice was stored right next to those of everyman, Einstein responded: "My political ideal is democracy! Everybody shall be respected as a person and no one shall be idolized!"

Ever since my youth, all my scientific striving has been directed toward penetrating deeper into the foundations of physics; philosophical aspects and requirements in a narrower sense had only a secondary influence on me. I give you here a brief report about this striving and its results up to now. I am omitting everything that has occupied me only occasionally or somewhat coincidentally.

My first problem lay in the apparent incongruity between the laws of the propagation of light, in particular Lorentz's theory,[a] and the equivalence of all inertial systems according to experience. After seven years of pondering in vain (1898–1905), the solution suddenly came to me with the thought that our concepts and laws about space and time may claim validity only insofar as they stand in clear relation to experiences, and that experience could very well lead to our modifying these concepts and laws. Through a revision of the concept of simultaneity and the character of rigid bodies, I thus arrived at the special theory of relativity, whose four-dimensional mathematical formulation, however, was only found three years later by Minkowski.[b]

In the attempt to incorporate the law of gravitation into special relativity, the conviction was impressed upon me at the end of 1907 that the spatial state of a gravitational field was identical with the state of a space free of the gravitational field if the latter is only referred to an accelerated coordinate system (in the sense of classical mechanics). This insight, briefly denoted as the principle of equivalence, in connection with the natural tendency to generalize the principle of relativity, led me to the general theory of relativity, the consistent foundation of which could only be laid at the end of 1915. The main difficulty resided in the failure of Euclidean geometry and the difficulty of still ascribing clear meaning to the laws of physics without it as a basis.

The other major problem that I have been grappling with since about 1900 is the theory of radiation and quanta. Inspired by Wien's and Planck's research, I recognized that mechanics and electrodynamics stood in unresolvable contradiction to observational facts and contributed toward creating that complex of ideas known under the name of quantum theory, which has been developed to great fruitfulness, especially by Bohr.[c] I shall probably devote the rest of my life to the fundamental clarification of this problem, slight though the prospects of reaching this goal may seem.

Editorial Notes

[a] See Herneck 1966, p. 134. The words "bezw. der Lorentzschen Theorie" are interlineated in the manuscript.

[b] Einstein 1905b (*Scientific Writings*, Doc. 4) and Minkowski 1908.

[c] Wilhelm Wien (1864–1928) was Professor of Physics at the University of Munich; Max Planck; Niels Bohr.

35. The Compton Experiment: Does Science Exist for Its Own Sake? (1924)

Berliner Tageblatt (20 April 1924), p. 1. [*Collected Papers* Vol. 14, Doc. 236].

In 1923, the American physicist Arthur Holly Compton (1892–1962) carried out his famous scattering experiment of X-rays by individual electrons. It was the first direct and incontrovertible proof for Einstein's light

> quantum theory. The dual particle and wave nature of light became the most significant research topic for many physicists, including Einstein, who encouraged various colleagues to carry out further experiments.

Does science exist for its own sake? Depending on how it is meant, this question must be answered in the affirmative and negative with equal determination. The scientist must be of service to science for its own sake, without regard for any practical results. Otherwise, science withers in that it loses sight of the general perspective; nor would it fulfill its major educational mission, which is to awaken and maintain the striving for causal insights in everyone. This grand mission of defending one of humanity's most valuable ideals also shows, however, the extent to which science is *not* allowed to exist for its own sake. Scientists, taken together, can be compared with one organ of the body of mankind that is nurtured by its blood, producing a vitally important secretion that must be conveyed to all its parts if it is not to waste away. This should not be understood as having to cram everyone full of learning and detailed knowledge, as unfortunately often happens to excess in schools; nor should the broader public make decisions on scientific issues. But every thoughtful person must be allowed to clearly experience the major scientific problems of his day, even when his social position does not permit him to devote a substantial part of his time and energy to pondering theoretical questions. Seen from the social perspective, only by doing justice to this important mission does science earn the right to exist.

In what follows, I would like to give an account, from this perspective, of an important experiment regarding light—or *electromagnetic radiation*—performed about one year ago by the American physicist Compton.[a] To recognize the full significance of this experiment, we must bring to mind the highly peculiar situation in which the theory of radiation finds itself.

Up to the first half of the nineteenth century, optics was mainly preoccupied with the reflection and refraction of light (concave mirrors, lens systems). Until then, one steadfastly held on to Newton's corpuscular theory and emission theory of light, according to which light was supposed to be composed of corpuscles that move uniformly through a homogeneous environment in straight lines, but generally experience a sudden change of direction when reaching surfaces. By using this fundamental notion, a quite complete theory was created for almost all phenomena known up to then, particularly those involving the telescope and the microscope.

But some one hundred years ago, when one became better acquainted with the phenomena of interference and diffraction (along with the polarization of light), it became necessary to replace Newton's fundamental assumption about the nature of light with the entirely different one of the *wave theory*, which had already been postulated by Huygens some one and a half centuries earlier. According to it, light would consist of elastic waves propagating outward through space (i.e., the ether) in all directions, similarly to the propagation of waves in two dimensions at the surface of water when set into oscillatory motion from a single point on that surface. It was only this theory that succeeded in clarifying why a ray of light propagates in all directions after passing through a very narrow opening. Only this theory can explain why there are dark spots in interference and diffraction effects in the middle of an area of space flooded with light, or why many bundles of light can locally cancel one another's effect. This wave theory was able to represent the complicated phenomena of diffraction and interference with downright astronomical precision, thus cementing the conviction of its correctness.

The wave theory was modified and, at the same time, given even firmer grounding through the investigations of Faraday and Maxwell,[b] according to which the wave field was divested of its mechanical character. Maxwell's theory of electricity and magnetism also encompasses the wave theory of light without altering any of its formal content. This theory establishes quantitative relationships between the optical and electrical behavior of empty space, as well as of ponderable bodies, and reduces the number of mutually independent hypotheses upon which wave optics is based. Thus, around the turn of the century, physics seemed to possess a foundation that, it was hoped, could forever serve as a basis for all its subfields, including mechanics.

But it happened otherwise. From Planck's[c] papers on the law governing the radiation emitted by hot bodies, it emerged that the theory was unable to explain this law, nor could one explain the general finding that the effects of the radiation depended qualitatively not on the latter's intensity, but only on its color. This is highly paradoxical and seems incompatible with the fundamental idea of the wave theory. Imagine huge waves generated somewhere on the open sea, spreading out in all directions from the center of excitation. Naturally, the wave peaks generated this way will not be as high the farther away they have propagated from the center of agitation. Now imagine equally sized ships had been distributed over that area of the ocean before the aforementioned waves were to be generated. What will happen when the waves start to mount? The ships positioned close to the source will keel over or be smashed, but no harm will come to the ships sufficiently far away from it; those ships will just rock harmlessly. One would think molecules struck by radiation would respond analogously to ships struck by ocean waves. Whether molecules are changed chemically ought to depend not just on the wavelength, but also on the intensity of the acting radiation; this is, in fact, what experience does not confirm.

Given this failure of the general theory, the *light quantum hypothesis*[d] was resorted to. Notwithstanding all due respect for the wave theory, a working hypothesis gained ground that radiation behaves in energetic respects as though it were composed of energy projectiles whose energy magnitude depends only on the radiation's frequency (color) and is proportional to it. Newton's corpuscular theory of light is being revived, even though it has failed completely with respect to the essentially geometrical properties of light.

Thus, there are now two theories of light, both indispensable and—as must be conceded today, despite twenty years of immense efforts by theoretical physicists—without any logical connection. Quantum theory has made Bohr's theory of the atom possible and has explained so many facts that there must be a great amount of truth to it. Given this situation, the question of the highest importance is: To what extent should a projectile-like quality be assigned to light corpuscles or quanta?

A projectile will transfer to the obstacle it has struck not only energy, but also an impulse in its direction of motion. Is it likewise for light quanta? Based on theoretical considerations, the response to this question has long been "yes," and Compton's experiment has demonstrated the correctness of this foresight. For a thorough grasp of this experimental method, one must take a closer look at the mechanism of a process known as "scattering" upon which is based, e.g., the blue color of the sky.

If an electromagnetic wave meets a free-charged elementary particle (electron) or one bound to an atom, then the particle is set into oscillatory motion by the wave's alternating electric fields. In turn, it thereby emits waves of the same frequency, whose energy is taken from the original wave, in every direction (like an antenna in wireless telegraphy). The effect of this is that the light irradiated through the medium containing such particles is scattered (at least in part) in every direction—that is, the more intense, the shorter the wavelength of the primary light. This is how scattering is interpreted according to the wave theory.

This process is interpreted differently according to quantum theory. According to it, a light quantum collides with the electron, whereupon it changes its direction and, at the same time, imparts a velocity to the electron. The kinetic energy transferred onto the electron in this collision must therefore be taken from the colliding quantum so that the scattered quantum therefore must have a lower energy—expressed, according to the wave theory, as a lower frequency—than the impinging radiation. More careful consideration shows that this frequency diminution of the scattered radiation is exactly calculable. The percentage change in frequency comes out very small for visible light; for harder Roentgen radiation, however,—which is, of course, nothing but very short-waved light—it is quite substantial. Compton has now found that the Roentgen light scattered by suitable substances does, in fact, manifest the change in frequency demanded by quantum theory (but not by the wave theory). This can be explained as follows: According to the Rutherford-Bohr theory, every atom possesses several electrons so loosely bound to it that they react to the quantum collision of the Roentgen ray roughly as though they were freely mobile. Hence, the above consideration applies to the light scattered by this collision. The positive outcome of Compton's experiment proves that radiation acts as though it were composed of discrete projectiles of energy—not just as regards energy transfer, but also as regards the collision effects.

Editorial Notes

[a] The results referred to were published as Compton 1923, dated 13 December 1922, published May 1923.
[b] Michael Faraday (1791–1867) and James Clerk Maxwell (1831–1879).
[c] Max Planck (1858–1947).
[d] Introduced in Einstein 1905a (*Scientific Writings*, Doc. 2).

36. On the League of Nations (1924)

Frankfurter Zeitung 69, no. 646 (29 August 1924), p. 1. [*Collected Papers* Vol. 14, Doc. 314].

Einstein was involved in discussions on establishing a postwar league of nations as early as 1918. In May 1922, he was invited to join the League of Nations' International Committee on Intellectual Cooperation (ICIC), but resigned shortly thereafter, most likely because of the French-Belgian occupation of the Ruhr. However, following the victory of the left in the French parliamentary elections in May 1924, Einstein reflected that his resignation had been caused by a moment of demoralization and rejoined the committee.

Having returned from the convention in Geneva of the League of Nations Committee on Intellectual Cooperation,[a] I have the desire to say something to the German public about the impressions I formed there. The goal of this Committee is to create or promote endeavors for the benefit of science and intellectual life, generally in the individual countries, through international cooperation and thus to draw cultural spheres separated by language and traditions closer together. Utopian plans are not the intention. Rather, more modest yet fruitful *small-scale missions* were immediately initiated: on the international organization of scientific reporting, exchange of publications, protection of intellectual property, exchanges of professors and students, between different countries, etc. Scientific reporting was the problem most extensively supported.

All these things are naturally of little interest to the general public. Of great significance to every man, however, is the question: What stance should Germans—what stance should the *German Reich*—take toward the *League of Nations*? About this important question, I shall say something here about my personal impressions. In all the issues that were addressed, all members endeavored to preserve a genuinely international character for the institution. Germany is viewed, on all issues, as though it were already a member of the League of Nations. Although there may also be a certain unconscious preponderance of the French viewpoint—which, given the history of the League of Nations' establishment and considering the hitherto lack of participation in the League of Nations by important nations,[b] is not otherwise conceivable at all—I was nevertheless happy to be able to perceive an *honest desire for objectivity*. Out of this spirit, only good can spring, and I may surely express the conviction that the League of Nations is a suitable instrument to gradually encourage an internal, and thereby also an external, *restoration to health for Europe*. The experience I had was that, with sensible reasons and unconditional openness, much can be achieved.

I also had occasion to discuss with well-informed people, particularly with French people,[c] the question of *Germany's entry* into the League of Nations. All, without exception, believed that Germany would receive a permanent seat in the Council of the League of Nations, like other large states. All were also of the opinion that, upon the successful conclusion of the London Conference, Germany should join the League of Nations. I myself am definitely also of this conviction, and believe I should add that other conditions or reservations should not be raised by Germany for it joining the League of Nations. With trust, one attracts trust, and *without* trust, fruitful cooperation is not at all possible. May all friends of international reconciliation in Germany use their influence toward disallowing the present favorable situation for Germany's entry into the League of Nations to pass by unused.

Editorial Notes

[a] The fourth session of the International Committee on Intellectual Cooperation took place 25–29 July. Before that, Einstein had attended the sessions of the sub-committee on bibliography, which met on 23–24 July.

[b] The United States and the Soviet Union had also not yet joined (see Hastedt 2004).

[c] Among them were Henri Bergson, Paul Langevin, Marie Curie, and Julien Luchaire.

37. Review of Alfred C. Elsbach's *Kant and Einstein* (1924)

Deutsche Literaturzeitung 1 (20 September 1924), cols. 1685–1692.
[*Collected Papers* Vol. 14, Doc. 321].

> The bicentennial of Immanuel Kant's birth was marked on 22 April 1924. Einstein was asked to review two books on (neo-)Kantianism and its relationship to relativity theory. In the course of these reviews, Einstein gave a detailed criticism of Kantianism. In particular, he explained why, in his view, the recent attempts by neo-Kantians to reconcile Kant with general relativity are ultimately futile.

Elsbach's book[1] distinguishes itself by its clarity and precision in how it forms concepts, by its honesty and thoroughness—the latter a bit too much, even. It first delivers an account of the epistemology of the Neo-Kantians (Cohen, Natorp, Cassirer)[a] and compares this doctrine to positivism on the one hand, and to realism on the other. This epistemological system had never been so clear to me as it became through this book. According to this system, reality and truth are nothing other than being subject to one unitary temporal-spatial-causal framework (chaps. 1 and 2). Thus, the problem of experienced reality (as opposed to merely dreamed experience) and thing reality (e.g., sun, hydrogen atom) is easily dealt with. Regarding these fundamental questions, the later thinkers mentioned have essentially adhered to Kant's opinion. The reviewer must confess that, although he found the argument on this as presented in the book logical and clear, it did not convince him. Doesn't an experienced reality that one senses directly exist, and is that indirectly also the source of what science denotes as "real"? Aren't, furthermore, the realists right, after all, along with all scientists (who don't happen to be philosophizing) when, by the highly astounding possibility to arrange experiences within a (temporal/spatial/causal) conceptual framework, they allow themselves to

---

[1] Alfred C. Elsbach [private lecturer at the University of Utrecht], *Kant and Einstein. Analyses on the Relation between Modern Epistemology and Relativity Theory.* Berlin and Leipzig: Walter de Gruyter & Co., 1924. VIII & 374 pp. 8°. 8 M[arks], bound 9.20 M[arks].

assume that real, existing things are independent of their own thinking and being? Isn't the incomprehensibility of being able to build a conceptual framework that connects experiences just as painful to the idealistic philosopher (from the logician's point of view) as accepting the reality hypothesis of the realistic philosopher and of the non-philosophizing person (and animal)? Is there indeed a difference at all between assuming that the totality of observations or experiences permits a logical conceptual framework that connects them with each other, versus accepting the reality hypothesis? That we can connect experiences with one another and order them through conceptual frameworks, that we can base correct predictions on them, is a miracle that is acknowledged by the reality hypothesis and cannot be eliminated from the world by any philosophical acumen.

In the third chapter, the Aristotelian theory of concepts (the "category" theory of concepts) is opposed and, following Cassirer closely, a new theory of concepts ("serial principle") is presented. One will surely, definitely have to concur with the author that concepts cannot be logically derived from sensory experiences through some method (abstraction), even though sensory experiences do psychologically point out the way for the formation of concepts. On the other hand, however, I don't think it has been successfully shown that the Aristotelian principle of categories does not correctly reflect the relationship between concepts. It definitely does not signify a contradiction if such a notion leads to the more general concept being poorer in some respects; it is simply poorer in characteristic traits, but richer for it with respect to the subsumed individual cases (concepts). The weakness in the deduction seems, to me, to be that no sharp distinction is drawn between a single sensory experience on the one hand, and the concept on the other. Logic generally only relates to the concepts and their mutual relations; and in this area, the Aristotelian approach does not appear to me persuasively refuted. On the contrary, I am certain that the relations between concepts and single sensory experiences cannot be grasped otherwise than intuitively. One can, for example, imagine the entire conceptual framework of a natural science be preserved, while the relations between its concepts and the sensory experiences fall into oblivion; this science would then not exist anymore in the sense of a natural science. It is different in mathematics, as usually conceived by mathematicians, where the relationship to experiences is forgone.

In general, it seems to be the weakness of positivists that the logical independence of concepts from sensory experiences is not clearly brought to the fore, while idealists like to forget that the dignity of concepts ultimately rests solely on the fact that they connect sensory experiences.

Chapter 4 deals with the "Structure of Theoretical Physics and the Task of Theoretical Philosophy." The content of this chapter is indicated by the author himself in the sentence: "In physics, by means of independently constructed concepts and laws, the manifoldness and heterogeneousness of individual observations are ordered and united to a firmly interrelated (and measurable) whole."[b] (The words "and measurable" are enclosed in parentheses by me as not logical, because the whole, composed of concepts and laws, cannot be called "measurable.") Most people would probably approve of the content of this chapter; it contains polemics against the purely positivist interpretation that are worth reading. I would just like to add the wish that the position of "conventionalism" (which probably refers to Poincaré's epistemological position on physics) be given more attention in the second edition.

In the fifth chapter, "Summary of Critical Philosophy and Its Relation to Relativity Theory," one first encounters a profoundly thought-through characterization of Kant's teachings as far it has been adopted by neo-Kantians. Then on pp. 281–301,[c] the author arrives at a point of crucial importance to an assessment of the relation between Kantian teachings and relativity theory—indeed, for an assessment of Kantian teachings overall. There is the thesis (not cited verbatim): Critical idealism can neither contradict science, nor anticipate any of its results. For this, philosophical scholarship accepts the essence of concepts and relations of science as the material it must deal with, just as natural science accepts the experimental data from the observation of nature. From this, the author concludes that the individual theses of critical epistemology cannot raise claim to being valid independently of the *respective state* of science at a given time. I think, here, that the author would be in agreement with neither Mohammed nor the Prophet. It is my conviction that it was the goal of Kant and all Kantians to locate those *a priori* (i.e., not deducible from experience) concepts and relations that *must* be the foundation of *every* science because, without them, science is not conceivable at all. Assuming this goal really is achievable and has been achieved, those *a priori* elements could not come into conflict with any future reasonable physical theory. Kant considered this goal achievable and believed he had achieved it. If one does not consider this goal achievable, one probably should not call oneself a "Kantian."

Until some time ago, it could be regarded as possible that Kant's system of *a priori* concepts and norms really could withstand the test of time. This was defensible as long as the content of later science held to be confirmed[2] did not violate those norms. This case occurred indisputably only with the theory of relativity. However, if one would rather not assert that relativity theory goes against reason, they cannot retain the *a priori* concepts and norms of Kant's system.

For starters, this at least does not exclude the retention of Kant's way of *posing the problem*, as Cassirer does, for instance. I am even of the opinion that this standpoint cannot be strictly refuted by any scientific development; for one will always be able to say that critical philosophers had hitherto erred in setting up the *a priori* elements, and one will always be able to set up a system of *a priori* elements that does not conflict with a given physical system. Surely, I may briefly indicate why I do not find this standpoint natural. Let a physical theory consist of the parts (elements) A, B, C, D, which together form a logical whole correctly connecting the pertinent experiments (sensory experiences). Then the tendency is that less than all four elements—e.g., A, B, D—still say nothing about the experiences without C, no more so A, B, C without D. One is then free to regard three of these elements— e.g., A, B, C—as *a priori* and only D as empirically determined. What always remains unsatisfactory in this is the *arbitrariness of the choice* of elements to be designated as *a priori*, even disregarding the fact that the theory could be replaced at some point by another theory that substitutes some of these elements (or all four of them) with others. However, one might think that, through direct analysis of human reason or thought, we would be in a position to recognize elements that would have to be present in *any* theory. But most researchers would probably agree that we lack a method for recognizing such elements,

---

[2]To refute Kant's system, it actually suffices to indicate a logically *conceivable* theory (corresponding to *conceivable* observational material) that conflicts with Kantian norms. Whether non-Euclidean geometries accomplished this remains controversial.

even if one were inclined to believe in their existence. Or should one imagine that the search for *a priori* elements was a kind of asymptotic process that advances along with the development of science?

Elsbach is fully aware of these problems and limits epistemology to the analytic examination of physical science as it presently exists. He rightfully admonishes this reviewer, as well as H. A. Lorentz, for having asserted "epistemological aspects" in the decision between physical theories where one ought just to have spoken of "logical arguments."

In chapter 6, "The Logical Standards of Judgment," it is set forth, with great acumen and clarity, that one should not set two theories in parallel by comparing their claims, but rather by juxtaposing the arguments supposedly supporting these claims. This analysis very much deserves to be noticed and taken to heart. Much has been perpetrated on this point, particularly in assessments of relativity theory from philosophical points of view.

Chapters 7 and 8 treat Cohen's and Natorp's views about time and space. I have not been able to make much out of Cohen's elaborations; Natorp's way of arguing is transparent; but in my view, it bypasses the essentials. Elsbach draws from his account the following arguments:

1. Wanting to determine the structure of space by measurement is based on circular reasoning; for in order to perform the measurement, Euclidean geometry is already presupposed.
2. If a deviation from Euclidean geometry were established by experiment, this would not necessarily need to lead to an alteration of the geometrical laws, as the agreement between theory and experience can also be produced by an alteration of the physical laws.
3. Even if non-Euclidean geometry were introduced, that would be based only on technical considerations about the simplicity of the calculation, whereby the real nature of space cannot be challenged. Nothing about reality can be found out by mere conversion. Observation and experiment decide on reality.
4. The metric of real space cannot be determined by experiment because space is not real.

How one responds to these theses depends on whether one grants reality to the practically rigid body. If so, then the concept of distance corresponds to something that can be experienced. Geometry, then, contains statements about possible experiments; it is a physical science directly subject to experimental testing (standpoint A). If no reality is granted to the practically rigid measuring-body, then geometry alone holds no statements about experiences (experiments); rather, only geometry together with the physical sciences (standpoint B). Physics has hitherto always used the simpler standpoint A and largely owes its productivity to it; it uses it in all its measurements. Seen from this standpoint, all of Natorp's aforementioned assertions are incorrect; this surely does not need to be elaborated further in detail. If, however, one assumes standpoint B,—which, at the current state of science, should be regarded as overly cautious—then geometry on its own is not experimentally testable. Then geometrical measurements don't exist at all. However, one does not have to speak of the "ideality of space" as a result. "Ideality" is ascribable to all concepts; this is no more and no less true with respect to space and time. Definite attribution to sensory experiences is possible only for a complete scientific system of concepts. In my view, Kant influenced developments unfavorably in that he granted a special place to spatio-temporal concepts and their relations compared to other concepts.

Seen from standpoint B, though, the choice of geometric concepts and relations is determined only by reasons of simplicity and practicality. Under no condition does the choice of a non-Euclidean geometry presuppose Euclidean geometry as its basis. But then nothing can be discerned empirically about the metric of space; not because "space is not real," but rather because, with this choice of standpoint, geometry is not a complete physical system of concepts, only part of one.

Elsbach certainly does not agree with this argument. Contrary to other representatives of critical epistemology, he advocates the view that the latter could neither contradict a physical theory, nor anticipate one. In the last two chapters of the book,—which endorse Cassirer but, in essential parts, contain the author's independent arguments—criticism is directed against representatives of critical philosophy who had earlier worked on the relation between philosophy and relativity theory. Based on philosophical arguments, they arrived partly at a rejection of relativity theory, partly at the result that the latter was merely an implementation of Kantian thought. In both cases, the error supposedly arose out of different usages of words in philosophy and physics. A philosopher calls space "absolute" in the sense that spatial concepts do not emerge logically out of observation; rather, they are taken as the foundation of science. On the other hand, when a physicist calls space "absolute," he means so "in its properties, independent and co-determinate of the phenomena." When a philosopher speaks of the ideality of space, he does not think of the equivalence of the coordinate systems, nor can the latter be concluded from the former. In what follows, the empiricist theory of space is rejected, following Cassirer; this reviewer must comment, however, that not enough light is shed on the relation between the logical system and experiences. But here, too, the reader finds much of merit. In particular, I would like to mention that criticism was rightly aimed against one statement by the reviewer: that a concept should only be permissible in physics when it can be established that it applies in concrete cases of observation; the objection is that, in general, experiences must correspond not merely to individual concepts, but to the system as a whole.

Elsbach's book offers much neat and honest work of thought and deserves to be studied by those interested in the relationship between philosophy and the natural sciences.

Editorial Notes

[a] Hermann Cohen (1842–1918), Professor of Philosophy at the University of Marburg; Paul Natorp (1854–1924), Professor of Philosophy and Pedagogy at the University of Marburg; Ernst Cassirer (1874–1945), Chair of Philosophy at University of Hamburg. All were prominent representatives of the so-called Marburg school of neo-Kantian philosophy.

[b] The quotation is from p. 123. Quotation marks are missing after "Ganzen."

[c] The page references should be to pp. 180–201.

38. Non-Euclidean Geometry and Physics (1925)

Die Neue Rundschau 36, no. 1 (1925), pp. 16–20. [*Collected Papers* Vol. 14, Doc. 220].

> Einstein completed this article in March 1924, several months before reviewing Elsbach's book *Kant and Einstein* (see Doc. 37). He had earlier published discussions on the historical foundations of geometry. In this popular exposition, Einstein displays his ambivalence regarding the concept of a rigid body, which is indispensable for the current state of knowledge (i.e., general relativity), but problematic as something whose existence has to be assumed rather than derived as holding approximately.

Thinking about the connection between non-Euclidean geometry and physics necessarily leads to the question of the connection between geometry and physics in general. I will consider this latter one first, and shall try to avoid controversial philosophical questions as much as possible.

In the most ancient eras, geometry was no doubt a semi-empirical science, a kind of primitive physics. A point was a body whose extension was ignored; a straight line was roughly defined by points that could be brought to optically coincide in the direction of one's gaze or by using a taut thread. Indeed, the concepts involved, as is always the case with concepts, did not originate directly from experience,— that is, they were not logically determined by it—but were nevertheless seen as directly related to the objects of experience. Propositions concerning points, lines, and the equality of lengths or angles were, to this mode of understanding, equivalent to propositions concerning certain experiences with natural objects.

Geometry as thus understood became a mathematical science after the recognition that most of its propositions could be derived by a purely logical path from fewer propositions,—the so-called axioms— as mathematics comprises every science concerning itself exclusively with logical relations between given entities according to given rules. The derivation of these relations then became the sole interest, as the independent construct of a logical system—uninfluenced by the uncertain external experience dependent on chance—has always possessed an irresistible charm for the human spirit.

The basic concepts—point, line, distance, etc.—and the so-called axioms remained the only ones logically irreducible in the geometrical system, or as witnesses of its empirical origins. One attempted to confine the number of these logically irreducible fundamental concepts and axioms to a minimum. The attempt to lift all of geometry out of the murky sphere of the empirical has now led imperceptibly to a mental reorientation somewhat analogous to the promotion of revered heroes of antiquity to gods. One gradually came to consider the fundamental concepts and axioms of geometry as "evident"—i.e., as objects and qualities of perception given with the human intellect itself—in such a way that inner objects of intuition correspond to the fundamental concepts of geometry and that negating a geometrical axiom is, in fact, actually inconceivable. From this perspective, the applicability of these fundamentals to real objects becomes a problem—precisely the problem from which Kant's conception of space emerged, we may add.

A second motive for the separation of geometry from its empirical foundation was furnished by physics. According to its more refined concepts of the nature of solid bodies and those of light, there exist no natural objects corresponding exactly in their properties to the foundations of Euclidean geometry. The solid body is not absolutely rigid, and the ray of light does not strictly embody a straight line, or even any one-dimensional structure. According to modern science, geometry does not correspond to any experiences, strictly speaking, but only together with mechanics, optics, etc. Since geometry must moreover precede physics,—the latter's laws cannot be expressed without the former—geometry appears to logically precede every experience and experiential science. Thus, at the beginning of the nineteenth century, it came about that the foundations of Euclidean geometry appeared as something absolutely immutable not only to mathematicians and philosophers, but to physicists as well.

One may add that, during the whole nineteenth century, this situation presented itself even more simply, schematically, and rigidly to the physicists who did not direct their attention particularly to epistemology. Their unconsciously maintained position corresponded to two propositions: The concepts and postulates of Euclidean geometry were evident. Under certain limitations, ruled solids realized the geometrical concept of length, and light rays that of the straight line.

Overcoming this situation was difficult work and required about a century. Strangely enough, it had its origin in purely mathematical investigations, existing long before the garments of Euclidean geometry became too tight for physics. One of the mathematicians' tasks was to establish geometry on as few axioms as possible. Now among the Euclidean axioms was one that seemed, to the mathematicians, less immediately obvious than the others and which they, therefore, tried for a long time to reduce to—i.e., to prove from—the others. This was the so-called axiom of parallels. Since all attempts to provide such a proof failed, the notion that such a proof is impossible had gradually developed—that is, that this axiom was independent of the rest. This could be proven by erecting a consistent logical structure that differed from Euclidean geometry in that—and only in that—it replaced the axiom of parallels with another. It remains the eternal achievement of Lobachevsky on the one hand, and of the Bolyai father and son[a] on the other, to independently conceive and convincingly accomplish this idea

The conviction that there exist still other geometries just as logically justifiable as Euclidean geometry thus became established among mathematicians, and this obviously led to the question whether no other geometry besides the Euclidean one could be the necessary basis of physics. The question was also posed in the specific form: Is it only Euclidean geometry that is valid in the physical world, or could another?

There has been much argument about whether this last question is meaningful. To see this clearly, one must logically adopt one of the two following viewpoints. In the first, one must assume the geometrical "body" is principally realized in the natural solids only when certain rules concerning temperature, mechanical stresses, etc., are observed. This is the practical physicist's perspective. A natural object, then, corresponds to geometrical "distance," and all propositions of geometry take on the character of statements about real bodies. Helmholtz[b] represented this perspective especially clearly, and one may add that the formulation of the theory of relativity would have been all but impossible without him.

Alternatively, one may deny in principle the existence of objects corresponding to the basic concepts of geometry. Then, geometry alone contains no statements about objects of reality unless it is paired with physics. This standpoint, which may be the more comprehensive for the systematic presentation of a complete physics, was especially clearly represented by Poincaré. From this perspective, the whole content of geometry is conventional; whether a geometry is to be preferred depends on how "simple" a physics in harmony with experience may be constructed with it.

Here, we shall choose the first perspective as the one better suited to the present state of our knowledge. From its vantage, our question as to the validity or invalidity of Euclidean geometry has a clear meaning. Euclidean geometry and geometry in general retain, as before, the character of a mathematical science inasmuch as the derivation of their theorems from the axioms remains a purely logical one; but at the same time, they become a physical science inasmuch as the axioms contain assertions about natural objects whose validity is only deducible by experiment. However, we must always be conscious that the idealization contained in the fiction of rigid (measuring) bodies actually existing as objects in nature may one day prove unjustified, or only justified with respect to certain natural phenomena. The general theory of relativity has already proven this concept is unjustified for spaces of small extension in an astronomical sense. The theory of elementary electrical quanta may prove the concept unjustified for an extension of atomic orders of magnitude. Riemann already realized both were possible.

Riemann's contribution to the development of our ideas about the relationship between geometry and physics is twofold. First, he developed spherical-elliptical geometry, which is the counterpart to Lobachevsky's hyperbolic geometry. He thus presented, for the first time, the possibility that geometric space could be of finite extent in a metrical sense. This idea was soon understood and has led to the oft-considered question of whether physical space might not be finite.

Second, Riemann had the daring thought of constructing a geometry incomparably more general than that of Euclid or the non-Euclidean geometries, in the narrower sense. He thus created "Riemannian geometry," which (like the non-Euclidean geometries in the narrower sense) is Euclidean only in the infinitely small; it represents the application of Gaussian surface theory to a continuum of arbitrarily many dimensions. According to this more general geometry, the metric properties of space—in particular, the possibility of stacking infinitely many, infinitely small rigid bodies over finite regions—are not determined by the axioms of geometry alone. Instead of letting himself be discouraged by this realization and concluding from it that his system was physically meaningless, Riemann had the daring thought that the geometric behavior of bodies might be determined by physical realities, particularly forces. He thus arrived, through purely mathematical speculation, at the idea of the inseparability of geometry and physics—an idea that actually prevailed seventy years later in the general theory of relativity, through which geometry and the theory of gravitation were fused into one.

After Levi-Civita brought Riemannian geometry to its simplest form by introducing the concept of infinitesimal parallel-shift,[c] Weyl and Eddington further generalized Riemannian geometry in the hope that the electromagnetic laws would also find a place in the thus-expanded conceptual system.[d] Whatever may be the result of these endeavors, one may reasonably say, without regard, that the ideas that developed from non-Euclidean geometry have proven themselves eminently fruitful in modern theoretical physics.

Editorial Notes

[a] Farkas Bolyai (1775–1856), his son János Bolyai (1802–1860), and Nikolai Lobachevsky (1792–1856).

[b] See Helmholtz 1868 and Helmholtz 1884. Einstein read the latter in his youth as part of the readings of the Olympia Academy.

[c] Levi-Civita 1917.

[d] See Weyl 1918a, Weyl 1919, Weyl 1923, and Eddington 1921.

39. Introductory Letter to *Letters from Russian Prisons* (1925)

A. Berkman, ed. *Letters from Russian Prisons*. New York: A. & C. Boni (1925), p. 7.
[*Collected Papers* Vol. 14, Doc. 423].

> This document is one of twenty-two prefatory letters by "Celebrated Intellectuals" that were solicited by Isaac Don Levine (1892–1981), a Russian-born American journalist and writer, on behalf of the International Committee for Political Prisoners for an anthology compiled by Alexander Berkman (1870–1936), a Russian-American anarchist. The introduction to the volume was written by Roger N. Baldwin on behalf of the committee. Other contributors to the prefatory letters were, among others, Karel Capek, Knut Hamsun, Gerhart Hauptmann, Selma Lagerlöf, Sinclair Lewis, Thomas Mann, Romain Rolland, Bertrand Russell, Arthur Schnitzler, Upton Sinclair, and H. G. Wells. The prisoners' letters had been smuggled out of the Soviet Union.

You of those who read these letters and live under a well-ordered government are not to believe that the people surrounding you are better than or different from those carrying on a regime of frightfulness in Russia. You will contemplate this tragedy of frightfulness in Russia with horror. You will contemplate this tragedy of human history, in which one murders from fear of being murdered, with horror. And it is just the best, the most altruistic individuals who are being tortured and slaughtered—but not in Russia alone—because they are feared as a potential political force.

All serious people should be obliged to the editor of these documents. Their publication should contribute to affecting a change in this terrible state of affairs; for the powers that be in Russia, if they desire to continue their attempt to acquire moral standing among the civilized peoples, will be compelled to alter their methods after the appearance of these letters in print. They will lose the last shred of sympathy they now enjoy if they cannot demonstrate through a great and courageous act of liberation that they do not require this bloody terror to put their political ideas in force.

40. Why Zionism/A Message (1925)

Jüdische Rundschau 30, no. 14 (17 February 1925), pp. 129–130. [*Collected Papers* Vol. 14, Doc. 438].

> Einstein began working on behalf of the establishment of the Hebrew University in 1921. During his visit to British Mandate Palestine in early 1923, he was greatly impressed with developments in the Jewish community, both in the towns and in the agricultural settlements. He was well aware of the diversity of opinions about Zionism among Jews, both in Europe and in the United States.

In my opinion, the presence of various nationalities—and, consequently, of mutually competing nationalisms, inside as well as outside Europe—must be regarded as a misfortune. Is it necessary to repeat that a certain nationalism poses a real danger to peace and constitutes an inexhaustible source of injustices and misery?

On the other hand, there is one fact that cannot be denied: Virtually everywhere, Jews are treated as members of a clearly characterized national group. This may seem regrettable to Jews who consider themselves, as I do, to be part solely of the nation of mankind, an admittedly hard to attain, yet possible ideal.

Nietzsche said that one of the peculiarities of the Jewish people consists of knowing and implementing "the subtle utility of misfortune."[a]

The Jews must make use of their nationality as well. May they do this for the benefit of the common good!

It is their duty to develop, of their own accord, those virtues and beliefs indispensable to one who wishes to serve humanity. Since the disappearance of Jewish nationality seems impossible, at least for the moment, the Israelites must justify their existence. And for this reason, they must regain awareness of the human value they represent without ridiculous arrogance. By studying their past, by a better understanding of their spirit in accordance with their race, they must acquaint themselves anew with the mission they must fulfill.

Zionism can help them in this way, as it recalls to their minds a past full of glory and pain and opens their eyes to a healthier, more dignified future. It guides them to view themselves less negatively and take courage. It gives back to them the moral energy that will allow them to live and act in dignity. It frees their souls of the inexcusable feeling of excessive modesty, which can only oppress them and make them unproductive. Finally, it reminds them that the centuries of commonly experienced suffering impose obligations of solidarity.

Enlivened by the mystique of Zionism, they will perhaps finally be able to fulfill the responsibilities that are their duty and that the honorable endeavor and united labors of Israel demand. Only at this price will those who view it as impossible for people of all nations to live without bonds of brotherhood be able to spread words of wisdom—more necessary today than ever before—in a useful way.

Therefore, I cannot regard the Zionist movement as an excrescence on the tree with poisonous fumes that destroy the joy of thinking and life.

A Jew who strives to steep his mind in the humanist ideal can declare himself a Zionist without contradiction.

One must be thankful to Zionism for the fact it is the only movement that has given many Jews justifiable pride, has given back to a race in despair the necessary faith, and—if I may express myself thus—has given a weakened nation new nourishment.

Zionism is in the process of creating a center of Jewish intellectual life in Palestine, and one will have to be forever grateful to its leaders for that. The existence of this moral homeland will have the result, I hope, of giving a nation that does not yet deserve to perish a boost in vitality. I have been able to observe the first manifestations of this moral restoration.[b]

That is why I believe that I can assert that Zionism, a seemingly nationalistic movement, is, in the final analysis, an important service to humanity.

Editorial Notes

[a] The Nietzsche quotation is an allusion to Nietzsche 2003, aphorism 205: "On the people of Israel ... Every Jew has in the history of his fathers and forefathers a treasure trove of examples of the coldest self-possession and perseverance in terrible circumstances, of the subtlest circumvention and exploitation of misfortune and coincidence; their bravery under the guise of wretched subjugation, their heroism in *spernere se sperni* (feeling contempt for being despised) exceeds the virtues of all the saints."

[b] Einstein visited Palestine in February 1923.

41. Pan-Europe (1925)

Das Junge Japan: Deutsch-Japanische Monatschrift 1, no. 10 (April 1925), pp. 369–372. [*Collected Papers* Vol. 14, Doc. 463].

> Einstein was invited to South America by wealthy Jewish communities and several universities. He embarked on a three-month tour of Argentina, Uruguay, and Brazil in March 1925. During his visits, he gave several lectures and published several articles and papers. This article was published in Spanish on the day of his arrival in Montevideo. Its pacifist ideals inflamed part of the German community in Argentina.

Irrespective of its political divisions, Europe forms a spiritual union going back to ancient times. An American, observing Europe in its intellectual and economic expressions, grasps particularly well that our continent is leading a common and international life despite its national peculiarities and selfishness; consequently, the belligerent entanglements must appear to him as something foolhardy, one would like to say suicidal. The history of European intellectual life shows that the different nations of this part of the Earth have always mutually stimulated each other. The powerful intellectual forces that influenced scientific and artistic life in various forms in the different centuries were not impeded by Europe's political boundaries. In every epoch, the multiplicity of national manifestations and their individual peculiarities found its epitome in the European idea, which continually renews itself.

It is, therefore, only too natural that those leading minds of Europe, who were very well aware of the bond existing between cultural life in their own countries and that of other countries of the continent, recognized the unreasonableness of situations that threatened and destroyed the highest values of this part of the world and had been born of an atavism difficult for the non-European to understand. Although the great European wars were caused by dynastic or commercial egoism, they touched neither the original essence, nor the value and real intellectual mission of the nations, nor their specific particularity within the European whole. Despite independence from any particular peace policy, it is therefore precisely that almost all important historical personalities of intellectual Europe are pacifists. They are that way not entirely by nature alone, but also because of the deep logic of their occupations. They are furthermore pacifists because they have a clear notion of the European spirit and the duties it imposes on them. Many of them did not even confine themselves to this natural pacifism alone, but, for that reason, engaged themselves in practical politics as well. I would like to recall the French abbot Saint-Pierre,[a] who proposed around 1715 that the European states must unite into a league that was supposed to create and guarantee perpetual peace. Kant, too, made the following suggestion in the second article of his work on Perpetual Peace: "International law must have a union of free states as its basis."[b]

Europe's intellectual life presents a wholly different picture from what we see when we observe the politics of the powers and interests of its various states. In most cases, particularly in those countries that granted free rein to the irresponsible ambitions of dynasties or some ruling castes, a striking contrast existed between those exercising power and the leading men of intellectual and artistic life. This contrast is explained not just by the diversity of interests; it also arises out of that fine psychological motive from which intellectual personalities who are "good Europeans"[c] in Goethe's and Nietzsche's sense emerge, who evade the absurdity of a pitiful patriotism that would want them to stop behind state borders.

In our epoch, especially after the terrible experiences of the world war, the idea of a European community has re-awoken to new life and perhaps with greater impact in Germany and France, the two European nations that have always mutually enriched each other in the intellectual area despite their centuries-long political feuding. A new point can be raised that permits a certain optimism about Europe's future. It is very well possible that a "good European's" psychology and mentality could give rise to a new policy. As questionable as the League of Nations has remained to this day in its structure and its work accomplished to date, it nevertheless represents the first attempt being made in European history to put into practice the idea of a European community of states. However, the League of Nations has not yet freed itself of the old power relations or the European power groupings. It will not be able to banish the danger a divided Europe holds for the world as long as Germany and Russia do not join it as well.[d] The League of Nations still has not found a method by which to unite the totality of Europe's most essential components. A short time ago, the idea of a League of Nations experienced a fruitful addition to the pan-European movement originating from Count Coudenhove-Kalergi.[e] This movement was inspired by the continent's common interests and attitudes. It almost seems to me that the program and methods of this movement's young founder are too artificial and overly simplified. When the program is put forward by idealistic politicians, the goal it strives for often lets the practical and theoretical problems appear to be already solved. A task as difficult as Europe's unification does not permit simplification to one design formula; rather, all the complicated preconditions must be cleared up and consolidated in advance.

All in all, though, the idea is striding forward. Not only are idealistic thinkers professing it, but also political personalities, as can be gathered from the example of various speeches by the French prime minister.[f] This idea must govern Europe's future—and it will govern it and prepare the way to the unification of Pan-Europe—if the continent does not want to dig its own grave.

It will appear strange to you Americans that I am always speaking of Europe, when it actually does involve a subject that concerns the whole world. It is easier to extend one's hand to one's faraway comrades concerning legal foundations, but it is far more difficult to harmonize with one's neighbors, where the recollections of countless earlier frictions and quarrels that arose out of close proximity act as an impediment. For that reason, conciliatory agreements between neighbors must prepare the regime of justice in the world as a whole.

One more observation must be added: Without a unification of Europe, there can never be a common League of Nations in reality. Europe's political union will be achieved entirely of itself when all nations of America have joined the League of Nations, as they do not need to struggle against any major obstacles in order to join that large community. When observed from this point of view, the fate of history lies in America's hands.

Editorial Notes

[a] Charles-Irénée Castel de Saint-Pierre (1658–1743) was a French abbot and author, member of the Académie française. He proposed his concept of a unified Europe in Saint-Pierre 1713.

[b] The exact quotation is: "International law should be based on a federalism of free

states" (see "Zweyter Definitivartikel" in Kant 1795, p. 30).

[c] The phrase was used by Friedrich Nietzsche in Nietzsche 1886, §377. This phrase was also used in October 1914 in the "Manifesto to the Europeans" (Doc. 3), co-signed by Einstein, but in that instance, it had been attributed to Johann Wolfgang von Goethe.

[d] Neither Germany nor the Soviet Union were members of the League of Nations.

[e] Richard Nikolaus von Coudenhove-Kalergi (1894–1972) was an Austrian political author. He presented his concept of a pan-European union in Coudenhove-Kalergi 1923.

[f] Édouard Herriot (1872–1957) was the French prime minister from the Radical Party.

42. The Mission of the Hebrew University/ A Word for the Journey (1925)

Jüdische Rundschau 30, no. 27/28 (3 April 1925), p. 244. [*Collected Papers* Vol. 14, Doc. 470].

The formal dedication of the Hebrew University took place on 1 April 1925 on Mount Scopus in Jerusalem. The ceremony was opened with a prayer by Abraham I. Kook, the Ashkenazi chief rabbi of Palestine. Major speeches were given by Chaim Weizmann, Sir Herbert Samuel, and Lord Balfour, the former British foreign secretary. The ceremony ended with a poem by Chaim N. Bialik and a prayer of dedication by Joseph Hertz, Chief Rabbi of the United Hebrew Congregations of the British Empire. Other distinguished guests present were Lord Allenby, High Commissioner of Egypt, and William Rappard, Permanent Secretary of the Mandates Commission of the League of Nations. The ceremony was held in an open-air amphitheater and attended by a crowd of approximately eight thousand people, including chief rabbis, Christian clergy, British officials, members of the consular corps, and representatives of nine foreign governments and forty-seven universities and other academic institutions.

The inauguration of our Hebrew University on Mount Scopus, Jerusalem, should not only fill us with justifiable pride, but also give occasion for thought.

A university is a place where the universality of the human mind is revealed. Truth alone is the aim of research and science. Therefore, it is natural for the institutions serving science to be part of the connection between nations and people. Universities in Europe are now, unfortunately, mostly places for fostering the mindless spirit of nationalism and blind intolerance for all that is alien or different from one's own people or race. Jews particularly suffer as a result, not only because they are being impeded from working or educating themselves freely, but also because most Jews perceive this nationalistic narrow-mindedness as particularly strange. On this birthday of our university, I would like to voice the wish that our university always be free of this evil, that the teachers and students always retain the awareness that they serve their nation best when they relate it to humanity and the highest human values beyond national boundaries.

Jewish nationalism is currently a necessity because we can eliminate the conflicts a Jew presently suffers only by consolidating our life as a nation. May the time soon come when this nationalism has become so self-evident that we no longer have any need to emphasize it in particular. The bond to our past and our nation's current endeavors gives us the security to stand confidently before the whole world. Yet, our educational institutions in particular must regard keeping our nation free from nationalistic self-conceit and aggressive intolerance to be one of their noblest missions.

Our university is a modest enterprise for now. In principle, it is entirely right to begin with some research institutes and let the university develop organically. I am convinced this development will make rapid progress and that, over time, this will be the first place to give full, clear evidence of the achievements of which the Jewish mind is capable.

One special responsibility to emerge for the university is the intellectual guidance and education of our nation's active labor forces in the countryside. We would rather not create in Palestine another populace of city-dwellers who live the way they do in Europe's cities, replete with the bourgeois mentality of Europe. We want a working people first of all,—a Jewish village—and we want the benefits of civilization to be accessible also to this working class, especially since we know that Jews set education above all else in any circumstances. It is incumbent on the university to create something entirely new here that would meet the specific needs of life for the developing nation in Palestine.

Let all of us work together so that the university can fulfill its mission. May a recognition of the importance of this cause make headway across broad sections of Jewry! Then our university will soon develop into a major intellectual center that will gain the respect of the entire civilized humankind.

43. On Ideals (1925)

La Prensa (28 April 1925), p. 10. [*Collected Papers* Vol. 14, Doc. 479].

> During his trip to South America, Einstein wrote several popular articles. In Japan, he had already expressed his opinion on religion and its relation to science, published in *Kaizo* in 1923: respectful of others' beliefs, he nevertheless wrote: "I can regard confessional traditions only historically and psychologically; I have no other relationship to them."

All lives, both individual and collective, after satisfying their most common material needs, yearn for a world of superior values that, because of its reactive effect, tends to make them nobler and more spiritual. All things begin with myth, primitive religion, pure animism, the deification of nature and the forces that control it. But in the subsequent development of the "European" people, the determination of life's values and ideals is in no sense limited to religion; rather, through constant growth as well as historical evolution, it becomes externalized, most notably in the literary, artistic, and philosophical life of nations. This fact—or rather, these fields of intellectual production that, for the most part, are essentially subjective—and perhaps the principal ideals of life differ markedly from Eastern thought,[a] wherein the great religious systems and equally religious compendiums of knowledge constitute almost all superior and superhuman truth and, with their pretensions of eternal validity and objectivity, exclude historical evolution, as well as a multiplicity of doctrines and those who teach them.

Linked to this formal distinction is also the essential one: the European ideal of life tends, first and foremost, to produce a "great and unique personality," set apart from the crowd and the present moment. The quintessential European ideal is that of "the hero," "the fighter," and its devotion to the world of ideals beyond the material is practically the equivalent of a "veneration of heroes" tinged with religious overtones. This explains the mythical character acquired by men like Caesar and Napoleon; but spiritual creators—a Dante, a Goethe, a Nietzsche—can also take on heroic proportions in the conscience of the people. The Asian ideal, unlike the European, entirely disregards the man of action and his culmination in the heroic. In this regard, Taoism, Buddhism, Judaism, and Christianity completely coincide. For them, doctrine—devotion to an idea that is valid for all, the recommendation of a morally pure life—comes first. The contradiction resulting from this contrast between East and West can be noted throughout the history of European Christianity so that, despite the widespread Christianization of the continent, the Christian ideal of life never managed to prevail to the exclusion of all others. This ideal has a passive character and is at odds with active Europeanism, which tends to create great personalities, heroism, and individual productivity when taken to an extreme.

This Europeanism is already clearly evident in Hellenism. Devotion to the ideals of beauty and truth is the manifestation of an active creative spirit, at the same time requiring the establishment of values that respond to such a psychology. Plato's world of ideas, with its closeness to the religious East, is absolutely European insofar as the establishment of different values is concerned. No matter how much the ideals and values of life may have changed in the history of Europe, they still retain this active, productive character. Moreover, European heroism is apparent even in those spiritual currents that oppose it by nature. (Here I see, for example, the difference in principle between Asian and European mysticism.)

The newest "American" ideal of life is as different from the Asian as it is from the European. It does not tend toward the creation of metapsychic moral values, nor is it hero worship or faith in personalities. Rather, its objective is economic power. The progressive Americanization of Europe gives this ideal a pronounced practical validity on our continent as well—though it may be denied and contested as an ideal value—because it is the opposite of the European ideal. For an American, the only criterion by which all values are judged is practical reality, a concept of life that had its philosophical expression in pragmatism, whose underlying idea goes like this: "The truth is whatever can be proved in praxis."[b]

Despite this contrast, as far as spiritual and artistic life is concerned, I do not see any antagonism between Europe and America. The psyche of both worlds tends toward productive growth. A high level of technical and economic acumen prevails in America; however, I do not believe this excludes all spiritual life. Technical and economic action, after all, also provide room for creative gifts, since genius can overcome mechanical rules and develop those gifts freely. Furthermore, the most rigorous organization of economic life creates possibilities for freeing spiritual creators from material concerns.

For the Western world (Europe and America), the religious ideal of the East calls for productivity. He who truly adopts this ideal in the spiritual, artistic realm of life does not, in my opinion, need any other ideal whatsoever and therefore cannot propose any objective for himself other than productive development. "He who possesses science and art also possesses religion; he who does not possess them needs religion" (Goethe).[c]

Editorial Notes

[a] See "Musings on My Impressions in Japan" (Doc. 30).

[b] He was reading Vaz Ferreira's criticism of pragmatism (Vaz Ferreira 1914) at the time of writing this article.

[c] A quotation from Goethe's poem "Zahme Xenien IX," Goethe 1952, p. 367. The original reads:

> He who has science and has art,
> He also has religion.
> But he who is devoid of both,
> He surely needs religion.

44. Space-Time, *Encyclopædia Britannica* (1926)

J. L. Garvin, ed. *Encyclopædia Britannica*, 13th ed., Supplementary Vol. 3, London and New York: The Encyclopædia Britannica Co., Ltd. (1926), pp. 608–611. [*Collected Papers* Vol. 15, Doc. 148].

> Einstein was invited by Thomas C. Hodson to contribute three articles by 1 January 1926 to a new edition of the *Encyclopædia Britannica*. Einstein sent the manuscript of "Raum" and "Zeit" on 3 January, and Hodson acknowledged their receipt on 5 January. The English translation was published as the following article. The third article, "Modern Conception of the Universe," was eventually written by Hermann Weyl.

Space-Time (*see* 25.525)—All our thoughts and concepts are called up by sense-experiences and have a meaning only in reference to these sense-experiences. On the other hand, however, they are products of the spontaneous activity of our minds; they are thus in no wise logical consequences of the contents of these sense-experiences. If, therefore, we wish to grasp the essence of a complex of abstract notions we must for the one part investigate the mutual relationships between the concepts and the assertions made about them; for the other, we must investigate how they are related to the experiences.

So far as the way is concerned in which concepts are connected to each other and with the experiences, there is no difference in principle between the concept-systems of science and those of daily life. The concept-systems of science have grown out of those of daily life and have been modified and completed according to the objects and purposes of the science in question.

The more universal a concept is the more frequently it enters into our thinking; and the more indirect its relation to sense-experience, the more difficult it is for us to comprehend its meaning; this is particularly the case with pre-scientific concepts that we have been accustomed to since childhood. Consider the concepts referred to in the words "where," "when," "why," and "being," to the elucidation of which innumerable volumes of philosophy have been devoted. We fare no better in our speculations than a fish which should strive to become as clear as to what is water.

§I. SPACE

In the present article, we are concerned with the meaning of "where," that is, of space. It appears that there is no quality contained in our individual primitive sense-experience that may be designated as spatial. Rather, what is spatial appears to be a sort of order of the material objects of experience. The concept of "material object" must therefore be available if concepts concerning space are to be possible. It is the logical primary concept. This is easily seen if we analyse the spatial concepts, for example, "next to," "touch," and so forth, that is, if we strive to become aware of their equivalents in experience. The concept of "object" is a means to take into account the persistence in time or the continuity, respectively, of certain groups of experience-complexes. The existence of objects is thus conceptual, and the meaning of the concepts of objects depends wholly on their being connected (intuitively) with groups of elementary sense-experiences. This connection is the basis of the illusion which makes primitive experience appear to inform us directly about the relation of material bodies (which exist, after all, only in so far as they are thought).

In the sense thus indicated we have (the indirect) experience of the contact of two bodies. We need do no more than call attention to this, as we gain nothing for our present purpose by singling out the individual experiences to which this assertion alludes. Many bodies can be brought into permanent contact with one another in manifold ways. We speak in this sense of the position-relationships of bodies (*Lagenbeziehugen*). The general laws of such position-relationships are essentially the concern of geometry. This holds, at least, if we do not wish to restrict ourselves to regarding the propositions that occur in this branch of knowledge merely as relationships between empty words that have been set up according to certain principles.

Pre-scientific Thought.—Now, what is the meaning of the concept "space" which we also encounter in pre-scientific thought? The concept of space in pre-scientific thought is characterized by the sentence: "We can think away things but not the space which they occupy." It is as if, without having had experience of any sort, we had a concept, nay even a presentation, of space and as if we ordered our sense-experiences with the help of this concept, present *a priori*. On the other hand, space appears as a physical reality, as a thing that exists independently of our thoughts, like material objects. Under the influence of this view of space, the fundamental concepts of geometry—the point, the straight line, the plane—were even regarded as having a self-evident character. The fundamental principles that deal with these configurations were regarded as being necessarily valid and as having, at the same time, an objective content. No scruples were felt about ascribing an objective meaning to such statements as "three empirically given bodies (practically infinitely small) lie on one straight line," without demanding a physical definition for such an assertion. This blind faith in evidence and the immediate real meaning of the concepts and propositions of geometry became uncertain only after non-Euclidean geometry had been introduced.

Reference to the Earth.—If we start from the view that all spatial concepts are related to contact-experiences of solid bodies, it is easy to understand how the concept "space" originated, namely, how a thing independent of bodies and yet embodying their position-possibilities (*Lagerungsmöglichkeiten*) was posited. If we have a system of bodies in contact and at rest relative to one another, some can be replaced by others. This property of allowing substitution is interpreted as "available space." Space denotes the property in virtue of which rigid bodies can occupy different positions. The view that space is something with a unity of its own is perhaps due to the circumstance that in pre-scientific thought all positions of bodies were referred to one body (reference body), namely the earth. In scientific thought the earth is represented by the coordinate system. The assertion that it would be possible to place an unlimited number of bodies next to one another denotes that space is infinite. In pre-scientific thought the concepts "space," "time," and "body of reference" are scarcely differentiated at all. A place or point in space is always taken to mean a material point on a body of reference.

Euclidean Geometry.—If we consider Euclidean geometry we clearly discern that it refers to the laws regulating the positions of rigid bodies. It turns to account the ingenious thought of tracing back all relations concerning bodies and their relative positions to the very simple concept of "distance" (*Strecke*). Distance denotes a rigid body on which two material points (marks) have been specified. The concept of the equality of distances (and angles) refers to experiments involving coincidences; the same remarks apply to the theorems on congruence. Now, Euclidean geometry, in the form in which it has been handed down to us from Euclid, uses the fundamental concepts "straight line" and "plane," which do not appear to correspond, or at any rate, not so directly, with experiences concerning the position of rigid bodies. On this, it must be remarked that the concept of the straight line may be reduced to that of the distance.[1] Moreover, geometricians were less concerned with bringing out the relation of their fundamental concepts to experience than with deducing logically the geometrical propositions from a few axioms enunciated at the outset.

Let us outline briefly how perhaps the basis of Euclidean geometry may be gained from the concept of distance.

We start from the equality of distances (axiom of the equality of distances). Suppose that of two unequal distances one is always greater than the other. The same axioms are to hold for the inequality of distances as hold for the inequality of numbers.

Three distances $\overline{AB^1}$, $\overline{BC^1}$, $\overline{CA^1}$ may, if $\overline{CA^1}$ be suitably chosen, have their marks BB^1, CC^1, AA^1 superposed on one another in such a way that a triangle ABC results. The distance CA^1 has an upper limit for which this construction is still just possible. The points A, (BB') and C then lie in a "straight line" (definition). This leads to the concepts: producing a distance by an amount equal to itself; dividing a distance into equal parts; expressing a distance in terms of a number by means of a measuring-rod (definition of the space-interval between two points).

When the concept of the interval between two points or the length of a distance has been gained in this way we require only the following axiom (Pythagoras' theorem) in order to arrive at Euclidean geometry analytically.

To every point of space (body of reference) three numbers (coordinates) x, y, z may be assigned—and conversely—in such a way that for each pair of points $A(x_1, y_1, z_1,)$ and $B(x_2, y_2, z_2)$ the theorem holds:

$$\text{measure-number } \overline{AB} = \sqrt{(x_2 - x_1)^2 + (y_2 - y_1)^2 + (z_2 - z_1)^2}$$

---

[1] A hint of this is contained in the theorem: "The straight line is the shortest connection between two points." This theorem served well as a definition of the straight line, although the definition played no part in the logical texture of the deductions.

All further concepts and propositions of Euclidean geometry can then be built up purely logically on this basis, in particular also the propositions about the straight line and the plane.

These remarks are not, of course, intended to replace the strictly axiomatic construction of Euclidean geometry. We merely wish to indicate plausibly how all conceptions of geometry may be traced back to that of distance. We might equally well have epitomised the whole basis of Euclidean geometry in the last theorem above. The relation to the foundations of experience would then be furnished by means of a supplementary theorem.

The coordinate may and must be chosen so that two pairs of points separated by equal intervals, as calculated by the help of Pythagoras' theorem, may be made to coincide with one and the same suitably chosen distance (on a solid). The concepts and propositions of Euclidean geometry may be derived from Pythagoras' proposition without the introduction of rigid bodies; but these concepts and propositions would not then have contents that could be tested. They are not "true" propositions but only logically correct propositions of purely formal content.

Difficulties.—A serious difficulty is encountered in the above represented interpretation of geometry in that the rigid body of experience does not correspond exactly with the geometrical body. In stating this I am thinking less of the fact that there are no absolutely definite marks than that temperature, pressure, and other circumstances modify the laws relating to position. It is also to be recollected that the structural constituents of matter (such as atom and electron, *q.v.*) assumed by physics are not in principle commensurate with rigid bodies, but that nevertheless the concepts of geometry are applied to them and to their parts. For this reason consistent thinkers have been disinclined to allow real contents of facts (*reale Tatsachenbestände*) to correspond to geometry alone. They considered it preferable to allow the content of experience (*Erfahrungsbestände*) to correspond to geometry and physics conjointly.

This view is certainly less open to attack than the one represented above; as opposed to the atomic theory, it is the only one that can be consistently carried through. Nevertheless, in the opinion of the author it would not be advisable to give up the first view, from which geometry derives its origin. This connection is essentially founded on the belief that the ideal rigid body is an abstraction that is well rooted in the laws of nature.

Foundations of Geometry.—We come now to the question: what is *a priori* certain or necessary, respectively in geometry (doctrine of space) or its foundations? Formerly we thought everything—yes, everything; nowadays we think—nothing. Already the distance-concept is logically arbitrary; there need be no things that correspond to it, even approximately. Something similar may be said of the concepts straight line, plane, of three-dimensionality and of the validity of Pythagoras' theorem. Nay, even the continuum-doctrine is in no wise given with the nature of human thought, so that from the epistemological point of view no greater authority attaches to the purely topological relations than to the others.

Earlier Physical Concepts.—We have yet to deal with those modifications in the space-concept which have accompanied the advent of the theory of relativity. For this purpose we must consider the space-concept of the earlier physics from a point of view different from that above. If we apply the theorem of Pythagoras to infinitely near points, it reads

$$\bar{ds}^2 = dx^2 + dy^2 + dz^2$$

where $\bar{ds}$[a] denotes the measurable interval between them. For an empirically-given ds the coordinate system is not yet fully determined for every combination of points by this equation. Besides being translated, a coordinate system may also be rotated.[2] This signifies analytically: the relations of Euclidean geometry are covariant with respect to linear orthogonal transformations of the coordinates.

In applying Euclidean geometry to pre-relativistic mechanics a further indeterminateness enters through the choice of the coordinate system: the state of motion of the coordinate system is arbitrary to a certain degree, namely, in that substitutions of the coordinates of the form

$$x' = x - vt$$
$$y' = y$$
$$z' = z$$

---

[2] Change of direction of the coordinate axes while their orthogonality is preserved.

also appear possible. On the other hand, earlier mechanics did not allow coordinate systems to be applied of which the states of motion were different from those expressed in these equations. In this sense we speak of "inertial systems." In these favoured-inertial systems we are confronted with a new property of space so far as geometrical relations are concerned. Regarded more accurately, this is not a property of space alone but of the four-dimensional continuum consisting of time and space conjointly.

Appearance of Time.—At this point time enters explicitly into our discussion for the first time. In their applications space (place) and time always occur together. Every event that happens in the world is determined by the space-coordinates x, y, z, and the time-coordinate t. Thus the physical description was four-dimensional right from the beginning. But this four-dimensional continuum seemed to resolve itself into the three-dimensional continuum of space and the one-dimensional continuum of time. This apparent resolution owed its origin to the illusion that the meaning of the concept "simultaneity" is self-evident, and this illusion arises from the fact that we receive news of near events almost instantaneously owing to the agency of light.

This faith in the absolute significance of simultaneity was destroyed by the law regulating the propagation of light in empty space or, respectively, by the Maxwell-Lorentz electrodynamics. Two infinitely near points can be connected by means of a light-signal if the relation

$$ds^2 = c^2 dt^2 - dx^2 - dy^2 - dz^2 = 0$$

holds for them. It further follows that ds has a value which, for arbitrarily chosen infinitely near space-time points, is independent of the particular inertial system selected. In agreement with this we find that for passing from one inertial system to another, linear equations of transformation hold which do not in general leave the time-values of the events unchanged. It thus became manifest that the four-dimensional continuum of space cannot be split up into a time-continuum and a space-continuum except in an arbitrary way. This invariant quantity ds may be measured by means of measuring-rods and clocks.

Four Dimensional Geometry.—On the invariant ds a four-dimensional geometry may be built up which is in a large measure analogous to Euclidean geometry in three dimensions. In this way physics becomes a sort of statics in a four-dimensional continuum.

Apart from the difference in the number of dimensions the latter continuum is distinguished from that of Euclidean geometry in that ds^2 may be greater or less than zero. Corresponding to this we differentiate between time-like and space-like line-elements. The boundary between them is marked out by the element of the "light-cone" $ds^2 = 0$ which starts out from every point. If we consider only elements which belong to the same time-value, we have

$$-ds^2 = dx^2 + dy^2 + dz^2$$

These elements ds may have real counterparts in distances at rest and, as before, Euclidean geometry holds for these elements.

Effect of Relativity, Special and General.—This is the modification which the doctrine of space and time has undergone through the restricted theory of relativity. The doctrine of space has been still further modified by the general theory of relativity, because this theory denies that the three-dimensional spatial section of the space-time continuum is Euclidean in character. Therefore it asserts that Euclidean geometry does not hold for the relative positions of bodies that are continuously in contact.

For the empirical law of the equality of inertial and gravitational mass led us to interpret the state of the continuum, in so far as it manifests itself with reference to a non-inertial system, as a gravitational field and to treat non-inertial systems as equivalent to inertial systems. Referred to such a system, which is connected with the inertial system by a non-linear transformation of the coordinates, the metrical invariant ds^2 assumes the general form:—

$$ds^2 = \sum_{\mu\nu} g_{\mu\nu} dx_\mu dx_\nu$$

where the $g_{\mu\nu}$'s are functions of the coordinates and where the sum is to be taken over the indices for all combinations 11, 12, ... 44. The variability of the $g_{\mu\nu}$'s is equivalent to the existence of a gravitational field. If the gravitational field is sufficiently general it is not possible at all to find an inertial system, that is, a coordinate system with reference to which ds^2 may be expressed in the simple form given above:—

$$ds^2 = c^2 dt^2 - dx^2 - dy^2 - dz^2 = 0$$

But in this case, too, there is in the infinitesimal neighbourhood of a space-time point a local system of reference for which the last-mentioned simple form for ds holds.

This state of the facts leads to a type of geometry which Riemann's genius created more than half a century before the advent of the general theory of relativity of which Riemann divined the high importance for physics.[b]

Riemann's Geometry.—Riemann's geometry of an n-dimensional space bears the same relation to Euclidean geometry of an n-dimensional space as the general geometry of curved surfaces bears to the geometry of the plane. For the infinitesimal neighbourhood of a point on a curved surface there is a local coordinate system in which the distance ds between two infinitely near points is given by the equation

$$ds^2 = dx^2 + dy^2$$

For any arbitrary (Gaussian) coordinate-system, however, an expression of the form[c]

$$ds^2 = g_{11}dx^2 + 2g_{12}dx_1 dx_2 + g_{22}dx_2^2$$

holds in a finite region of the curved surface. If the $g_{\mu\nu}$'s are given as functions of x_1 and x_2 the surface is then fully determined geometrically. For from this formula we can calculate for every combination of two infinitely near points on the surface the length ds of the minute rods connecting them; and with the help of this formula all networks that can be constructed on the surface with these little rods can be calculated. In particular, the "curvature" at every point of the surface can be calculated; this is the quantity that expresses to what extent and in what way the laws regulating the positions of the minute rods in the immediate vicinity of the point under consideration deviate from those of the geometry of the plane.

This theory of surfaces by Gauss has been extended by Riemann to continua of any arbitrary number of dimensions and has thus paved the way for the general theory of relativity. For it was shown above that corresponding to two infinitely near space-time points there is a number ds which can be obtained by measurement with rigid measuring-rods and clocks (in the case of time-like elements, indeed, with a clock alone). This quantity occurs in the mathematical theory in place of the length of the minute rods in three-dimensional geometry. The curves for which

$\int ds$ has stationary values determine the paths of material points and rays of light in the gravitational field, and the "curvature" of space is dependent on the matter distributed over space.

Just as in Euclidean geometry the space-concept refers to the position-possibilities of rigid bodies, so in the general theory of relativity the space-time-concept refers to the behaviour of rigid bodies and clocks. But the space-time-continuum differs from the space-continuum in that the laws regulating the behaviour of these objects (clocks and measuring-rods) depend on where they happen to be. The continuum (or the quantities that describe it) enters explicitly into the laws of nature, and conversely these properties of the continuum are determined by physical factors. The relations that connect space and time can no longer be kept distinct from physics proper.

Nothing certain is known of what the properties of the space-time-continuum may be as a whole. Through the general theory of relativity, however, the view that the continuum is infinite in its time-like extent but finite in its space-like extent has gained in probability.

§II. TIME

The physical time-concept answers to the time-concept of the extra-scientific mind. Now, the latter has its root in the time-order of the experiences of the individual, and this order we must accept as something primarily given.

I experience the moment "now," or, expressed more accurately, the present sense-experience (*Sinnen-Erlebnis*) combined with the recollection of (earlier) sense-experiences. That is why the sense-experiences seem to form a series, namely the time-series indicated by "earlier" and "later." The experience-series is thought of as a one-dimensional continuum. Experience-series can repeat themselves and can then be recognised. They can also be repeated inexactly, wherein some events are replaced by others without the character of the repetition becoming lost for us. In this way we form the time-concept as a one-dimensional frame which can be filled in by experiences in various ways. The same series of experiences answer to the same subjective time-intervals.

The transition from this "subjective" time (*Ich-Zeit*) to the time-concept of pre-scientific thought is connected with the formation of the idea that there is a real external world independent of the subject. In this sense the (objective) event is made to correspond with the subjective experience. In the same sense there is attributed to the "subjective" time of the experience a "time" of the corresponding "objective" event. In contrast with experiences, external events and their order in time claim validity for all subjects.

This process of objectification would encounter no difficulties were the time-order of the experiences corresponding to a series of external events the same for all individuals. In the case of the immediate visual perceptions of our daily lives, this correspondence is exact. That is why the idea that there is an objective time-order became established to an extraordinary extent. In working out the idea of an objective world of external events in greater detail, it was found necessary to make events and experiences depend on each other in a more complicated way.

This was at first done by means of rules and modes of thought instinctively gained, in which the conception of space plays a particularly prominent part. This process of refinement leads ultimately to natural science.

The measurement of time is effected by means of clocks. A clock is a thing which automatically passes in succession through a (practically) equal series of events (period). The number of periods (clock-time) elapsed serves as a measure of time. The meaning of this definition is at once clear if the event occurs in the immediate vicinity of the clock in space; for all observers then observe the same clock-time simultaneously with the event (by means of the eye) independently of their position. Until the theory of relativity was propounded it was assumed that the conception of simultaneity had an absolute objective meaning also for events separated in space.

This assumption was demolished by the discovery of the law of propagation of light. For if the velocity of light in empty space is to be a quantity that is independent of the choice (or, respectively, of the state of motion) of the inertial system to which it is referred, no absolute meaning can be assigned to the conception of the simultaneity of events that occur at points separated by a distance in space. Rather, a special time must be allocated to every inertial system. If no coordinate system (inertial system) is used as a basis of reference there is no sense in asserting that events at different points in space occur simultaneously.

It is in consequence of this that space and time are welded together into a uniform four-dimensional continuum.

Editorial Notes

[a] $\bar{d}s$ here and in the preceding equation occur like this in the published version but were written as $d\sigma$ in the German manuscript version.

[b] In 1854, Bernhard Riemann (1826–1866) developed the geometry of general curvature in arbitrary dimensions named after him. It was only published posthumously as Riemann 1867.

[c] Equation is correctly written $ds^2 = g_{11} dx_1^2 + 2g_{12} dx_1 dx_2 + g_{22} dx_2^2$ in the German manuscript.

45. New Experiments on the Effect of the Earth's Motion on Light Velocity with Respect to the Earth (1927)

Forschungen und Fortschritte 3 (1927), p. 36. [*Collected Papers* Vol. 15, Doc. 478].

For many years, the American physicist Dayton C. Miller carried out experiments at Case School of Applied Science in Cleveland, Ohio, and on Mt. Wilson in California aimed at detecting an ether drift. Early on, Einstein cautiously abstained from public criticism and explained that, if Miller's results should be confirmed, "then the special relativity theory, and with it the general theory in its present form, falls. Experiment is the supreme judge. Only the equivalence of inertia and weight remain, which would lead to an essentially different theory." Gradually, however, he was encouraged by others' observations to express his doubts more freely. He was even quoted in the press advising the public not to bet on confirmation of Miller's results.

It is well known that the outcome of the interference experiment by Michelson—that is, by Michelson and Morley[a]—has been a particularly important factor in the genesis of relativity theory; for the negative outcome of this experiment has led to the conviction that the propagation of light in empty space with respect to an inertial system proceeds at a constant velocity independent of the velocity of the respective inertial system. More precisely, the experiment shows slightly less; namely, the statement that the time needed by light to move back and forth along a rigid rod at relative rest to Earth is independent of the rod's spatial orientation. The present version of relativity theory stands or falls with this outcome. That is why it was an exciting event for theoreticians when Mr. Dayton Miller, a professor in Cleveland, arrived at a differing result based on scrupulous experiments lasting years, the most important of which were carried out on Mount Wilson.[b]

Miller found that the time it takes for light to go there and back again depends on the spatial orientation of the path with respect to the fixed stars. His experimental setup was, in itself, more precise than the one by Michelson and Morley, whose experiment compared light paths about 60 m long. To explain his experimental findings, Mr. Miller formed an opinion that had already been considered before the theory of relativity was proposed—namely, that the luminous ether was carried along by the Earth in its translatory motion, but to a degree diminishing with height above sea level. This was supposed to explain the positive outcome of the experiments conducted at sites located higher above sea level.

During the last few months, the experiments were independently repeated from two quarters and with different apparatuses—that is, by Mr. R. J. Kennedy from the California Institute of Technology, as well as by Messrs. A. Piccard and E. Stahel in Brussels.[c]

The weakest aspect of Miller's experiments had already become apparent to physicists beforehand, namely that the considerable size of his apparatus did not permit sufficient constancy in the temperature of the air through which the interfering light rays moved; local systematic temperature differences of a few hundredths of a degree could simulate the observed positive effect. Kennedy and Piccard avoided this drawback by using substantially smaller apparatuses than Miller; the required precision was attained by improvements in the optical equipment while employing special means to achieve constancy of temperature.

Mr. Kennedy worked with a light path of barely 5 m. The optical paths were arranged inside a sealed metal casing filled with helium at atmospheric pressure. The result of the experiments was definitely negative, at such an accuracy that the existence of an effect four times smaller than the one found by Miller could be ruled out.

Whereas Dr. Kennedy's experiments were conducted in the laboratory, Piccard and Stahel successfully carried out the bold plan of performing those extraordinarily subtle experiments in a free balloon. The major difficulty of using an apparatus of relatively low weight in a small space was offset by the advantage that the apparatus could rotate by having the entire balloon set into slow rotation through two small ventilators. Most importantly, however, the experiment could be performed at various altitudes, and thus the postulated dependence on height could be tested. Unfortunately, the attainable temperature constancy in the apparatus was not large enough to eliminate the existence of a positive effect of the order of magnitude posited by Mr. Miller with certainty. However, it could be shown that the observed effects did not increase with altitude as would have been expected, given Miller's results.

Without a doubt, it was meritorious of Prof. Miller to, with his experiments, begin a scrupulous re-examination of Michelson's important experiment. However, his result must be seen as refuted by Kennedy's and Piccard's experiments.

Editorial Notes

[a] Michelson and Morley 1887.
[b] See Miller 1925a, Miller 1925b, and Miller 1925c.
[c] Kennedy 1926 and Piccard and Stahel 1927.

46. Speech at Rally for the Keren Ha-Yesod in Berlin (1926)

Keren Hayesod. *Das Palästinawerk: eine Kundgebung deutscher Juden im ehemaligen Herrenhaus zu Berlin am 4. März 1926.* Berlin: Keren Hayesod (1926), p. 18. [*Collected Papers* Vol. 15, Doc. 208].

On 4 March 1926, Einstein participated in a rally at the *Preussisches Herrenhaus* in Berlin organized by the executive committee of the German *Keren Hayesod* (the Palestine Foundation Fund). According to the *Jüdische Rundschau*, the rally was notable for the participation of many non-Zionist Jews prominent in Germany.

Fundamentally, it does not comport with my ideal for race-based and tradition-based communities to specifically cultivate and emphasize their distinctiveness. However, if such a community is embattled, it must defend itself as a community so that its individual members can assert themselves physically and spiritually. Through association, it must be accomplished that the individual escapes spiritual dangers that necessarily arise out of isolation. Whoever has come to clearly understand this must approve and encourage the association of all Jews in the shared work, even though he is, in principle, opposed to any nationalistic attitude.

For me, there is no doubt that, with the current situation, the development of Palestine is the only objective with the appealing strength necessary to achieve an effective association of Jews. It is Herzl's[a] immortal service to have been the first to clearly recognize this situation and draw the practical consequences from this recognition.

Therefore, in my opinion, every Jew who is deeply concerned about the health of the Jewish community as a whole and the dignity of Jews in general must cooperate with all his might in the realization of Herzl's ideal.

The Jew who works for the Jewish community and homeland in Palestine ceases to be German no more than the Jew ceases to be a Jew as a result of baptism and name change. The two affiliations are based on realities of different kinds. The distinction is not between Jew and German, but rather between upstanding and spineless. Whoever remains loyal to his origin, race, and tradition will also remain loyal to the state of which he is a citizen; whoever is disloyal in the first case is also disloyal in the other.

Editorial Note

[a] Theodor Herzl (1860–1904) was the founder of modern political Zionism.

47. Newton's Mechanics and Its Influence on the Formation of Theoretical Physics (1927)

Die Naturwissenschaften 15 (25 March 1927), pp. 273–276. [*Collected Papers* Vol. 15, Doc. 503].

> The bicentenary of Isaac Newton's death was to be commemorated across the world on 31 March 1927. Einstein was overwhelmed by requests to write addresses for various events and journals. In the end, he sent a message for a gathering at Newton's birthplace in Grantham, England. He wrote this longer article for *Die Naturwissenschaften*, published in English in the *Manchester Guardian*, and another for the journal *Nord und Süd* which constituted the manuscript for a German radio broadcast read by Einstein on the day of the bicentenary.

The two-hundredth anniversary of the death of Newton falls at this time. One's thoughts cannot help but turn to this shining spirit who pointed out the path of Western thought, research, and practical construction as none had before or after him. He was not only an inventor of genius in respect of particular guiding methods, but also showed a unique mastery of the empirical material known in his time and was marvelously inventive in special mathematical and physical demonstrations. For all these reasons, he deserves our deep veneration. He is, however, a yet more significant figure than his own mastery made him, as he was placed by fate at a turning point in the world's intellectual development. This is brought home to us vividly when we recall that, before Newton, there was no comprehensive system of physical causality that could render the deeper characters of the world of concrete experience in any way.

The great materialists of ancient Greek civilization had indeed postulated the reference of all material phenomena to a process of atomic movements controlled by rigid laws, without appealing to the will of living creatures as an independent cause. Descartes, in his own fashion, revived this ultimate conception. But it remained a bold postulate, the problematic ideal of a school of philosophy. In the way of actual justification of our confidence in the existence of an entirely physical causality, virtually nothing had been achieved before Newton.

Newton aimed to find an answer to the question: Does there exist a simple rule by which the motion of our planetary system's heavenly bodies can be completely calculated, if the state of motion of all these bodies at a single moment is known? Kepler's empirical laws of the motion of the planets, based on Tycho Brahe's observations, were already enunciated and demanded an interpretation.[1] These laws gave a complete answer to how the planets moved around the sun (elliptical orbit, equal areas described by the radius vector in equal periods, relation between semi-major axis and period of revolution). However, these rules do not satisfy the requirement of causality. The three rules are logically independent of one another and show no sign of any interconnection. The third law cannot be extended numerically, as it stands, from the sun to another central body; there is, for instance, no relation between a planet's period of revolution around the sun and that of a moon around its planet. But the principal thing is that these laws have reference to motion as a whole, not to the question of *how there is developed, from one condition of motion of a system, that which immediately follows it in time*. They are integral laws, in our phraseology of today, not differential laws.

The differential law is the form that entirely satisfies the modern physicist's requirement of causality alone. The clear conception of the differential law is one of the greatest of Newton's intellectual achievements. What was needed was not only the idea, but a formal

---

[1] Everyone knows, today, what gigantic efforts were needed to discover these laws from the empirically ascertained orbits of the planets. But few reflect on the genius of the method by which Kepler ascertained the true orbits from the apparent ones, i.e., their directions as observed from the earth.

mathematical method that was, indeed, extant in rudiment, but had yet to gain a systematic shape. Newton found this in differential and integral calculus as well. It is unnecessary to consider whether Leibniz arrived at these same mathematical methods independently of Newton or not; in any case, their development was a necessity for Newton, as they were required in order to give Newton the means of expressing his thoughts.

Galileo had already made a significant first step in the recognition of the law of motion. He discovered the laws of inertia and of free-falling in the earth's field of gravitation: a mass (or, more accurately, a material point) that is uninfluenced by other masses moves uniformly in a straight line; the vertical velocity of a free body increases in the field of gravity in proportion to the time. To us today, it may seem only a small step from Galileo's observations to Newton's laws of motion. But it must be observed that the two propositions above relate to motion as a whole in the form in which they are given, whereas Newton's law of motion answers the question: How does the condition of motion of a point-mass change *in an infinitely small period* under the influence of an external force? Only after proceeding to consider the phenomenon during an infinitely short period (differential law) does Newton arrive at a formula applicable to all motions. He takes the conception of force from the already highly developed theory of statics. He can only connect force with acceleration by introducing the new conception of mass, which is indeed supported, curiously enough, by a pseudo-definition. Today, we are so accustomed to forming conceptions corresponding to differential quotients, we can hardly realize any longer how great a capacity for abstraction was needed to pass across a double barrier to the general differential laws of motion, with the further need to evolve the conception of mass.

But this was still a long way from the causal comprehension of the phenomena of motion, for the motion was only determined by the equation of motion if the force was given. Newton had the idea, to which he was probably led by the laws of the planetary motions, that the force acting on a mass is determined by the position of all masses at a sufficiently small distance from the mass in question. It was not until this connection was realized that a completely causal comprehension of the phenomenon of motion could be obtained. How Newton, proceeding from Kepler's laws of the motion of planets, solved this problem for gravitation and so discovered the identity of the nature of gravity with the motive forces acting on the stars is common knowledge. It is only the combination of

(Law of motion) + (Law of attraction)

which constitutes that wonderful thought structure enabling the earlier and later conditions of a system to be calculated from the conditions ruling at one particular time, in so far as the phenomena occur under the sole influence of the forces of gravitation. The logical completeness of Newton's system of ideas lay in the fact that the sole causes of the acceleration of the masses of a system prove to be the masses themselves. On the basis sketched, Newton succeeded in explaining the motions of the planets, moons, and comets down to fine details, as well as ebb and flow and the precessional movement of the earth—this last was a deductive achievement of peculiar brilliance. It was, no doubt, especially impressive to learn that the cause of the movements of the heavenly bodies is identical to the force of gravity so familiar to us from everyday experience.

The significance, however, of Newton's achievement not only lay in its provision of a serviceable and logically satisfactory basis for mechanics proper; up to the end of the nineteenth century, it formed the program of all theoretical physical research. All physical phenomena were to be referred to as masses subject to Newton's law of motion. Only the law of force had to be amplified and adapted to the type of phenomenon considered. Newton himself tried to apply this program in optics, on the hypothesis that light consisted of inert corpuscles. The optics of the undulatory theory also made use of Newton's law of motion, applying it to continuously diffused masses. The kinetic theory of heat rested solely on Newton's formulae of motion, and this theory not only prepared people's minds for the recognition of the law of the conservation of energy, but also supplied a theory of gases confirmed in its smallest details, as well as a deepened conception of the nature of the second law of thermodynamics. The theory of electricity and magnetism also developed down to modern times entirely under the guidance of Newton's basic ideas (electric and magnetic substance, forces at a distance). Even Faraday and Maxwell's revolution in electrodynamics and optics, the first great advance in the fundamental principles of theoretical physics since Newton, was achieved still entirely under the guidance of Newton's ideas. Maxwell, Boltzmann, and Lord Kelvin never tired of trying, again and again, to reduce electromagnetic fields and their dynamical reciprocal action to mechanical processes occurring in continuously distributed hypothetical masses. But owing to the barrenness—or, at least, fruitlessness—of these efforts, there gradually occurred a turnaround in fundamental conceptions at the end of the nineteenth century; theoretical physics outgrew Newton's framework, which had provided fixity and intellectual guidance for science for nearly two centuries.

Newton's basic principles were so satisfying from a logical standpoint that the impulse to fresh departures could only come from the pressure of the facts of experience. Before I enter into this, I must emphasize that Newton himself was better aware of the weak sides of his thought structure than the succeeding generations of students. This fact has always excited my reverent admiration; I should like, therefore, to dwell a little on it.

1. Although everyone has remarked on how Newton strove to represent his thought system as necessarily subject to the confirmation of experience and to introduce the minimum of conceptions not directly referable to matters of experience, he made use of the conceptions of absolute space and absolute time. In our own day, he has often been criticized for this. But it is in this very point that Newton is particularly consistent. He recognized that the observable geometrical magnitudes (distances of material points from one another) and their change in the process of time do not completely determine movements in a physical sense. He shows this in the famous bucket experiment. There is therefore, in addition to masses and their distances varying with time, something else that determines what happens; he conceives of this "something" as the relation to "absolute space." He recognizes that space must possess a sort of physical reality if his laws of motion are to have a meaning, a reality of the same sort as the material points and their distances.

   This clear recognition shows both Newton's wisdom and a weakness of his theory, for a logical construction of the theory would certainly be more satisfactory without this shadowy conception; only those objects (point-masses, distances), then, would come into the laws whose relation to our perceptions is perfectly clear.

2. The introduction of direct, instantaneously acting forces at a distance into the exposition of the effects of gravitation does not correspond to the character of most of the phenomena familiar to us in our daily experience. Newton meets this objection by pointing out that his law of reciprocal gravitation is not to be taken as an ultimate explanation, but as a rule induced from experience.
3. Newton's theory did not explain the very remarkable fact that the weight and inertia of a body are determined by the same magnitude (the mass). The remarkable nature of this fact struck Newton as well.

None of these three points can rank as logical objections against the theory. They form, as it were, merely unsatisfied needs of the scientific spirit in its effort to penetrate the processes of nature with a complete and unified set of ideas. Newton's theory of motion, considered as a program for the whole field of theoretical physics, suffered its first shock from Maxwell's theory of electricity. It was found that the reciprocal action between bodies through electrical and magnetic bodies does not take place through instantaneously acting forces at a distance, but through processes transmitted with finite velocity through space. Alongside the point-mass and its movements, there arose, in Faraday's conception, a new sort of physically real thing—the "field." It was first sought to conceive this, with the aid of mechanical modes of thought, as a mechanical condition (of movement or strain) of a hypothetical space-filling medium (the ether). However, when this mechanical interpretation refused to work despite the most obstinate efforts, students slowly accustomed themselves to the concept of the "electromagnetic field" as the ultimate irreducible foundation-stone of physical reality. We owe to H. Hertz the deliberate liberation of the concept of the field from all the scaffolding of the concepts of mechanics; and to H. A. Lorentz, the liberation of the concept of the field from a material bearer. According to Lorentz, the physical empty space (or ether) alone figured as the bearer of the field; in Newton's mechanics, indeed, space had not been devoid of all physical functions. When this development had been completed, no longer did anyone believe in directly acting, instantaneous forces at a distance, even in connection with gravitation—though a field theory for gravitation was not unmistakably indicated, lacking sufficient known facts. The development of the theory of the electromagnetic field also led, after Newton's hypothesis of action at a distance had been abandoned, to the attempt to find an electromagnetic explanation for Newton's law of motion, or to replace that law with a more accurate law based on the field theory. These efforts were not crowned with full success, but the mechanical basic conceptions ceased to be regarded as foundation-stones of the physical conception of the universe.

The Maxwell-Lorentz theory led inevitably to the special theory of relativity, which, by destroying the conception of absolute simultaneity, negated the existence of forces at a distance. Under this theory, mass is not an unalterable magnitude, but a magnitude dependent on (and, indeed, identical with) the amount of energy. The theory also showed that Newton's law of motion could be considered valid as a limiting law only for small velocities, and the theory substituted for it a new law of motion in which the velocity of light in a vacuum appears as the limiting velocity.

The last step in the development of the program of the field theory was the general theory of relativity. It made little modification in Newton's theory quantitatively, but a deep-seated one qualitatively. Inertia, gravitation, and the metrical behavior of bodies and clocks were reduced to the single quality of a field, and this field, in turn, was made dependent on the bodies (generalization of Newton's law of gravitation, or the corresponding field law as formulated by Poisson). Space and time were thus divested not of their reality, but of their causal absoluteness (absoluteness—influencing, that is, not influenced), which Newton was compelled to attribute to them in order to give expression to the laws then known. The generalized law of inertia takes over the role of Newton's law of motion. From this short characterization, it will be clear how the elements of Newton's theory passed over into the general theory of relativity, the three defects above-mentioned being overcome at the same time. It appears that, within the framework of the general theory of relativity, the law of motion can be deduced from the law of the field, which corresponds to Newton's law of force. Only when this aim has been fully attained can we speak of a pure theory of fields.

Newton's mechanics prepared the way for the theory of fields in a yet more formal sense. The application of Newton's mechanics to continuously distributed masses led necessarily to the discovery and application of partial differential equations, which in turn supplied the language in which the laws of the theory of fields alone could be expressed. In this formal connection, Newton's conception of the differential law also forms the first decisive step in the subsequent development.

The whole development of our ideas concerning natural phenomena described above may be conceived as an organic development of Newton's thought. But while the construction of the theory of fields was still actively in progress, the facts of heat radiation, spectra, radioactivity, and so on revealed a limit to the employment of the whole system of thought which, despite gigantic successes in detail, seems completely insurmountable for us today. Many physicists maintain, not without weighty arguments, that not only the differential law, but the law of causality itself—hitherto the ultimate basic postulate of all natural science—fails in the face of these facts. The very possibility of a spatio-temporal construction that can be clearly brought into consonance with physical experience is denied. That a mechanical system should permanently admit only discrete values of energy or discrete states—as experience, so to say, directly shows—seems, at first, hardly deducible from a theory of fields working with differential equations. The method of De Broglie and Schrödinger—which has, in a certain sense, the character of a theory of fields—does deduce the existence of purely discrete states and their transition into one another, based on differential equations, from a sort of consideration of resonance, in amazing agreement with the facts of experience, but it has to dispense with a localization of the mass-particles and strictly causal laws.[a] Who would be so venturesome, today, as to decide whether causal law and differential law—these ultimate premises of Newton's treatment of nature—must definitely be abandoned?

Editorial Note

[a] This refers to the probabilistic interpretation of Schrödinger's wave function.

48. Review of Émile Meyerson's *La Déduction Relativiste* (1928)

Revue philosophique de la France et de l'étranger 105, nos. 3–4 (January–June 1928), pp. 161–166.[a]
[*Collected Papers* Vol. 16, Doc. 6].

> Einstein had met the French philosopher, chemist, and historian of science Émile Meyerson (1859–1933) for the first time in 1922. They had bonded over their opposition to Machian positivism and Kantianism and corresponded over epistemological issues posed by the theory of relativity. Einstein first read Meyerson's book reviewed here while on his trip to South America, when he noted in his diary that Meyerson unjustly made a comparison between relativistic physics and the German idealist Hegel because he took the attempts at a unified field theory by Hermann Weyl and Arthur S. Eddington to be an essential part of the theory. In this review, Einstein formulates his most detailed opposition to the idea that general relativity reduces physics, and gravity in particular, to space-time geometry.

What makes this book so extraordinary can be readily stated: It was written by a man who has encompassed the thought processes of modern physics, as well as penetrated deeply into the history of philosophy and the exact sciences, with an authoritative view of psychological mainsprings and interrelations. Logical precision, psychological instinct, multifaceted knowledge, and clear and simple expression meet here in a happy conjunction.

Meyerson's guiding principle seems, to me, to be the following: A theory of knowledge cannot be arrived at through an analysis of the mind and logical speculations, but rather only by the consideration and intuitive assessment of empirical data. Here, "empirical data" refers to the totality of scientific results currently at hand and the history of their accrual. To the author, the following appeared to be the main problem: What is the relationship between scientific knowledge and the embodiment of empirical facts? To what extent can one speak of inductive and deductive methodologies of science?

Pure positivism and pragmatism are rejected, indeed, even passionately opposed. Experience and empirical facts admittedly form the basis of every science, but they are neither its content nor essence; instead, they are only the givens on which the science is based. Merely determining the empirical interrelations between experimental facts cannot be claimed as the unique goal of science, according to this author. First of all, interrelations of such a general nature as expressed by our natural laws are not simply determinations of what can be experienced at all; they can be formulated and derived only based on a conceptual construction that itself cannot be obtained from experience as such. Secondly, however, science does not content itself with formulating empirical laws. Instead, it attempts to construct a logical system based on as few premises as possible and containing all the natural laws as logical consequences. This system or the entities that occur within it are attributed to the objects of experience; reason tries to establish this system, which is taken to correspond to the world of the real, existing things in a pre-scientific worldview, in such a way that it encompasses the totality of empirical facts and experiences. All the natural sciences are thus based upon philosophical realism. The attribution of all empirical laws to logically deducible theorems is the final goal of all scientific research, according to Meyerson—a goal toward which we continually strive, but are murkily aware we can achieve only incompletely.

In this sense, Meyerson is a rationalist, not an empiricist. His thoughts are, however, also distinct from critical idealism in the Kantian sense, for we do not know *a priori* of any feature of the system sought that is necessarily a part of the nature of our thought. This also holds for the forms of logic and causality. We can only ask how the system of the sciences (in its present state of development) is constituted, not how it must be constituted. The logical fundamentals of the system and its structure are thus conventions (as seen from the standpoint of logic); their justification is to be found solely in the ability of the system to describe observed facts in their unity and in the small number of premises it requires.

In relativity theory, Meyerson sees a new deductive system of physics with respect to previously established physics; he calls it "relativism" to formally point out its novelty. Here, in my opinion, he has gone too far.[b] Relativity theory by no means claims to be a new system of physics. Starting from the idea suggested by our experiences with light, inertia, and gravity that there is no physically preferred state of motion (the principle of relativity), it establishes the formal principle that the equations of physics should be covariant with respect to arbitrary point transformations within four-dimensional space-time. The basic laws of physics—as they were previously known—are adapted to this principle with a minimal number of modifications. The principle of relativity or covariance principle alone would provide a much too weak foundation for constructing the edifice of theoretical physics. Thus, one could preferably speak of "physics that has been adapted to the principle of relativity" rather than of "relativism" as the new system of physics. By identifying the less assertive claims to the generality of the "principle of relativity" with the much more extensive claims of "relativism," Meyerson adopts, in my opinion, a not completely relevant point of view with respect to the novelty or claims of the theory to long-term significance. The theory as a whole is not new, merely its adaptation to the principle of relativity.[c] On the other hand, the principle of relativity in itself would appear to be much more clearly justified by the character of experience than the formal structure of the theory that has been adapted to our current state of physical knowledge. We cannot be certain today, but we fear that the concepts of the "metric field" and "electromagnetic field" will prove inadequate with respect to the facts of quantum theory. But the opinion that the principle of relativity as such could be brought down by this inadequacy need hardly be considered seriously.

We mention all of this only in passing.[d] For Meyerson, it was important to point out that the rational system of physics, through its adaptation to the principle of relativity, has attained the character of a logically complete deductive system to a much higher degree than it had previously. Meyerson does not criticize this strongly deductive-constructive, highly abstract character of the theory, but rather finds that this character corresponds to the tendency of the whole development of the exact sciences; this tendency relinquishes, more and more, the convenience of fundamentals and methods (in the psychological sense) to achieve a unity of the whole system in a logical sense.

This deductive-constructive character impels Meyerson to compare the theory of relativity, in a rather ingenious manner, with the systems of Hegel and Descartes. The success of all three theories with contemporary thinkers is attributed to their logical completeness and deductive character; the human mind strives not only to establish relationships, but also to comprehend them. The advantage of the theory of relativity, with respect to the other two theories mentioned, is seen by Meyerson to lie in its quantitative structure and compatibility with many empirical facts. Meyerson sees another essential correspondence between Descartes's theory of physical events and the theory of relativity, namely the reduction of all concepts of the theory to spatial—or rather, geometrical—concepts; however, this is presumed to apply completely to the theory of relativity only after the electric field has been subsumed into the theory in the manner of Weyl's or Eddington's theory.

I would like to consider this latter point in some more detail, as my opinion differs strongly here. To be specific, I cannot admit that the assertion relativity theory traces physics back to geometry has a clear meaning.[e] It would be more correct to say the theory of relativity has been accompanied by a loss of (metric) geometry's special status relative to those regularities that were always considered to be physical.[f] Even before the proposal of the theory of relativity, it was not justifiable to consider geometry as an *a priori* doctrine as compared to physics. This occurred simply because one usually forgets that geometry is the study of the possible positional properties of rigid bodies. According to the general theory of relativity, the metric tensor determines the behavior of the measuring rods and clocks, as well as the motion of free bodies in the absence of electrical effects.

The fact the metric tensor is denoted as "geometrical" is due simply to the fact this formal structure first appeared in the area of study denoted as "geometry." However, this by no means justifies calling every science in which that formal structure plays a role a "geometry," not even when, for the sake of illustration, one makes use of concepts familiar from geometry. With similar arguments, Maxwell and Hertz could have termed the electromagnetic equations for the vacuum "geometric" because the geometrical concept of the vector occurs in them. It is, by the way, gratifying to me to emphasize that Meyerson himself states expressly, in his last chapter, that applying the terminology "geometric" to the theory of relativity is, in fact, meaningless; one could just as well say the metric tensor describes the "state of the ether."[g]

The essential aspects of the theories of Weyl and Eddington for representing the electromagnetic field thus do not lie in the fact that these theories have incorporated the theory of this field into geometry, but rather that they have shown a possible way to represent gravitation and electromagnetism from a unified point of view when those fields had previously entered the theory as logically independent entities. Meyerson furthermore correctly emphasizes that, in many descriptions of relativity theory, one speaks incorrectly of a "spatialization of time." Space and time are admittedly fused into a unified continuum, but that continuum is not isotropic. The character of a spatial neighborhood remains distinct from that of a temporal neighborhood, particularly with respect to the signs within the squared distance between two neighboring world-points.[h]

Meyerson's book belongs, in my opinion, among the most worthwhile of those written about relativity from the point of view of the theory of knowledge. My only regret is that Meyerson has not taken M. Schlick's work into account. He would most certainly have been able to appreciate its value.

Editorial Notes

[a] The document is a translation of Einstein's original German manuscript, written before 15 June 1927, rather than the final French version published.

[b] Einstein does not allege that Meyerson has "gone too far" in the French publication of this review.

[c] This criticism is not mentioned in the French publication, which instead claims Einstein and Meyerson agree on this point.

[d] This sentence is not present in the French publication.

[e] The first two sentences in this paragraph differ significantly from the French publication, which instead reads as follows: "Here again there is confusion to be avoided in the interpretation of certain assertions of Mr. Meyerson, and in particular this one: 'Relativity brings back physics to geometry.'"

[f] A sentence is added here in the French publication: "Et M. Meyerson est à même de citer (*La Déduction relativiste*, p. 137) un passage d'Eddington où il est question d'une 'théorie géométrique' de l'univers."

[g] The final sentence in this paragraph is not present in the French publication.

[h] The French publication contains an additional sentence following this paragraph: "But the tendency that he denounces, for being often latent in the minds of physicists, is no less real and profound, and the excesses of popularizers and even many scientists in their presentations of relativity are unequivocal manifestations of it."

49. The New Field Theory (1929)

New York Times, 3 February 1929, reprinted in *The Observatory* 52 (1930), pp. 82–86; 114–118.
[*Collected Papers* Vol. 16, Doc. 387].

> This piece was prefaced in the journal by the following note: "The following, the first of two articles by Professor Einstein, is written as an explanation of his thesis for readers who do not possess an expert knowledge of mathematics. The first article is a historical sketch leading up to the introduction of the relativity theory. The second is an account of the new work which describes the mathematical methods which led to the general theory of relativity as well as the new unified field theory. The translation is by Mr. L. L. Whyte."

I. MATTER AND SPACE

The chain of discovery

While physics wandered exclusively on the paths prepared by Newton, the following conception of physical reality prevailed. Matter is real, and matter undergoes only those changes which we conceive as movements in space. Motion, space, and also time are real forms. Every attempt to deny the physical reality of space collapses in the face of the law of inertia. For if acceleration is to be taken as real, then that space must also be real within which bodies are conceived as accelerated. Newton saw this with perfect clarity, and consequently he called space "absolute." In his theoretical system, there was a third constituent of independent reality: the motive forces acting between material particles, such forces being considered to depend only on the position of the particles. These forces between particles were regarded as unconditionally associated with the particles themselves and as distributed spatially according to an unchanging law.

The physicists of the 19th century considered that there existed two kinds of such matter—namely, ponderable matter and electricity. The particles of ponderable matter were supposed to act on each other by gravitational forces under Newton's law, the particles of electrical matter by Coulomb forces also inversely proportional to the square of the distance. No definite views prevailed regarding the nature of the forces acting between ponderable and electrical particles.

Mere empty space was not admitted as a carrier of physical changes and processes. It was only, one might say, the stage on which the drama of material happenings was played. Consequently, Newton dealt with the fact that light is propagated in empty space by making the hypothesis that light also consists of material particles, interacting with ponderable matter through special forces. To this extent, Newton's view of Nature involved a third type of material particle, although this certainly had to have properties very different from those of the particles of the other forms of matter. Light particles had, in fact, to be capable of being formed and of disappearing. Moreover, even in the 18th century, it was already clear from experience that light traveled in empty space with a definite velocity, a fact which obviously fitted badly into Newton's theoretical system. For why on earth should the light particles not be able to move through space with any arbitrary velocity?

Newton's successors

It need not, therefore, surprise us that this theoretical system, built up by Newton with his powerful and logical intellect, should have been overthrown precisely by a theory of light. This was brought about by the Huygens-Young-Fresnel wave theory of light, which the facts of interference and diffraction forced on stubbornly resisting physicists. The great range of phenomena which could be calculated and predicted to the finest detail by the use of this theory delighted physicists and filled many fat and learned books. No wonder, then, that the learned men failed to notice the crack which this theory made in the statue of their eternal goddess. For in fact, this theory upset the view that everything real can be conceived as the motion of particles in space. Light waves were, after all, nothing more than undulated states of empty space, and space thus gave up its passive *rôle* as a mere stage for physical events. The aether hypothesis patched up the crack and made it invisible. The aether was invented, penetrating everything, filling the whole of space, and was admitted as a new kind of matter. Thus it was overlooked that, by this procedure, space itself had been brought to life. It is clear that this had really happened, since the aether was considered to be a sort of matter which could nowhere be removed. It was thus, to some degree, identical with space itself, i.e., something necessarily given with space.

Light was thus viewed as a dynamical process undergone, as it were, by space itself. In this way, the field theory was born as an illegitimate child of Newtonian physics, though it was cleverly passed off at first as legitimate.

To become fully conscious of this change in outlook was a task for a highly original mind, whose insight could go straight to essentials, a mind that never got stuck in formulas. Faraday was this favored spirit. His instinct revolted at the idea of forces acting directly at a distance, which seemed contrary to every elementary observation. If one electrified body attracts or repels a second body, this was for him brought about, not by a direct action from the first body to the second but through an intermediary action. The first body throws the space immediately around it into a certain condition which spreads itself into more distant parts of space, according to a certain spatio-temporal law of propagation. This condition of space was called "the electric field." The second body experiences a force because it lies in the field of the first, and *vice versa*. The "field" thus provides a conceptual apparatus which rendered unnecessary the idea of action at a distance. Faraday also had the bold idea that, in appropriate circumstances, fields might detach themselves from the bodies producing them and speed away through space as free fields; this was his interpretation of light. Maxwell then discovered the wonderful group of formulas, seeming so simple to us nowadays, which finally built the bridge between the theory of electromagnetism and the theory of light. It appeared that light consists of rapidly oscillating electromagnetic fields.

The revolution in physics

After Hertz in the eighties of the last century had confirmed the existence of the electromagnetic waves and displayed their identity with light by means of his magnificent experiments, the great intellectual revolution in physics gradually became complete. People slowly accustomed themselves to the idea that the physical states of space itself were the final physical reality, especially after Lorentz had shown in his penetrating theoretical researches that, even inside ponderable bodies, the electromagnetic fields are not to be regarded as states of the matter, but essentially as states of the empty space in which the material atoms are to be considered as loosely distributed.

At the turn of the century, physicists began to be dissatisfied with the dualism of a theory admitting two kinds of fundamental physical reality: on the one hand, the field, and on the other hand, the material particles. It is only natural that attempts were made to represent the material particles as structures in the field—i.e., as places where the fields were exceptionally concentrated. Any such representation of particles on the basis of the field theory would have been a great achievement, but in spite of all the efforts of science, it has not been accomplished. It must even be admitted that this dualism is, today, sharper and more troublesome than it was ten years ago. This fact is connected with the latest impetuous developments in the Quantum theory, where the theory of the continuum (field theory) and the essentially discontinuous interpretation of the elementary structures and processes are fighting for supremacy.

We shall not here discuss questions concerning molecular theory, but shall describe the improvements made in the field theory during this century.

These all arise from the theory of relativity, which has, in the last six months, entered its third stage of development. Let us briefly examine the chief points of view belonging to these three stages and their relation to field theory.

The first stage, the Special Theory of Relativity, owes its origin principally to Maxwell's theory of the electromagnetic field. From this, combined with the empirical fact that there does not exist any physically distinguishable state of motion which may be called "absolute rest," arose a new theory of space and time. It is well known that this theory discarded the absolute character of the conception of the simultaneity of two spatially separated events. Well known also is the courage of despair with which some philosophers still defend themselves in a profusion of proud but empty words against this simple theory. On the other hand, the services rendered by the special theory of relativity to its parent, Maxwell's field equations, are less adequately recognized. Up to that time, the electric field and the magnetic field were regarded as existing separately, even if a close causal correlation between the two types of field was provided by Maxwell's field equations. But the special theory of relativity showed that this causal correlation corresponds to an essential identity of the two types of field. In fact, the same condition of space, which, in one coordinate system, appears as a pure magnetic field, appears simultaneously in another coordinate system in relative motion as an electric field, and *vice versa*. Relationships of this kind, displaying an identity between different conceptions, which therefore reduce the number of independent hypotheses and concepts of field theory and heighten its logical self-containedness, are a characteristic feature of the theory of relativity. For instance, the special theory also indicated the essential identity of the conceptions, inertial mass and energy. This is all generally known and is only mentioned here in order to emphasize the unitary tendency which dominates the whole development of the theory.

Theory of gravitation

We now turn to the second stage in the development of the theory of relativity, the so-called General Theory of Relativity. This theory also starts from a fact of experience which, till then, had received no satisfactory interpretation: the equality of inertial and gravitational mass, or, in other words, the fact, known since the days of Galileo and Newton, that all bodies fall with equal acceleration in the earth's gravitational field. The theory uses the special theory as its basis and, at the same time, modifies it. The recognition that there is no state of motion whatever which is physically privileged, i.e., that not only velocity but also acceleration are without absolute significance, forms the starting point of the theory. It then compels a much more profound modification of the conceptions of space and time than was involved in the special theory. For even if the special theory forced us to fuse space and time together to an invisible four-dimensional continuum, the Euclidean character of the continuum remained essentially intact in this theory. In the general theory of relativity, this hypothesis regarding the Euclidean character of our space-time continuum had to be abandoned and the latter given the structure of a so-called Riemannian space. Before we attempt to understand what these terms mean, let us recall what this theory accomplished.

It furnished an exact field theory of gravitation and brought gravitation into a fully determinate relationship to the metrical properties of the continuum. The theory of gravitation, which until then had not advanced beyond Newton, was thus brought within Faraday's conception of the field in a necessary manner, i.e., without any essential arbitrariness in the selection of the field laws. At the same time, gravitation and inertia were fused into an essential identity. The confirmation which this theory has received in recent years through the measurement of the deflection of light rays in a gravitational field and the spectroscopic examination of binary stars is well known.

II. STRUCTURE OF SPACE-TIME

Dualism overcome

The characteristics which especially distinguish the General Theory of Relativity, and even more the new third stage of the theory, the Unitary Field Theory, from other physical theories, are the degree of formal speculation, the slender empirical basis, the boldness in theoretical construction, and finally the fundamental reliance on the uniformity of the secrets of natural law and their accessibility to the speculative intellect. It is this feature which appears as a weakness to physicists who incline toward realism or positivism, but is especially attractive, nay, fascinating, to the speculative mathematical mind. Meyerson,[a] in his brilliant studies on the theory of knowledge, justly draws a comparison of the intellectual attitude of the relativity theoretician with that of Descartes, or even of Hegel, without thereby implying the censure which a physicist would read into this.

However that may be, in the end, experience is the only competent judge. Yet in the meantime, one thing may be said in defense of the theory. Advances in scientific knowledge must bring about the result that an increase in formal simplicity can only be won at the cost of an increased distance or gap between the fundamental hypothesis of the theory on the one hand, and the directly observed facts on the other. Theory is compelled to pass, more and more, from the inductive to the deductive method, even though the most important demand to be made of every scientific theory will always remain that it must fit the facts.

Search for simplicity

We now reach the difficult task of giving to the reader an idea of the methods used in the mathematical constructions which led to the General Theory of Relativity and to the New Unitary Field Theory. The general problem is: Which are the simplest formal structures that can be attributed to a four-dimensional continuum, and which are the simplest laws which may be conceived to govern these structures? We then look for the mathematical expression of the physical fields in these formal structures, and for the field laws of physics—already known to a certain approximation from earlier researches—in the simplest laws governing these structures.

The conceptions which are used in this connection can be explained just as well in a two-dimensional continuum (a surface) as in the four-dimensional continuum of space and time. Imagine a piece of paper ruled in millimeter squares. What does it mean if I say that the printed surface is two-dimensional? If any point P is marked on the paper, one can define its position by using two numbers. Thus, starting from the bottom left-hand corner, move a pointer toward the right until the lower end of the vertical through the point P is reached. Suppose that, in doing this, one has passed the lower ends of x vertical (millimeter) lines. Then move the pointer up to the point P, passing y horizontal lines. The point P is then described without ambiguity by the numbers x, y (coordinates). If one had used, instead of a ruled millimeter paper, a piece which had been stretched or deformed, the same determination could still be carried out: but in this case, the lines passed would no longer be horizontals or verticals, or even straight lines. The same point would then, of course, yield different numbers, but the possibility of determining a point by means of two numbers (Gaussian coordinates) still remains. Moreover, if P and Q are two points which lie very close to one another, then their coordinates only differ very slightly. When a point can be described by two numbers in this way we speak of a two-dimensional continuum (surface).

Now consider two neighboring points P, Q on the surface; and a little way off, another pair of points P', Q'. What does it mean to say that the distance PQ is equal to the distance P'Q'? This statement only has a clear meaning when we have a small measuring rod which we can take from one pair of points to the other, and if the result of the comparison is independent of the particular measuring rod selected. If this is so, the magnitudes of the tracts PQ, P'Q' can be compared. If a continuum is of this kind, we say it has a metric. Of course, the distance of the two points P, Q must depend on the coordinate differences (dx, dy). But the form of this dependence is not known *a priori*. If it is of the form:

$$ds^2 = g_{11} \cdot dx^2 + 2g_{12} \cdot dx \cdot dy + g_{22} \cdot dy^2,$$

then it is called a Riemannian metric. If it is possible to choose the coordinates so that this expression takes the form:

$$ds^2 = dx^2 + dy^2 \text{ (Pythagoras's theorem)},$$

then the continuum is Euclidean (plane). Thus it is clear that the Euclidean continuum is a special case of the Riemannian. Inversely, the Riemannian continuum is a metric continuum which is Euclidean in infinitely small regions, but not in finite regions. The g_{11}, g_{12}, g_{22} describe the metrical properties of the surface, i.e., the metrical field.

By making use of empirically known properties of space, especially the law of the propagation of light, it is possible to show that the space-time continuum has a Riemannian metric. The quantities g_{11}, etc., appertaining to it determine not only the metric of the continuum, but also the gravitational field. The law governing the gravitational field is found in answer to the question: Which are the simplest mathematical laws to which the metric (i.e., the g_{11}, etc.) can be subjected? The answer was given by the discovery of the field laws of gravitation, which have proved themselves more accurate than the Newtonian law. This rough outline is intended only to give a general idea of the sense in which I have spoken of the "speculative" methods of the General Theory of Relativity.

Two fields as one

This theory, having brought together the metric and gravitation, would have been completely satisfactory if the world had only contained gravitational fields and no electromagnetic fields. Now, it is true that the latter can be included within the General Theory of Relativity by taking over and appropriately modifying Maxwell's equations of the electromagnetic field, but they do not then appear like the gravitational fields, as structural properties of the space-time continuum, but as logically independent constructions. The two types of field are causally linked in this theory, but still not fused to an identity. It can, however, scarcely be imagined that empty space has conditions or states of two essentially different kinds, and it is natural to suspect that this only appears to be so because the structures of the physical continuum are not completely described by the Riemannian Metric.

The new Unitary Field Theory removes this fault by displaying both types of field as manifestations of one comprehensive type of spatial structure in the space-time continuum. The stimulus to the new theory arose from the discovery that there exists a structure between the Riemannian space structure and the Euclidean which is richer in formal relationships than the former, but poorer than the latter. Consider a two-dimensional Riemannian space in the form of the surface of a hen's egg. Since this surface is embedded in our (accurately enough) Euclidean space, it possesses a Riemannian metric. In fact, it has a perfectly definite meaning to speak of the distance of two neighboring points, P, Q, on the surface. Similarly, it has, of course, a meaning to say of two such pairs of points $(PQ)(P'Q')$ on separate parts of the surface of the egg that the distance PQ is equal to the distance $P'Q'$. On the other hand, it is impossible to now compare the *direction PQ* with the direction $P'Q'$. In particular, it is meaningless to demand that $P'Q'$ shall be chosen parallel to PQ. In the corresponding Euclidean geometry of two dimensions, the Euclidean geometry of the plane, directions can be compared and the relationship of parallelism can exist between lines in regions of the plane at any distance from one another (distant parallelism). To this extent, the Euclidean continuum is richer in relationships than the Riemannian.

The new Unitary Field Theory is based on the following mathematical discovery: There are continua with a Riemannian metric *and* distant parallelism which nevertheless are not Euclidean. It is easy to show, for instance, in the case of three-dimensional space, how much a continuum differs from a Euclidean.

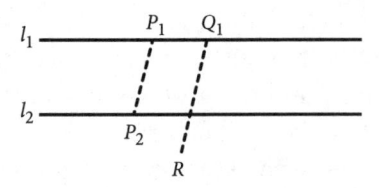

First of all, in such a continuum, there are lines whose elements are parallel to one another. We shall call these "straight lines." It also has a definite meaning to speak of two parallel straight lines, as in the Euclidean case. Now choose two such parallels l_1 and l_2 and mark on each a point (P_1, P_2). On l_1, choose in addition a point Q_1. If we now draw through Q_1 a straight line Q_1R parallel to the straight line P_1, P_2, then, in Euclidean geometry, this will cut the straight line l_2; in the geometry now used, the line Q_1R and the line l_2 do not in general cut one another. To this extent, the geometry now used is not only a specialization of the Riemannian, but also a generalization of the Euclidean geometry. My opinion is that our space-time continuum has a structure of the kind here outlined.

The mathematical problem whose solution, in my view, leads to the correct field laws is to be formulated thus: Which are the simplest and most natural conditions to which a continuum of this kind can be subjected? The answer to this question, which I have attempted to give in a new paper, yields unitary field laws for gravitation and electromagnetism.

Editorial Note

[a] Émile Meyerson (1859–1933); see Doc. 48.

50. Einstein Believes in Spinoza's God (1929)

New York Times (25 April 1929), p. 6. [*Collected Papers* Vol. 16, Doc. 508].

Professor Albert Einstein, the author of the theory of relativity, professed belief in "Spinoza's God"[a] in a radiogram received here yesterday from Dusseldorf, Germany, by Rabbi Herbert S. Goldstein[b] of the Institutional Synagogue 37 West 116th Street. The message came in response to a cablegram to the scientist asking him in German: "Do you believe in God?"

Professor Einstein's reply read:

Ich glaube an Spinozas Gott der Sich mit Schicksalen und Handlungen der Menschen abgibt.

(I believe in Spinoza's God, who reveals Himself in the orderly harmony of what exists, not in a God who concerns Himself with fates and actions of human beings.)

Rabbi Goldstein declared in an interview yesterday that Professor Einstein's message substantiates the reply the rabbi is now making to the recent attack by Cardinal O'Connell[c] of Boston on the Einstein theory, wherein the prelate declared that it "was cloaking the ghastly apparition of atheism."[d]

Rabbi Goldstein said that Einstein very clearly disproves by his own words the charge of atheism made against him. "In fact," he continued, "quite the reverse is true. Spinoza, who is called 'the God intoxicated man' and who saw God manifest in all of nature, certainly could not be called an atheist."

The rabbi was of the opinion that Einstein by his positive acceptance of God surely could not be classified either as an atheist or even as an agnostic. "Einstein points to a unity," he declared. "Einstein's theory if carried out to its logical conclusion would bring to mankind a scientific formula for monotheism. He does away with all thought of dualism or pluralism. There can be no room for any aspect of polytheism. This latter thought perhaps may have caused the Cardinal to speak out. Let us call a spade a spade."

Rabbi Goldstein added that "if Einstein's notion of relativity and unity are only vague theories, it scarcely behooves a man of the Cardinal's standing to speak and write as he did."

The address by Cardinal O'Connell referred to by Rabbi Goldstein was delivered April 7 in Boston to the members of the New England Province of the Catholic Clubs of America, in which he warned his audience against Professor Einstein's theory of relativity as "befogged speculation producing universal doubt about God and His Creation."[e]

Informed of Cardinal O'Connell's assertion, Professor Einstein later said, speaking through his wife, that the attack left him "cold" and was "devoid of interest." He said he was wholly disinclined to enter into a controversy with the Cardinal.[f]

Editorial Notes

[a] Baruch Spinoza (1632–1677) was a Dutch-Jewish rationalist philosopher and theologian. In his *Theologico-Political Treatise*, he argued for a pantheistic,

non-intervening god, stating, "Nature herself is the power of God under another name, and our ignorance of the power of God is co-extensive with our ignorance of Nature."

[b] Herbert S. Goldstein (1890–1970) was the founder of the Orthodox Institutional Synagogue in New York.

[c] Cardinal William Henry O'Connell (1859–1944) was the archbishop of Boston. He attacked Einstein's theories at a communion breakfast for members of the New England Province of College Catholic Clubs of America in Boston on 7 April 1929.

[d] Full quotation: "I mean that while I do not wish to accuse Einstein at present of deliberately wishing to destroy the Christian faith and the Christian basis of life, I half suspect that if we wait a little longer we will find he unquestionably will ultimately reveal himself in this attitude. In a word, the outcome of this doubt and befogged speculation about time and space is a cloak beneath which lies the ghastly apparition of atheism." (*Jewish Daily Bulletin*, 9 April 1929).

[e] Full quotation: "Now, I have my own ideas about the so-called theories of Einstein, with his relativity and his utterly befogged notions about space and time. It seems nothing short of an attempt at muddying the waters so that without perceiving the drift innocent students are led away into a realm of speculative thought, the sole basis of which, so far as I can see, is to produce a universal doubt about God and His creation" (*Jewish Daily Bulletin*, 9 April 1929).

[f] Quoted from *Jewish Daily Bulletin*, 10 April 1929.

51. The Palestine Troubles (1929)

Manchester Guardian Weekly (12 October 1929), p. 314. [*Collected Papers* Vol. 17, Doc. 77].

In August 1929, right-wing Jewish groups organized marches and demonstrations in Jerusalem, claiming the Western Wall as "theirs." The tense atmosphere escalated in both Jewish and Arab circles. A fatal stabbing of a Jewish youth led to riots on 23 August, and a day later Arab mobs attacked the Jewish quarter of Hebron. Acts of horrendous violence were perpetrated. More than 60 Jews were killed. Further attacks followed during the next five days. It is estimated that more than 130 Jews and more than 116 Arabs were killed in the riots. Einstein wrote this letter of protest against the British press's reaction to recent Arab attacks on Jews in Palestine. His letter was accompanied by an editorial urging the British government to uphold its commitments to the Jewish people that it had adopted upon assuming the responsibilities of the Mandate in Palestine. Einstein's letter appeared subsequently in other European and Hebrew newspapers.[a]

51. The Palestine Troubles

To the Editor of the *Manchester Guardian*,

Sir, I have been following with anxious concern the comments in the British press on the recent events in Palestine.[b] What I have read has so deeply affected me that, despite my general reluctance to enter the political arena, I feel impelled to ask for the hospitality of your columns for the following observations.

It was with a wonderful enthusiasm and a deep sense of gratitude that the Jews, afflicted more than any other people by the chaos and horror of the war, obtained from Great Britain a pledge to support the re-establishment of the Jewish national home in Palestine.[c] The Jewish people, beset with a thousand physical wrongs and moral degradations, saw in the British promise the sure rock on which it could recreate a Jewish national life in Palestine, which, by its very existence as well as by its material and intellectual achievements, would imbue the Jewish masses, dispersed all over the world, with a new sense of hope, dignity, and pride. Jews of all lands gave of their best in man-power and in material wealth in order to fulfil the inspiration that had kept the race alive through a martyrdom of centuries. Within a brief decade, some £10,000,000 were raised by voluntary contributions, and 100,000 picked Jews entered Palestine to redeem by their physical labour the almost derelict land.[d] Deserts were irrigated, forests planted, swamps drained, and their crippling diseases subdued. A work of peace was created which, although still perhaps small in size, compelled the admiration of every observer.

Has the rock on which we have built begun to shake? A considerable section of the British press now meets our aspirations with lack of understanding, with coldness, and with disfavour. What has happened?

Arab mobs, organised and fanaticised by political intriguers working on the religious fury of the ignorant, attacked scattered Jewish settlements and murdered and plundered wherever no resistance was offered. In Hebron, the inmates of a rabbinical college, innocent youths who had never handled weapons in their lives, were butchered in cold blood; in Safed, the same fate befell aged rabbis and their wives and children. Recently, some Arabs raided a Jewish orphan settlement where the pathetic remnants of the great Russian pogroms had found a haven of refuge. Is it not then amazing that an orgy of such primitive brutality upon a peaceful population has been utilised by a certain section of the

British press for a campaign of propaganda directed, not against the authors and instigators of these brutalities, but against their victims?

No less disappointing is the amazing degree of ignorance of the character and the achievement of Jewish reconstruction in Palestine displayed in many organs of the press. A decade has elapsed since the policy of the establishment of a Jewish national home in Palestine was officially endorsed by the British Government, with the almost unanimous support of the entire British press and of the leaders of all political parties. On the basis of that official recognition, which was approved by almost every civilized Government, and which found its legal embodiment in the Palestine Mandate, Jews have sent their sons and daughters and have given their voluntary offerings for this great work of peaceful reconstruction. I think it may be stated without fear of exaggeration that, except for the war efforts of the European nations, our generation has seen no national effort of such spiritual intensity and such heroic devotion as that which the Jews have displayed during the last ten years in favour of a work of peace in Palestine. When one travels through the country, as I had the good fortune to do a few years ago, and sees young pioneers, men and women of magnificent intellectual and moral calibre, breaking stones and building roads under the blazing rays of the Palestinian sun; when one sees flourishing agricultural settlements shooting up from the long-deserted soil under the intensive efforts of the Jewish settlers; when one sees the development of water-power and the beginnings of an industry adapted to the needs and possibilities of the country, and, above all, the growth of an educational system ranging from the kindergarten to the university, in the language of the Bible—what observer, whatever his origin or faith, can fail to be seized by the magic of such amazing achievement and of such almost superhuman devotion? Is it not bewildering that, after all this, brutal massacres by a fanaticised mob can destroy all appreciation of the Jewish effort in Palestine and lead to a demand for the repeal of the solemn pledges of official support and protection?

Zionism has a two-fold basis. It arose on the one hand from the fact of Jewish suffering. It is not my intention to paint here a picture of the Jewish martyrdom throughout the ages which has arisen from the homelessness of the Jew. Even to-day there is an intensity of Jewish suffering throughout the world of which the public opinion of the civilised West never obtains a comprehensive view. In the whole of

Eastern Europe the danger of physical attack against the individual Jew is constantly present. The degrading disabilities of old have been transformed into restrictions of an economic character, while restrictive measures in the educational sphere, such as the "numerus clausus" at the universities, seek to suppress the Jew in the world of intellectual life.[e] There is, I am sure, no need to stress at this time of the day that there is a Jewish problem in the Western world also. How many non-Jews have any insight into the spiritual suffering and distortion, the degradation and moral disintegration engendered by the mere fact of the homelessness of a gifted and sensitive people? What underlies all these phenomena is the basic fact, which the first Zionists recognised with profound intuition, that the Jewish problem cannot be solved by the assimilation of the individual Jew to his environment. Jewish individuality is too strong to be effaced by such assimilation, and too conscious to be ready for such self-effacement. It is, of course, clear that it will never be possible to transplant to Palestine anything more than a minority of the Jewish people, but it has for a long time been the deep conviction of enlightened students of the problem, Jews and non-Jews alike, that the establishment of a National Home for the Jewish people in Palestine would raise the status and the dignity of those who would remain in their native countries and would thereby materially assist in improving the relations between non-Jews and Jews in general.

But Zionism springs from an even deeper motive than Jewish suffering. It is rooted in a Jewish spiritual tradition, whose maintenance and development are for Jews the *raison d'être* of their continued existence as a community. In the re-establishment of the Jewish nation in the ancient home of the race, where Jewish spiritual values could again be developed in a Jewish atmosphere, the most enlightened representatives of Jewish individuality see the essential preliminary to the regeneration of the race and the setting free of its spiritual creativeness.

It is by these tendencies and aspirations that the Jewish reconstruction in Palestine is informed. Zionism is not a movement inspired by Chauvinism or by a *sacro egoismo*. I am convinced that the great majority of the Jews would refuse to support a movement of that kind. Nor does Zionism aspire to divest anyone in Palestine of any rights or possessions he may enjoy. On the contrary, we are convinced that we shall be able to establish a friendly and constructive co-operation with the kindred Arab race which will be a blessing to both sections of the population materially and spiritually. During the whole of the work of Jewish colonisation, not a single Arab has been dispossessed; every acre of land acquired by the Jews has been bought at a price fixed by buyer and seller.[f] Indeed, every visitor has testified to the enormous improvement in the economic and sanitary standard of the Arab population resulting from the Jewish colonisation. Friendly personal relations between the Jewish settlements and the neighbouring Arab villages have been formed throughout the country. Jewish and Arab workers have associated in the trade unions of the Palestine railways, and the standard of living of the Arabs has been raised. Arab scholars can be found working in the great library of the Hebrew University, while the study of the Arabic language and civilisation forms one of the chief subjects of study at this university.[g] Arab workmen have participated in the evening courses conducted at the Jewish Technical Institute at Haifa. The native population has come to realise in an ever-growing measure the benefits, economic, sanitary and intellectual, which the Jewish work of reconstruction has bestowed on the whole country and all its inhabitants. Indeed, one of the most comforting features in the present crisis has been the reports of personal protection afforded by Arabs to their Jewish fellow-citizens against the attacks of the fanaticised mob.[h]

I submit, therefore, that the Zionist movement is entitled, in the name of its higher objectives and on the strength of the support which has been promised to it most solemnly by the civilized world, to demand that its unprecedented reconstructive effort—carried out in a country which still largely lies fallow, and in which by methods of intensive cultivation such as the Jews have applied room can be found for hundreds of thousands of new settlers without detriment to the native population— shall not be defeated by a small clique of agitators, even if they wear the garb of ministers of the Islamic religion. Does public opinion in Great Britain realize that the Grand Mufti of Jerusalem, who is the centre of all the trouble and speaks so loudly in the name of all the Moslems, is a young political adventurer of not much more, I understand, than thirty years of age, who in 1920 was sentenced to several years' imprisonment for his complicity in the riots of that year, but was pardoned under the terms of an amnesty?[i] The mentality of this man may be gauged from a recent statement he gave to an interviewer accusing me, of all men, of having demanded the rebuilding of the Temple on the site of the Mosque of Omar.[j] Is it tolerable that, in a country where ignorant fanaticism can so easily be incited to rapine and murder by interested agitators, so utterly irresponsible and unscrupulous a politician should be enabled to continue to exercise his evil influence garbed in all the spiritual sanctity of religion and invested with all the temporal powers that this involves in an Eastern country?

The realisation of the great aims embodied in the Mandate for Palestine depends to a very large degree on the public opinion of Great Britain, on its press, and on its statesmen. The Jewish people is entitled to expect that its work of peace shall receive the active and benevolent support of the Mandatory Power. It is entitled to demand that those found guilty in the recent riots shall be adequately punished, and that the men in whose hands is laid the responsible task of the administration of a country of such a unique past and such unique potentialities for the future shall be so instructed as to ensure that this great trust, bestowed by the civilized world on the Mandatory Power, is carried out with vision and courage in the daily tasks of routine administration. Jews do not wish to live in the land of their fathers under the protection of British bayonets; they come as friends of the kindred Arab nation. What they expect of Great Britain is that it shall promote the growth of friendly relations between Jews and Arabs, that it shall not tolerate poisonous propaganda, and that it shall create such organs of security in the country as will afford adequate protection to life and peaceful labour.

The Jews will never abandon the work of reconstruction which they have undertaken. The reaction of all Jews, Zionist and non-Zionist alike, to the events of the last few weeks has shown this clearly enough. But it lies in the hands of the Mandatory Power materially to further or materially to hamper the progress of the work. It is of fundamental importance that British public opinion and the Governments of Great Britain and of Palestine shall feel themselves responsible for this great trust, not because Great Britain once undertook this responsibility in legal form, but because they are deeply convinced of the significance and importance of the task and believe that its realisation will tend to promote the progress and the peace of mankind, and to right a great historic wrong. I cannot believe that the greatest colonial Power in the world will fail when it is faced with the task of placing its unique colonising experience at the service of the reconstruction of the ancient home of the People of the Bible. The task may not be an easy one for the Mandatory Power, but for the success it will attain it is assured of the undying gratitude not only of the Jews but of all that is noblest to mankind.

Yours, be,

A. Einstein

Berlin, October 9.

Editorial Notes

[a] The editors of the *Guardian* wrote: "Dr. Einstein's letter should make the coldest reader understand that the future of the Palestine State is closely and inextricably associated with the happiness of the Jewish race. But it should make the coldest reader understand that it is also inextricably associated with the self-respect of the British people" (*Manchester Guardian*, 12 October 1929).

Einstein consistently advocated cooperation: "Should we be unable to find a way to honest cooperation and honest pacts with the Arabs, then we have learned absolutely nothing during our 2,000 years of suffering and will deserve our fate" (letter to Chaim Weizmann, 25 November 1929).

[b] In the immediate aftermath of the riots, most major British newspapers blamed both sides in the conflict, describing "chronic feuds" between Jews and Arabs, and holding responsible both "Zionist arrogance and Moslem fanaticism" for leading the disturbances (*The Times*, 26 and 28 August 1929).

[c] A reference to the Balfour Declaration of 1917.

[d] The actual amount was $25,000,000 or the equivalent of about £5,000,000. In the previous decade, some 104,000 Jewish immigrants arrived in Palestine.

[e] Refers to the quotas imposed on the admission of Jewish students at Eastern European universities.

[f] During the 1920s, there was no centrally coordinated plan by Jews to acquire lands. The two major Zionist institutions favored land acquisitions that could easily accommodate Jewish immigrants, and therefore prioritized purchasing land that was free of Arab inhabitants.

[g] Palestine was the only country in the Middle East without a university aimed at enrolling Arab students. The Hebrew University opened in 1925, and did not specifically aim to educate Arab students.

[h] Around 435 Jews were saved by their Arab neighbors.

[i] Mohammed Ain al-Husayni (1897–1974) had been imprisoned by the British in 1920 for instigating an attack by Arabs on Jews. The following year, he was pardoned by High Commissioner Sir Herbert Samuel and appointed mufti of Jerusalem.

[j] Al–Husayni was quoted as charging that "Lord Melchett (Alfred Mond, president of the British Zionist Foundation) and Prof. Albert Einstein are responsible for revealing designs on the Mosque area" in Jerusalem (*Jewish Daily Bulletin*, 26 September 1929).

52. To a Young Scholar (1929)

Berliner Tageblatt (25 December 1929), p. 1. [*Collected Papers* Vol. 17, Doc. 173].

> This article was most likely prompted by inquiries Einstein received from a young correspondent a few days earlier.

Dear Mr. X,

You have reservations about starting a course of study in mathematics because, in today's difficult economic conditions, the choice of profession ought to be entirely adapted to external exigencies. Yet from our previous conversations, I know you are imbued with a passionate urge to understand all things, especially the simplest, logically ascertainable ones.

I understand your hesitation, but advise not to lose faith through such anxious deliberations. Only he who remains true to himself is a man in full, and only then will he be one to others as well. But this is ultimately what everyone must aspire to.

"What shall become of me if I study mathematics or physics?" I hear you ask. "Does the teaching and research profession offer one sufficient prospects?"

My answer will seem a little peculiar to you: It is totally unnecessary for you to become a teacher or researcher. This does not depend on your qualifications alone, but also on demand. If you spend your money, you expect something in return, something you really need. Out of necessity, society functions similarly; it is prepared to provide you sustenance if you do what it needs in exchange. If you now wonder: "Why should I study mathematics?" then it would really be better not to study. But if you think: "According to his view, I could certainly study mathematics without neglecting my obligations to myself and my family," then you ought to study.

For this optimistic view, I give you two reasons which you should not disdain. Namely, I studied mathematics and physics in Switzerland and was then useless as a teacher.[a] Although lacking any technical skills, I was hired as an examiner at the Swiss patent office.[b] And it soon became apparent that a general grasp of physical connections was often more valuable than specialized knowledge and experience. That which is acquired through the simple joy of understanding is a nifty, useful instrument in the hands of a fully alive human being.

There is, however, a second point to be considered. It is still unclear if the professional researcher has more advantages in research than he who earns his bread as a "shoemaker." How often does one feel pushed into shallow prolixness and busyness to "get ahead"!

It may be that science is like those delicate women who prefer being a lover rather than a housewife. But on that, I dare not decide.

Good luck,
Yours,

Editorial Notes

[a] At age twenty-one, Einstein earned a degree from the ETH in Zurich that accredited him to teach mathematics and physics at the secondary-school level. He was subsequently unsuccessful in obtaining a university position, but was employed as a substitute teacher for a few months in 1901.

[b] Einstein was hired as Technical Expert, third class, at the Swiss Patent Office in Bern in June 1902.

53. What I Believe: Living Philosophies XIII (1930)[a]

Forum and Century 84, no. 4 (October 1930), pp. 193–194. [Collected Papers Vol. 17, Docs. 424 and 425].

The *Forum* was a monthly publication known as the "magazine of controversies." Founded in New York in 1885, it addressed issues in politics, art, religion, and science. It reached its greatest popularity in the late 1920s through the mid-1940s under the editorship of Henry Goddard Leach (1880–1970). The series of articles titled "What I Believe—Living Philosophies" was initiated in September 1929 with an essay by Bertrand Russell. The editor had "invited some of the most distinguished and representative thinkers of the modern world to formulate their personal credos, to set forth the intimate views of life which they have built up out of their own experiences" during an "age of transition," because, "while many of the accepted beliefs of the past no longer command respect, we have not yet been able to agree upon new beliefs to take their place." Einstein's was the thirteenth in a series of twenty-two essays that were subsequently published as a book (Einstein et al. 1931). Among the other authors were John Dewey, Sir James Jeans, H. G. Wells, H. L. Mencken, Beatrice Webb, Fridtjof Nansen, Lewis Mumford, Robert A. Millikan, and J. B. S. Haldane.

Strange is our situation here upon earth. Each of us comes for a short visit, not knowing why, yet sometimes seeming to divine a purpose.

From the standpoint of daily life, however, there is one thing we do know: one is here for the sake of other people—above all for those upon whose smile and well-being our own happiness depends, and also for the countless, unknown souls with whose fate we are connected by a bond of sympathy. Many times a day, I realize how much my own outer and inner life is built upon the labors of other people, both living and dead, and how earnestly I must exert myself in order to give, in return, as much as I have received. My peace of mind is often troubled by the depressing sense that I have borrowed too heavily from the work of others.[b]

I do not believe we can have any freedom at all in the philosophical sense, for we act not only under external compulsion but also by inner necessity. Schopenhauer's saying—"A man can surely do what he wills to do, but he cannot determine what he wills"—impressed itself upon me in youth and has always consoled me when I have witnessed or suffered life's hardships.[c] This conviction is an inexhaustible source of tolerance, for it does not allow us to take ourselves or others too seriously; it leads rather to a world view that makes room especially for humor.

To ponder interminably over the reason for one's own existence or the meaning of life in general seems to me, from an objective point of view, senseless. And yet everyone holds certain ideals by which he guides his aspiration and his judgment. The ideals which have always shone before me and filled me with the joy of living are goodness, beauty, and truth. To make a goal of comfort or happiness has never appealed to me; a system of ethics built on this basis would be sufficient only for a herd of cattle.

Without the sense of collaborating with like-minded beings in the pursuit of the ever unattainable in art and scientific research, my life would have been empty. Ever since childhood, I have scorned the commonplace limits so often set upon human ambition. Possessions, outward success, publicity, luxury—to me these have always been contemptible. I believe that a simple and unassuming manner of life is best for everyone, best both for the body and the mind.

My passionate interest in social justice and social responsibility has always stood in curious contrast to a marked lack of desire for direct association with men and women. I am a horse for single harness, not cut out for tandem or team work. I have never belonged wholeheartedly to country or state, to my circle of friends, or even to my own family. These ties have always been accompanied by a vague aloofness, and the wish to withdraw into myself increases with the years.

Such isolation is sometimes bitter, but I do not regret being cut off from the understanding and sympathy of other men. I lose something by it, to be sure, but I am compensated for it in being rendered independent of the customs, opinions, and prejudices of others, and am not tempted to rest my peace of mind upon such shifting foundations.

My political ideal is democracy. Everyone should be respected as an individual, but no one idolized. It is an irony of fate that I should have been showered with so much uncalled-for and unmerited admiration and esteem. Perhaps this adulation springs from the unfulfilled wish of the multitude to comprehend the few ideas which I, with my weak powers, have advanced.

Full well do I know that in order to attain any definite goal, it is imperative that one person should do the thinking and commanding and carry most of the responsibility. But those who are led should not be driven, and they should be allowed to choose their leader. It seems to me that the distinctions separating the social classes are false; in the last analysis they rest on force. I am convinced that degeneracy follows every autocratic system of violence, for violence inevitably attracts moral inferiors. Time has proved that illustrious tyrants are succeeded by scoundrels.

For this reason, I have always been passionately opposed to such régimes as exist in Russia and Italy today.[d] The thing which has discredited the European forms of democracy is not the basic theory of democracy itself, which some say is at fault, but the instability of our political leadership, as well as the impersonal character of party alignments.

I believe that those in the United States have hit upon the right idea. A President is chosen for a reasonable length of time and enough power is given to him to acquit himself properly of his responsibilities. In the German Government, on the other hand, I like the state's more extensive care of the individual when one is ill or unemployed. What is truly valuable in our bustle of life is not the nation, I should say, but the creative and impressionable individuality, the personality—who produces the noble and sublime while the common herd remains dull in thought and insensible in feeling.

This subject brings me to that vilest offspring of the herd mentality—the odious military. The man who enjoys marching in line and file to the strains of music falls below my contempt; he received his large brain by mistake—the spinal cord would have been amply sufficient. This heroism at command, this senseless violence, this accursed bombast of patriotism—how intensely I despise them! War is low and despicable, and I had rather be smitten to shreds than participate in such doings.

Such a stain on humanity should be erased without delay. I think well enough of human nature to believe that it would have been wiped out long ago had not the common sense of nations been systematically corrupted through school and press for business and political reasons.

The most beautiful thing we can experience is the mysterious. It is the source of all true art and science. He to whom this emotion is a stranger, who can no longer pause to wonder and stand rapt in awe, is as good as dead: his eyes are closed. This insight into the mystery of life, coupled though it be with fear, has also given rise to religion. To know that what is impenetrable to us really exists, manifesting itself as the highest wisdom and the most radiant beauty which our dull faculties can comprehend only in their most primitive forms—this knowledge, this feeling, is at the center of true religiousness. In this sense, and in this sense only, I belong in the ranks of devoutly religious men.

I cannot imagine a God who rewards and punishes the objects of his creation, whose purposes are modeled after our own—a God, in short, who is but a reflection of human frailty. Neither can I believe that the individual survives the death of his body, although feeble souls harbor such thoughts through fear or ridiculous egotism. It is enough for me to contemplate the mystery of conscious life perpetuating itself through all eternity, to reflect upon the marvelous structure of the universe which we can dimly perceive, and to try humbly to comprehend even an infinitesimal part of the intelligence manifested in nature.

Editorial Notes

[a] Einstein's draft manuscript for this article is titled: "How I Look Into the World."
[b] The English translators repeatedly replaced "people" or "others" with "men."
[c] The closest formulation by Schopenhauer is: "You can do what you will, but at any given moment of your life you can will only one definite thing and absolutely nothing else but this one thing." See Schopenhauer 1999, p. 21.
[d] Einstein's criticism of the Soviet Union became outspoken in 1925 (see Doc. 39).

54. Religion and Science (1930)

New York Times Magazine (9 November 1930), p. 1.[a] [*Collected Papers* Vol. 17, Doc. 462].

This essay caused much discussion, praise, and criticism. The *New York Times* sent a questionnaire to various ministers, asking them whether or not they agree with Einstein and replies were printed the following Sunday, November 16. The *Jewish Daily Bulletin* wrote: "Rev. Cox, a Catholic and professor of philosophy at Fordham University, finds that the cosmic religion of Einstein is no religion at all. All the others, while quarreling with Einstein on certain points, nevertheless find his views on religion stimulating." Felix Adler, in a sermon of the same day, declared that "Einstein's article revealed the mind of a mathematical physicist and not of a religious philosopher, and that for this reason the article was unconvincing."

Everything that men do or think concerns the satisfaction of the needs they feel or the escape from pain. This must be kept in mind when we seek to understand spiritual or intellectual movements and the way in which they develop. For feeling and longing are the motive forces of all human striving and productivity—however nobly these latter may display themselves to us.

What, then, are the feelings and the needs which have brought mankind to religious thought and to faith in the widest sense? A moment's consideration shows that the most varied emotions stand at the cradle of religious thought and experience.

In primitive peoples it is, first of all, fear that awakens religious ideas—fear of hunger, of wild animals, of illness and of death. Since the understanding of causal connections is usually limited on this level of existence, the human soul forges a being more or less like itself, on whose will and activities depend the experiences which it fears. One hopes to win the favor of this being by deeds and sacrifices, which, according to the tradition of each lineage, are supposed to appease the being or to make it well-disposed to man. I call this the religion of fear.

This religion is considerably stabilized—though not caused—by the formation of a priestly caste which claims to mediate between the people and the being they fear, and so attains a position of power. Often a leader or despot, or a privileged class whose power is maintained in other ways, will combine the function of the priesthood with its own temporal rule for the sake of greater security, or an alliance may exist between the interests of the political power and the priestly caste.

* * *

A second source of religious development is found in the social feelings. Fathers and mothers, as well as leaders of great human communities, are fallible and mortal. The longing for guidance, for love and succor, provides the stimulus for the growth of a social or moral conception of God. This is the God of Providence, who protects, decides, rewards and punishes. This is the God who, according to man's widening horizon, loves and provides for the life of the race, or of mankind, or who even loves life itself. He is the comforter in unhappiness and in unsatisfied longing, the protector of the souls of the dead. This is the social or moral idea of God.

It is easy to follow in the sacred writings of the Jewish people the development of the religion of fear into the moral religion, which is carried further in the New Testament. The religions of all the civilized peoples, especially those of the Orient, are principally moral religions. An important advance in the life of a people is the transformation of the religion of fear into the moral religion. But one must avoid the prejudice that regards the religions of primitive peoples as pure fear religions and those of the civilized peoples as pure moral religions. All are mixed forms, though the moral element predominates in the higher levels of social life. Common to all these types is the anthropomorphic character of the idea of God.

Only exceptionally gifted individuals or especially noble communities rise *essentially* above this level; in these there is found a third level of religious experience, even if it is seldom found in a pure form. I will call it the cosmic religious sense. This is hard to make clear to those who do not experience it, since it does not involve an anthropomorphic idea of God; the individual feels the vanity of human desires and aims, and the nobility and marvelous order which are revealed in nature and in the world of thought. He feels the individual destiny as an imprisonment and seeks to experience the totality of existence as a unity full of significance. Indications of this cosmic religious sense can be found even on earlier levels of development—for example, in the Psalms of David and in the Prophets. The cosmic element is much stronger in Buddhism, as, in particular, Schopenhauer's magnificent essays have shown us.[b]

The religious geniuses of all times have been distinguished by this cosmic religious sense, which recognizes neither dogmas nor God made in man's image. Consequently, there cannot be a church whose chief doctrines are based on the cosmic religious experience. It comes about, therefore, that we find precisely among the heretics of all ages men who were inspired by their highest religious experience; often they appeared to their contemporaries as atheists, but sometimes also as saints. Viewed from this angle, men like Democritus, Francis of Assisi and Spinoza are near to one another.

How can this cosmic religious experience be communicated from man to man, if it cannot lead to a definite conception of God or to a theology? It seems to me that the most important function of art and of science is to arouse and keep alive this feeling in those who are receptive.

Thus, we reach an interpretation of the relation of science to religion which is very different from the customary view. From the study of history, one is inclined to regard religion and science as irreconcilable antagonists, and this for a perspicuous reason. For anyone who is pervaded with the sense of causal law in all that happens, who accepts in real earnest the assumption of causality, the idea of a Being who interferes with the sequence of events in the world is absolutely impossible. Neither the religion of fear nor the social-moral religion can have any hold on him. A God who rewards and punishes is for him unthinkable, because man acts in accordance with an inner and outer necessity, and would, in the eyes of God, be as little responsible as an inanimate object is for the movements which it makes.

* * *

Science, in consequence, has been accused of undermining morals—but certainly wrongly. The ethical behavior of man is better based on sympathy, education and social relationships, and requires no support from religion. It would indeed be sad for mankind to be tamed by fear of punishment and hope of rewards after death.

It is, therefore, quite natural that the churches have always fought against science and have persecuted its supporters. But, on the other hand, I assert that the cosmic religious experience is the strongest and the noblest driving force behind scientific research. No one who does not appreciate the terrific exertions, and above all, the devotion without which groundbreaking creations in scientific thought cannot come about, can judge the strength of the feeling from which alone such work, turned away as it is from immediate practical life, can emerge. What a deep faith in the rationality of the structure of the world and what a longing to grasp even a small glimmer of this rationality revealed in the world there must have been in Kepler and Newton to enable them to unravel the mechanism of the heavens in long years of lonely work!

Anyone who only knows scientific research in its practical applications may easily come to a wrong interpretation of the state of mind of those who, surrounded by skeptical contemporaries, have shown the way to kindred spirits scattered over all countries in all centuries. Only those who have dedicated their lives to similar ends can have a living conception of the inspiration which gave them the power to remain loyal to their purpose in spite of countless failures. It is the cosmic religious sense which grants this power.

A contemporary has rightly said that the only deeply religious people of our largely materialistic age are the earnest researchers.

Editorial Notes

[a] A German iteration was also published in *Berliner Tageblatt*, No. 532 (11 November 1930), p. 1.

[b] Arthur Schopenhauer (1788–1860) was the first major Western philosopher to declare himself a "Buddhist."

55. Militant Pacifism/The Two Percent Speech (1930)

World Tomorrow 14, no. 1 (January 1931), p. 9.[a]

> Address delivered at the Ritz-Carlton Hotel in New York on 14 December 1930 before the New History Society, an organization advocating the Baha'i Faith founded in 1929. Translated by Rosika Schwimmer (1877–1948), a Hungarian-born pacifist, feminist, women's suffragist, and social activist who participated in establishing the Women's International League for Peace and Freedom.

When pacifists come together they usually have the feeling that they are the sheep and the wolves are outside. The trouble is that pacifists generally contact only their own groups—that is to say, those who are already convinced. They make no effort to go afield and convert others. Serious-minded pacifists should try to actually do something instead of contenting themselves with idle dreams or merely talking about their pacifism. Our next step is to act—to *do something*. We must realize that when war comes, everyone considers it his duty to commit a crime—the crime of killing. People must be made to understand the immorality of war. They must do everything in their power to disentangle themselves from this antiquated, barbarous institution, and to free themselves from the shackles of slavery.

For this I have two suggestions. One of them has already been tried and found practical. It is the refusal to engage in war service of any kind, under any circumstances. Even at the risk of great personal sacrifice and hardship, all who wish to do something concrete toward world pacification must refuse war service. Pacifists who mean what they say should adopt this position in time of peace even in countries where there is compulsory military service. In other countries, where there is no such service, pacifists should declare openly that they will never bear arms or take part in any military service whatsoever. I advise the recruiting of people for this idea all over the world. And for the timid who say, "What is the use of our trying, we are so few in number," my answer is: "If you can get only two percent of the population of the world to assert in time of peace that they will not fight, you will have the solution for international troubles." Even so small a proportion as two percent will accomplish the desired result, for they could not be put in jail. There are not enough jails in the world to accommodate them!

The second suggestion I offer appears less illegal. International legislation should be attuned to the idea that those who declare themselves against war should, in time of peace, be allowed to take up some kind of difficult or even dangerous work either for their country or for the international benefit of mankind. In this way, they can prove that their opposition to war is not prompted by selfish or cowardly motives.

I feel confident that whoever adopts this program will eventually succeed in establishing international legislation, either by legal or other methods. I advise all war resisters to organize and to internationalize; I also advise them to collect money so that they may reinforce war resisters in other countries who have no means with which to carry on their work. Let all those who wish to promote pacifism, who have the courage to suffer, devote their energy to the initiation of these activities and stand firm so that the whole world may see and respect them for what they are doing.

Editorial Note

[a] Also in: *The Fight Against War*, ed. Alfred Lief. New York: John Day, 1933. The text above differs from the one in Lief 1933, which is presented by Rowe and Schulmann 2007, p. 240.

56. Some Remarks Concerning My American Impressions (1931)

A. Einstein. *Mein Weltbild*. Amsterdam: Querido Verlag (1934), pp. 54–60.[a]

> This document was written by Einstein during his first visit to the California Institute of Technology in Pasadena in January-March 1931 and was first published in the *Los Angeles Times* on 29 March. It was published three years later in a collection of essays titled first *The World As I See It*, compiled by his son-in-law Rudolf Kayser, and later as *Ideas and Opinions*, edited by Carl Seelig. Einstein expressed displeasure at the German title *Mein Weltbild*.

I must keep my promise to say something about my impressions of this country. This is no easy task for me, as it is not so easy to assume the attitude of an objective observer when one has been received with so much love and exaggerated worship as I was in these United States.

A few words in this respect:

The worshiping of personalities is, in my eyes, always something unjustifiable. Of course, nature distributes her gifts quite diversely among her children. But there are, thank God, many well-bred among them, and I am convinced most of them lead a quiet and unnoticed existence. It seems to me unjust—not even in good taste—if only a very few of these were admired beyond measure, attributed almost superhuman powers of mind and character. This has become precisely my fate, and there exists a grotesque contrast between the capabilities and achievements attributed to me by my contemporaries and my actual abilities. The consciousness of this strange fact would be unbearable were it not for *one* beautiful consolation: It is a gratifying sign that our so-much decried materialistic age makes heroes out of human beings whose aims are exclusively spiritual and moral. This fact proves knowledge and justice are placed above wealth and power by a great part of humanity, and according to my experiences, this idealistic attitude seems especially prevalent in the United States, which is denounced so much as being a most materialistic country. After this digression, I return to my topic in the hope that my modest remarks will not be given more weight than they deserve.

What first strikes the newcomer is the superiority of this country in things technical and organizational. The essential tools of daily use are more solidly built than those in Europe. Household appliances are incomparably more practical. Everything is designed to save human labor. Labor is expensive because the country is sparsely populated proportionally to its natural resources. The high price of labor has brought about the marvelous development of technical machinery and labor-saving devices. Compare the all-too-densely populated China and India, in contrast, where the cheap price of labor has prevented the development of technical machinery. Europe stands between them. When the machine is fully developed, it becomes cheaper, in the end, than the most inexpensive manpower. The fascists in Europe who advocate increasing population density in their respective homelands for narrow political reasons should bear this in mind. This seems to contrast with the impression of anxiety with which the United States halts the importation of goods through protective tariffs. But you must not ask a harmless visitor to worry too much, and in the last analysis, it is not quite certain that every question admits of a reasonable answer.

The second thing that strikes the visitor is a joyful, positive attitude to life. The smiling faces in photographs are one of the strongest traits of the American. He is friendly, unself-conscious, optimistic and unenvious. The European who comes in contact with Americans feels at ease and perfectly at home. The European, on the other hand, is more critical, more self-conscious, less good-hearted, less ready to render assistance, more isolated, more fastidious in his diversions and in his reading. In comparison with the American, the European is inclined to pessimism.

The amenities and comforts of life play an important role. Calm, carefreeness, and security are sacrificed to this end. The American lives more for the goal—the future—than the European. Life, for him, is a constant becoming, never a being. In this respect, he is still more dissimilar from the Russian and Asian than from the European. There is, however, one point in which the American resembles the Asian more than does the European. Looked upon from the psychological and not from the economic angle, he is less of an individualist than the European. The "we" is more strongly emphasized than the "I." This peculiarity makes for very powerful customs and conventions, and the philosophy of life of the individuals, as well as their moral and aesthetic preferences, are much more standardized than in Europe. To this circumstance America owes her largely economic superiority over Europe. Cooperation and effective division of labor are accomplished with more ease and less friction than in Europe, be it in the factory, university, or in private charity. In a certain sense this social attitude may be ascribed to English tradition.

In apparent contradiction to this is the relatively small field of activity of the nation as compared to that of Europe. The European wonders why the telegraph, telephone, railways, and schools are mostly in the hands of private companies. This is possible here through the above-mentioned more social attitude of the individual. This also accounts for the fact that the extremely uneven distribution of property does not entail intolerable bitterness. The feeling of social responsibility is much more developed among the propertied than it is in Europe. The propertied person considers it self-evident that a large part of his possessions, and even of his labor, must be placed at the disposal of the community. Public opinion, all-powerful in America, demands this of him. Thus it is that the most important cultural functions can be entrusted to private initiative, and the role of government in this country is proportionately a very limited one.

The reputation of state power has, without any doubt, been diminished by the Prohibition law. There is nothing more dangerous for the reputation of the state and the law than enacting laws beyond its power to enforce them. It is an open secret that the threatening development of criminality in the United States is closely connected to this fact.

Prohibition, in my opinion, weakens the state in yet another respect. The tavern is a place where people have an opportunity to exchange their thoughts and opinions on public affairs. I feel that such an opportunity seems to be missing in this country, so that the press, controlled chiefly by interest groups, wields too much influence on public opinion.[b]

The overestimation of money in this country is still greater than in Europe, but it seems to me to be waning. There is certainly a gradual realization that not many possessions are necessary for a happy and blessed life.

In artistic matters I have sincerely admired the high taste expressed in the modern buildings and in the objects of daily use; on the other hand, I find that the fine arts and music are little alive in the soul of the people in comparison with Europe.

I have great admiration for the achievements of the scientific research institutes. It is wrong to attribute the growing superiority of American research work solely to greater wealth; dedication, patience, a spirit of camaraderie, and a tendency to cooperate play an important role in these successes.

One final remark. The United States of America is the most powerful, technologically advanced nation on earth. Her influence on shaping international relations is almost incalculable. But America is big, and its inhabitants have so far not shown much interest in the major international problems, at the top of which is the problem of disarmament. This must change, if only in America's own interests. The last war showed that the continents are no longer separated, but that the destinies of all countries are now closely intertwined. The conviction must prevail in this country that its inhabitants bear a great responsibility in the field of international politics. The role of an impassive spectator is not worthy of this country and would be disastrous for everyone in the long run.

Editorial Notes

[a] This is is a translation of Einstein's 1931 pre-published manuscript [AEA 72-530].

[b] The two paragraphs on the Prohibition law were not printed in the *Los Angeles Times*.

57. The 1932 Disarmament Conference (1931)

Nation 133, no. 3455 (23 September 1931), p. 300.

Written on 22 August 1931 from Berlin to *The Nation*, New York, in regards to the World Disarmament Conference to be held in Geneva in February 1932.

What the inventive genius of mankind has bestowed upon us in the last hundred years could have made human life carefree and happy if the development of the organizing power of man had been able to keep step with his technical advances. As it is, the hard-won achievements of the machine age in the hands of our generation are as dangerous as a razor in the hands of a three-year-old child. The possession of wonderful means of production has not brought freedom—only care and hunger.

Worst of all is the technical development which produces the means for the destruction of human life and the dearly created products of labor. We older people lived through that shudderingly in the World War. But even more terrible than this destruction seems to me the unworthy servitude into which the individual is swept by war. Is it not terrible to be forced by the community to deeds which every individual feels to be most despicable crimes? Only a few have had the moral greatness to resist; they are, in my eyes, the true heroes of the World War.

There is one ray of hope. It seems to me that today the responsible leaders of the several peoples have, in the main, the honest will to abolish war. The opposition to this unquestionably necessary advance lies in the unhappy traditions of the people which are passed on like an inherited disease from generation to generation because of our faulty educational machines. Of course, the main supports of this tradition are military training and its glorification, and not less important, the press which is so dependent upon the military and the larger industries. Without disarmament, there can be no lasting peace. On the contrary, the continuation of military armaments in their present extent will with certainty lead to new catastrophes.

Hence, the Disarmament Conference in Geneva in February, 1932, will be decisive for the fate of the present generation and the one to come. If one thinks back to the pitiful results achieved by the international conferences thus far held, it must be clear that all the thoughtful and responsible human beings must exercise all their powers again and again to inform public opinion of the vital importance of the conference of 1932. Only if the statesmen have, to urge them forward, the will to peace of a decisive majority in their respective countries, can they arrive at their important goal. For the creation of this public opinion in favor of disarmament, every person living shares the responsibility, through every deed and every word.

The failure of the conference would be assured if the delegates were to arrive in Geneva with fixed instructions and aims, the achievement of which would at once become a matter of national prestige. This seems to be universally recognized, for the meetings of the statesmen of any two states, of which we have seen a number of late, have been utilized for discussions of the problem of disarmament in order to clear the ground for the conference. This procedure seems to me a very happy one, for two persons, or two groups, ordinarily conduct themselves most sensibly, most honorably, and with the greatest freedom from passion if no third person listens in, whom the others believe they must consider or conciliate in their speeches. We can only hope for a favorable outcome in this most vital conference if the meeting is prepared for exhaustively in this way by advance discussions in order that surprises shall be made impossible, and if, through honest good will, an atmosphere of mutual confidence and trust can be effectively created in advance.

Success in such great affairs is not a matter of cleverness, or even shrewdness, but instead a matter of honorable conduct and mutual confidence. You cannot substitute intellect for moral conduct in this matter—I should like to say, thank God that you cannot!

It is not the task of the individual who lives in this critical time merely to await results and to criticize. He must serve this great cause as well as he can. For the fate of all humanity will be that fate which it honestly earns and deserves.

58. To American Negroes (1932)

The Crisis 39, no. 2 (February 1932), p. 45.[a]

On 14 October 1931, W.E.B. Dubois (1868–1963), editor of the National Association for the Advancement of Colored People's monthly magazine *The Crisis*, both founded in 1910, invited Einstein to contribute a short statement on the "evil of race prejudice in the world," given that the magazine addressed the "defense of the citizenship rights of 12 million people descended from the former slaves of this country."[b] Einstein sent the text below on 29 October 1931 with an expression of regret that he did not have time for a longer essay. The magazine issue carried a portrait of Einstein on its cover, and the subtitle in the table of contents reads: "A word of advice by one of the world's greatest citizens."

It seems to be a universal fact that the minorities, especially when their individuals are recognizable because of physical differences, are treated by the majorities among whom they live as an inferior class. The tragic part of such a fate, however, lies not only in the automatically realized disadvantages suffered by these minorities in economic and social relations, but also in the fact that those who meet such treatment themselves, for the most part, acquiesce in this prejudiced estimate because of the suggestive influence of the majority, and come to regard people like themselves as inferior. This second and more important aspect of the evil can be met through closer union and conscious educational enlightenment among the minority, and so an emancipation of the soul of the minority can be attained.[c]

The determined effort of the American negroes in this direction deserves every recognition and assistance.

Editorial Notes

[a] Einstein did not provide a title in his original statement. Also in *Mein Weltbild*, edited by Rudolph Kayser, and its translations (1934), titled "On Minorities" (Von den Minderheiten), written in Kayser's hand and numbered (19) in the German manuscript [AEA 28-207].

[b] The total population of the U.S. in 1932 was about 124 million.

[c] Einstein advocated the same closer union and enlightenment for Jews (see, e.g., Document 32).

59. Why War? (1932)

International Institute of Intellectual Cooperation, League of Nations, *Open Letters* (1933), no. 2, pp. 11–20.

In late 1931, the International Institute of Intellectual Cooperation, of which Einstein was a member representing Germany, suggested that Einstein engage in a literary exchange on a significant topic of the day, to be published with other similar letters. At first, Einstein agreed, suggesting Sigmund Freud as a possible correspondent. Although he retracted his willingness sometime later, the exchange of letters eventually took place between July and December 1932. They were published in early 1933. The booklet was banned in Germany.

Dear Professor Freud,

The proposal of the League of Nations and its International Institute of Intellectual Co-operation at Paris that I should invite a person, to be chosen by myself, to a frank exchange of views on any problem that I might select affords me a very welcome opportunity of conferring with you upon a question which, as things now are, seems the most insistent of all the problems civilization has to face. This is the problem: Is there any way of delivering mankind from the menace of war? It is common knowledge that, with the advance of modern science, this issue had come to mean a matter of life and death for civilization as we know it; nevertheless, for all the zeal displayed, every attempt at its solution has ended in a lamentable breakdown.

I believe, moreover, that those whose duty it is to tackle the problem professionally and practically are growing only too aware of their impotence to deal with it, and have now a very lively desire to learn the views of men who, absorbed in the pursuit of science, can see world-problems in the perspective distance lends. As for me, the normal objective of my thought affords no insight into the dark places of human will and feeling. Thus, in the enquiry now proposed, I can do little more than seek to clarify the question at issue and, clearing the ground of the more obvious solutions, enable you to bring the light of your far-reaching knowledge of man's instinctive life to bear upon the problem. There are certain psychological obstacles whose existence a layman in the mental sciences may dimly surmise, but whose interrelations and vagaries he is incompetent to fathom; you, I am convinced, will be able to suggest educative methods, lying more or less outside the scope of politics, which will eliminate these obstacles.

As one immune from nationalist bias, I personally see a simple way of dealing with the superficial (i.e., administrative) aspect of the problem: the setting up, by international consent, of a legislative and judicial body to settle every conflict arising between nations. Each nation would undertake to abide by the orders issued by this legislative body, to invoke its decision in every dispute, to accept its judgments unreservedly and to carry out every measure the tribunal deems necessary for the execution of its decrees. But here, at the outset, I come up against a difficulty; a tribunal is a human institution which, in proportion as the power at its disposal is inadequate to enforce its verdicts, is all the more prone to suffer these to be deflected by extrajudicial pressure. This is a fact with which we have to reckon; law and might inevitably go hand in hand, and juridical decisions approach more nearly the ideal justice demanded by the community (in whose name and interests these verdicts are pronounced) in so far as the community has effective power to compel respect of its juridical ideal. But at present we are far from possessing any supranational organisation competent to render verdicts of incontestable authority and enforce absolute submission to the execution of its verdicts. Thus I am led to my first axiom: the quest of international security involves the unconditional surrender by every nation, in a certain measure, of its liberty of action, its sovereignty that is to say, and it is clear beyond all doubt that no other road can lead to such security.

The ill-success, despite their obvious sincerity, of all the efforts made during the last decade to reach this goal leaves us no room to doubt that strong psychological factors are at work, which paralyze these efforts. Some of these factors are not far to seek. The craving for power which characterises the governing class in every nation is hostile to any limitation of the national sovereignty. This political power-hunger is wont to batten on the activities of another group, whose aspirations are on purely mercenary, economic lines. I have specially in mind that small but determined group, active in every nation, composed of individuals who, indifferent to social considerations and restraints, regard warfare, the manufacture and sale of arms, simply as an occasion to advance their personal interests and enlarge their personal sphere of power.

But recognition of this obvious fact is merely the first step toward an appreciation of the actual state of affairs. Another question follows hard upon it: How is it possible for this small clique to bend the will of the majority, who stand to lose and suffer by a state of war, to the service of their ambitions? (In speaking of the majority, I do not exclude soldiers of every rank who have chosen war as their profession, in the belief that they are serving to defend the highest interests of their people, and that attack is often the best method of defense.) An obvious answer to this question would seem to be that the minority, the ruling class at present, has the schools and press, usually the religious organizations as well, under its thumb. This enables it to organize and sway the emotions of the masses, and make its tool of them.

Yet even this answer does not provide a complete solution. Another question arises from it: How is it these devices succeed so well in rousing men to such wild enthusiasm, even to sacrifice their lives? Only one answer is possible. Because man has within him a lust for hatred and destruction. In normal times this passion exists in a latent state, it emerges only in unusual circumstances; but it is a comparatively easy task to call it into play and raise it to the power of a collective psychosis. Here lies, perhaps, the crux of all the complex of factors we are considering, an enigma that only the expert in the lore of human instincts can resolve.

And so we come to our last question. Is it possible to guide people's psychological development so that it becomes more resistant to the psychoses of hatred and destructiveness? Here I am thinking by no means only of the so-called uneducated. According to my own experiences, it is rather the so-called "Intelligentsia" that is most easily subject to the seductive mass suggestions, since the intellectual has no direct contact with life in the raw, but captures it most conveniently and completely on the printed page.

To conclude: I have so far spoken only of wars between nations; what are known as international conflicts. But I am well aware that the aggressive instinct operates under other forms and in other circumstances. (I am thinking of civil wars, for instance, due in earlier days to religious zeal, but nowadays to social factors; or, again, the persecution of racial minorities.) But my insistence on what is the most typical, most cruel and extravagant form of conflict among human communities was deliberate, for here we have the best occasion of discovering ways and means to render all armed conflicts impossible.

I know that in your writings we may find answers, explicit or implied, to all the issues of this urgent and absorbing problem. But it would be of the greatest service to us all were you to present the problem of world peace in the light of your most recent discoveries, for such a presentation might well blaze the trail for new and fruitful modes of action.

Yours very sincerely,

A. Einstein.

60. Is There a Jewish Philosophy? (1932)[a]

The Jewish Review 2 (September–December 1932), pp. 14–16.

In May 1932, Norman Bentwich (1883–1971), British legal scholar and former attorney general of Mandatory Palestine, solicited a contribution from Einstein for the new London *Jewish Review*. The journal would first address the question "Is there a Jewish *Weltanschauung*." Einstein also sent the essay to James W. Wise, the editor of the American journal *Opinion: A Journal of Jewish Life and Letters*.

In my opinion there is no such thing as a Jewish Weltanschauung in the philosophical sense, i.e., a Jewish conception of the external world. Judaism seems to me to be concerned solely with the moral attitude in life, and toward life.

I consider the content of Judaism to be rather the attitude to life inherent in the Jewish people than the laws laid down in the *Torah* and interpreted in the *Talmud*. *Torah* and *Talmud* are to me only the most important testimonies to the prevalence of the Jewish conception of life in earlier times.

The essence of this conception seems to me to lie in the affirmation of the life of all creatures. The life of the individual has a meaning only in the service of beautifying and ennobling of the life of all living things. Life is holy, i.e., the highest value, on which all other values depend. The sanctification of the supra-individual life entails the veneration of everything spiritual, a particularly characteristic trait of Jewish tradition.

Judaism is not a creed. The Jewish God is only a negation of superstition, an imaginary result of its elimination. He is also an attempt to base the moral law on fear, a regrettable, inglorious attempt. But it seems to me that the strong moral tradition among the Jewish people has largely detached itself from this fear. It is also clear that "serving God" was equated with "serving the living." For this end, the best spirits of the Jewish people, especially the prophets along with Jesus, have struggled untiringly.

Thus Judaism is essentially not a transcendent religion; it only has to do with the life we experience, to a certain extent tangible, and nothing else. It seems to me, therefore, questionable whether it can be called a "religion" in the ordinary sense of the word, especially as no "belief" is required of the Jew, but only the sanctification of life in the suprapersonal sense (*im überpersönlichen Sinne*).

There is, however, another element in Jewish tradition, which reveals itself so arrestingly in many of the Psalms, namely, a kind of drunken joy and wonder at the beauty and incomprehensible sublimity of this world, of which man can only gain but a faint inkling. This is the feeling from which true inquiry also draws its spiritual strength, but which also seems to express itself in the song of birds. This seems to me to be the most sublime content of the idea of God.[b]

Is all this characteristic of Judaism? Is it, perhaps, to be found elsewhere under another name? In the pure state it is to be found nowhere, not even in Judaism, where much literalism often obscures the pure doctrine. Nevertheless I see in Judaism one of its most vivid and pure realizations. This applies especially to the principle of the sanctification of life.

It is remarkable that animals were also explicitly included in the commandment to keep the Sabbath holy, such was the ideal of solidarity with the living. The command for solidarity between all people is expressed even more strongly, and it is no coincidence that the socialist demands originated mostly from Jews.

How strong the feeling of the sanctity of life is in the Jewish people may be illustrated by a remark which Walther Rathenau[c] once made to me in the course of a conversation: There is no simpler way to express the awareness of the sacredness and connectedness of life as it lives in the Jewish people.

Editorial Notes

[a] This translation by the editors hones closely to the German typed text sent by Einstein. At least three English-language translations of this text have been published in the past. They all have virtues and defects. Two German-language original versions are extant, both titled "Gibt es eine Jüdische Weltanschauung?": a manuscript in Einstein's hand [AEA 28-197] and a typed transcription with Einstein's handwritten emendations [AEA 28-196]. The latter was sent to the journal *Opinion* on 3 August 1932. A third English version appeared in *Ideas and Opinions*, pp. 185–187, with the title "Is There a Jewish Point of View?."

[b] In the manuscript [AEA 28-197], this last sentence reads: "To associate it with the God idea seems to be childish simplicity."

[c] On Rathenau, see Document 28.

61. Statement on Hitler upon Leaving Pasadena (1933)

New York World Telegram (11 March 1933).

On 30 January 1933, during Einstein's third and final visit to the California Institute of Technology in Pasadena, Adolf Hitler was appointed Chancellor of Germany. On 4 February, his cabinet restricted the press and political meetings. After the Reichstag fire on 27 February, the Decree of the Reich President for the Protection of the People and the State gave Hitler emergency powers. The decree suspended many articles of the constitution. It restricted freedom of the press and freedom of speech. The police received authority to dissolve political organizations and arrest and incarcerate political opponents of the regime.

As long as I have any choice, I will only stay in a country where political liberty, tolerance, and equality of all citizens before the law prevail. Political liberty implies the freedom to express one's political opinions orally and in writing; tolerance implies respect for any and every individual opinion.

These conditions do not obtain in Germany at the present time. Those who have done most for the cause of international understanding, among them some of the leading artists, are being persecuted there.

Any social organism can become physically distempered just as any individual can, especially in times of difficulty. Nations usually survive these distempers. I hope that healthy conditions will soon supervene in Germany and that in future her great men like Kant and Goethe will not merely be commemorated from time to time but that the principles which they taught will also prevail in public life and in the general consciousness.

62. Letter of Resignation from the Prussian Academy of Sciences (1933)

28 March 1933, from Antwerp, on board S. S. Belgenland.

Following Einstein's statement to the press on his departure from Pasadena (Doc. 61), widely reported in Germany as well, Einstein received a letter sent on 18 March 1933 by the secretary of the Prussian Academy, inquiring as to its veracity. The letter also stated that Max Planck, permanent secretary of the academy, viewed Einstein's "public declaration against the government" as an unfortunate development to which the Academy would need to respond.

To the Prussian Academy of Sciences, Berlin:

The current conditions in Germany prompt me to hereby resign my position at the Prussian Academy of Sciences.

For nineteen years, the Academy has given me the possibility to devote myself to scientific work, free of any professional obligations. I know how much I am indebted to it. Grudgingly, I also abandon the stimulation and the beautiful human relationships I enjoyed during this long period as its member, and which I still value highly.

Yet in the current conditions, I find my dependence on the Prussian government entailed by this position unbearable.

With all respect,

Albert Einstein

63. Militant Pacifism No More (1933)

La patrie humaine (18 August 1933), 81: 1.[a]

> An open letter to Belgian pacifist Alfred Nahon clarifying Einstein's apparent reversal of his militant pacifist views in light of the rearmament of Germany under Hitler. The letter was written on 20 July 1933, while Einstein was in exile at Le Coq-sur-mer (De Haan), Belgium. Already before its publication, the letter caused dismay and bitter criticism in pacifist circles.[b]

What I shall tell you will greatly surprise you. Until quite recently, we in Europe could assume that personal war resistance constituted an effective attack on militarism.

Today, we face an altogether different situation. In the heart of Europe lies a power, Germany, that is obviously pushing toward war with all available means. This has created such a serious danger to the Latin countries, especially Belgium and France, that they have come to depend completely on their armed forces.

As for Belgium, surely so small a country cannot possibly misuse its armed forces; rather, it needs them desperately to protect its very existence. Imagine Belgium occupied by present-day Germany! Things would evidently be far worse than in 1914, and they were bad enough even then.

Hence, I must tell you candidly: Were I a Belgian, I should not, in the present circumstances, refuse military service; rather, I should enter such service gladly in the belief that I would thereby be helping to save European civilization.

This does not mean that I am surrendering the principle for which I have stood heretofore. I have *no greater* hope than that the time may not be far off when refusal of military service will once again be an effective method of serving the cause of human progress.

Editorial Notes

[a] The bi-weekly French journal was established in 1931 by Victor Méric (1876–1933), an anarchist, socialist, and communist absolute pacifist, under the auspices of Victor Margueritte (1866–1942), pacifist essayist and novelist.

[b] Einstein's text was presented as an insert within a long diatribe by Gustave Dupin titled "Akeldama" (Potters' field, or Field of blood), referring to the Christian tradition according to which the land was bought with the thirty pieces of silver that Judas received for his betrayal of Jesus. Dupin wrote: "[Through this betrayal] Einstein has, unconsciously, placed himself on the same pillory as his tribesman Judas Iscariot."

64. On the Method of Theoretical Physics/The Herbert Spencer Lecture (1933)

On the Method of Theoretical Physics: The Herbert Spencer Lecture, Delivered at Oxford 10 June 1933.
Oxford: Clarendon Press (1933), pp. 5–16.[a]

A printing of Einstein's Herbert Spencer Lecture delivered 10 June 1933 at Oxford. It was preceded by the following note:

> I wish to preface what I have to say by expressing to you the great gratitude which I feel to the University of Oxford for having given to me the honor and privilege of delivering the Herbert Spencer Lecture. May I say that the invitation makes me feel that the links between this University and myself are becoming progressively stronger?

> I want also to render thanks to my colleagues at Christ Church, Mr. Ryle, Mr. Page, and Dr. Hurst, who have helped me—and perhaps a few of you—by translating into English the lecture which I wrote in German.

If you wish to learn from the theoretical physicist anything about the methods which he uses, I would give you the following piece of advice: Don't listen to his words, examine his achievements. For to the discoverer in that field, the constructions of his imagination appear so necessary and so natural that he is apt to treat them not as the creations of his thoughts but as given realities.

This statement may seem to be designed to drive my audience away without more ado. For you will say to yourselves, "The lecturer is himself a constructive physicist; on his own showing, therefore he should leave the consideration of the structure of theoretical science to the epistemologist."

So far as I personally am concerned, I can defend myself against an objection of this sort by assuring you that it was no suggestion of mine but the generous invitation of others which has placed me on this dais, which commemorates a man who spent his life in striving for the unification of knowledge.

But even apart from that, I have this justification for my pains, that it may possibly interest you to know how a man thinks about his science after having devoted so much time and energy to the clarification and reform of its principles.

Of course, his view of the past and present history of his subject is likely to be unduly influenced by what he expects from the future and what he is trying to realize to-day. But this is the common fate of all who have adopted a world of ideas as their dwelling-place.

He is in just the same plight as the historian, who also, even though unconsciously, disposes events of the past around ideals that he has formed about human society.

I want now to glance for a moment at the development of the theoretical method, and while doing so especially to observe the relation of pure theory to the totality of the data of experience. Here is the eternal antithesis of the two inseparable constituents of human knowledge, Experience and Reason, within the sphere of physics. We honor ancient

Greece as the cradle of western science. She for the first time created the intellectual miracle of a logical system, the assertions of which followed one from another such rigor that not one of the demonstrated propositions admitted of the slightest doubt—Euclid's geometry. This marvelous accomplishment of reason gave to the human spirit the confidence it needed for its future achievements. The man who was not enthralled in youth by this work was not born to be a scientific theorist. But the time for a science that could comprehend reality was not ripe until a second elementary truth had been realized, which only became the common property of philosophers after Kepler and Galileo. Pure logical thinking can give us no knowledge whatsoever of the world of experience; all knowledge about reality begins with experience and terminates in it.

Conclusions obtained by purely rational processes are, so far as Reality is concerned, entirely empty. It was because he recognized this, and especially because he impressed it upon the scientific world, that Galileo became the father of modern physics and in fact of the whole of modern natural science.

But if experience is the beginning and end of all our knowledge about reality, what role is there left for reason in science? A complete system of theoretical physics consists of concepts and basic laws to interrelate those concepts and of consequences to be derived by logical deduction. It is these consequences to which our particular experiences are to correspond, and it is the logical derivation of them which in a purely theoretical work occupies by far the greater part of the book. This is really exactly analogous to Euclidean geometry, except that in the latter the basic laws are called "axioms"; and, further, that in this field there is no question of the consequences having to correspond with any experiences. But if we conceive Euclidean geometry as the science of the possibilities of the relative placing of actual rigid bodies and accordingly interpret it as a physical science, and do not abstract from its original empirical content, the logical parallelism of geometry and theoretical physics is complete.

We have now assigned to reason and experience their place within the system of theoretical physics. Reason gives the structure to the system; the data of experience and their mutual relations are to correspond exactly to consequences in the theory. On the possibility alone of such a correspondence rest the value and the justification of the whole system, and especially of its fundamental concepts and basic laws. But for this, these latter would simply be free inventions of the human mind which admit of no *a priori* justification either through the nature of the human mind or in any other way at all.

The basic concepts and laws which are not logically further reducible constitute the indispensable and not rationally deducible part of the theory. It can scarcely be denied that the supreme goal of all theory is to make the irreducible basic elements as simple and as few as possible without having to surrender to the adequate representation of a single datum of experience.

The conception here outlined of the purely fictitious character of the basic principles of physical theory was in the eighteenth and nineteenth centuries still far from being the prevailing one. But it continues to gain more and more ground because of the ever-widening logical gap between the basic concepts and laws on the one side and the consequences to be correlated with our experiences on the other—a gap which widens progressively with the developing unification of the logical structure, that is, with the reduction of the number of the logically independent conceptual elements required for the basis of the whole system.

Newton, the first creator of a comprehensive and workable system of theoretical physics, still believed that the basic concepts and laws of his system could be derived from experience; his phrase "*hypotheses non fingo*" can only be interpreted in this sense. In fact, at that time it seemed that there was no problematical element in the concepts, Space and Time. The concepts of mass, acceleration and force, and the laws connecting them, appeared to be directly borrowed from experience. But if this basis is assumed, the expression for the force of gravity seems to be derivable from experience; and the same derivability was to be anticipated for the other forces.

One can see from the way he formulated his views that Newton felt by no means comfortable about the concept of absolute space, which embodied that of absolute rest; for he was alive to the fact that nothing in experience seemed to correspond to this latter concept. He also felt uneasy about the introduction of action at a distance. But the enormous practical success of his theory may well have prevented him and the physicists of the eighteenth and nineteenth centuries from recognizing the fictitious character of the principles of his system.

On the contrary, the scientists of those times were for the most part convinced that the basic concepts and laws of physics were not in a logical sense free inventions of the human mind, but rather that they were derivable by abstraction, i.e., by a logical process, from experiments. It was the general Theory of Relativity which showed in a convincing manner the incorrectness of this view. For this theory revealed that it was possible for us, using basic principles very far removed from those of Newton, to do justice to the entire range of the data of experience in a manner even more complete and satisfactory than was possible with Newton's principles. But quite apart from the question of comparative merits, the fictitious character of the principles is made quite obvious by the fact that it is possible to exhibit two essentially different bases, each of which in its consequences leads to a large measure of agreement with experience. This indicates that any attempt logically to derive the basic concepts and laws of mechanics from the ultimate data of experience is doomed to failure.

If, then, it is the case that the axiomatic basis of theoretical physics cannot be an inference from experience, but must be free invention, have we any right to hope that we shall find the correct way? Still more—does this correct approach exist at all, save in our imagination? Have we any right to hope that experience will guide us aright, when there are theories (like classical mechanics) which agree with experience to a very great extent, even without comprehending the subject in its depths? To this I answer with complete assurance, that in my opinion there is the correct path and, moreover, that it is in our power to find it. Our experience up to date justifies us in feeling sure that in Nature is actualized the ideal of mathematical simplicity. It is my conviction that pure mathematical construction enables us to discover the concepts and the laws connecting them which give us the key to the understanding of the phenomena of Nature. Experience can of course guide us in our choice of serviceable mathematical concepts; it cannot possibly be the source from which they are derived; experience of course remains the sole criterion of the serviceability of a mathematical construction for physics, but the truly creative principle resides in mathematics. In a certain sense, therefore, I hold it to be true that pure thought is competent to comprehend the real, as the ancients dreamed.

To justify this confidence of mine, I must necessarily avail myself of mathematical concepts. The physical world is represented as a four-dimensional continuum. If in this I adopt a Riemannian metric, and look for the simplest laws which such a metric can satisfy, I arrive at the relativistic gravitation-theory of empty space. If I adopt in this space a vector-field, or the anti-symmetrical tensor-field derived from it, and if I look for the simplest laws which such a field can satisfy, I arrive at the Maxwell equations for free space.

Having reached this point we have still to seek a theory for those parts of space in which the electrical density does not vanish. De Broglie surmised the existence of a wave-field, which could be used to explain certain quantum properties of matter. Dirac found in the "spinors" field-quantities of a new kind, whose simplest equations make it possible to deduce a great many of the properties of the electron. I and my colleague[b] discovered that these "spinors" constitute a special case of a field of a new sort which is mathematically connected with the metrical continuum of four dimensions, and it seems that they are naturally fitted to describe important properties of the electrical elementary particles.

It is essential for our point of view that we can arrive at these constructions and the laws relating them one with another by adhering to the principle of searching for the mathematically simplest concepts and connections of them. In the paucity of the mathematically existent simple field-types and of the relations between them lies the justification for the theorist's hope that he may comprehend reality in its depths.

The most difficult point for such a field-theory at present is how to include the atomistic structure of matter and energy. For the theory in its basic principles is not an atomic one in so far as it operates exclusively with continuous functions of space, in contrast to classical mechanics whose most important feature, the material point, agrees with the atomic structure of matter.

The modern quantum theory, as associated with the names of De Broglie, Schrödinger, and Dirac, which of course operates with continuous functions, has overcome this difficulty by means of a daring interpretation, first given in a clear form by Max Born:—the space functions which appear in the equations make no claim to be a mathematical model of atomic objects. These functions are only supposed to determine in a mathematical way the probabilities of encountering those objects in a particular place or in a particular state of motion if we make a measurement. This conception is logically unexceptionable, and has led to important successes. But unfortunately, it forces us to employ a continuum of which the number of dimensions is not that of previous physics, namely 4, but which has dimensions increasing without limit as the number of the particles constituting the system under examination increases. I cannot help confessing that I myself accord to this interpretation no more than a transitory significance. I still believe in the possibility of giving a model of reality, a theory, that is to say, which shall represent events themselves and not merely the probability of their occurrence.

On the other hand, it seems to me certain that we have to give up the notion of an absolute localization of the particles in a theoretical model. This seems to me to be the correct theoretical interpretation of Heisenberg's indeterminacy relation. And yet a theory may perfectly well exist, which though it is in a genuine sense an atomistic one (and not merely on the basis of a particular interpretation), nevertheless involves no localizing of the particles in a mathematical model. For example, in order to include the atomistic character of electricity, the field equations only need imply that a three-dimensional volume of space, on whose boundary the electrical density vanishes everywhere, contains a total electrical charge of an integral amount. Thus in a continuum theory, the atomistic character could be satisfactorily expressed by integral propositions without localizing the particles which constitute the atomic system.

Only if this sort of representation of the atomistic structure be obtained could I regard the quantum problem within the framework of a continuum theory as solved.

Editorial Notes

[a] Also printed in *Philosophy of Science* 1, No. 2 (April 1934), pp. 163–169, without Einstein's preface.

[b] Einstein is here referring to his collaborator Walther Mayer.

65. Science and Civilization/The Albert Hall Speech (1933)[a]

Royal Albert Hall Programme (3 October 1933).

> Einstein, supported by Lord Rutherford, was the chief figure at the Royal Albert Hall at a meeting organized for the purpose of raising funds for the Refugee Assistance Committee. Lord Rutherford said that four bodies were concerned in relief work for academic and professional workers, and in order to avoid overlapping and to avoid confusing the public they were making a joint appeal. Those bodies were the Academic Assistance Council, of which he was president, the International Students Service, the Refugee Professionals Committee, and the Germany Emergency Committee of the Society of Friends. Their object was to collect a fund to be employed for the relief of students, university teachers, and members of the professional classes.

I am glad that you have given me the opportunity of expressing to you here my deep sense of gratitude as a man, as a good European,[b] and as a Jew. Through your well-organized work of relief, you have done a great service not only to innocent scholars who have been persecuted, but also to humanity and science. You have shown that you and the British people have remained faithful to the traditions of tolerance and justice, which for centuries you have upheld with pride. It is in times of economic distress such as we experience everywhere today, one sees very clearly the strength of the moral forces that live in a people. Let us hope that a historian delivering judgement in some future period when Europe is politically and economically united, will be able to say that in our days the liberty and honor of this Continent was saved by its Western nations, which stood fast in hard times against the temptations of hatred and oppression; and that Western Europe defended successfully the liberty of the individual which has brought us every advance of knowledge and invention—liberty without which life to a self-respecting man is not worth living.

It cannot be my task to-day to act as judge of the conduct of a nation which for many years has considered me as her own; perhaps it is an idle task to judge in times when action counts.

Today, the questions which concern us are: How can we save mankind and its spiritual acquisitions, of which we are the heirs? How can one save Europe from a new disaster?

It cannot be doubted that the world crisis and the suffering and privations of the people resulting from the crisis are in some measure responsible for the dangerous upheavals of which we are the witnesses. In such periods, discontent breeds hatred, and hatred leads to acts of violence and revolution, and often even to war. Thus, distress and evil produce new distress and new evil. Again, the leading statesmen are burdened with tremendous responsibilities, just as twenty years ago. May they succeed through timely agreement to establish a condition of unity and clarity of international obligations in Europe, so that for every State a war-like adventure must appear as utterly hopeless. But the work of statesmen can succeed only if they are backed by the serious and determined will of the people.

We are concerned not merely with the technical problem of securing and maintaining peace, but also with the important task of education and enlightenment. If we want to resist the powers which threaten to suppress intellectual and individual freedom, we must keep clearly before us what is at stake, and what we owe to that freedom which our ancestors have won for us after hard struggles.

Without such freedom, there would have been no Shakespeare, no Goethe, no Newton, no Faraday, no Pasteur and no Lister. There would be no comfortable houses for the mass of people, no railway, no wireless, no protection against epidemics, no cheap books, no culture and no enjoyment of art for all. There would be no machines to relieve the people from the arduous labor needed for the production of the essential necessities of life. Most people would lead a dull life of slavery, just as under the ancient despotisms of Asia. It is only men who are free who create the inventions and intellectual works which to us moderns make life worthwhile.

Without doubt, the present economic difficulties will eventually bring us to the point where the balance between the supply and demand of labor, between production and consumption, will be enforced by law. But even this problem we shall solve as free men, and we shall not allow ourselves for its sake to be driven into a slavery which ultimately would bring with it stagnation of every healthy development.

In this connection, I should like to give expression to an idea which has occurred to me recently. I lived in solitude in the country and noticed how the monotony of a quiet life stimulates the creative mind. There are certain callings in our modern organization which entail such an isolated life without making a great claim on bodily and intellectual effort. I think of such occupations as the service in lighthouses and lightships. Would it not be possible to fill such places with young people who wish to think out scientific problems, especially of a mathematical or philosophical nature? Very few of such people have the opportunity during the most productive period of their lives to devote themselves undisturbed for any length of time to scientific problems. Even if a young person is lucky enough to obtain a scholarship for a short period, he must endeavor to arrive as quickly as possible at definite conclusions. That cannot be of advantage in the pursuit of pure science. The young scientist who carries on an ordinary practical profession which maintains him is in a much better position—assuming of course that this profession leaves him with sufficient spare time and energy. In this way, perhaps a greater number of creative individuals could be given an opportunity for mental development than is possible at present. In these times of economic depression and political upheaval, such considerations seem to be worth attention.

Shall we worry over the fact that we are living in a time of danger and want? I think not. Man, like every animal, is by nature indolent. If nothing spurs him on, then he will hardly think, and will behave from habit like an automaton. I am no longer young and can therefore say that, as a child and as a young man, I experienced that phase—when a young man thinks only about the trivialities of personal existence, and talks like his fellows and behaves like them. Only with difficulty can one see what is really behind such a conventional mask. For owing to habit and speech, his real personality is, as it were, wrapped in cotton wool.

How different it is to-day! In the lightning flashes of our tempestuous times, one sees human beings and things in their nakedness. Every action and every human being reveal clearly their aims, powers and weaknesses, and also their passions. Routine becomes no avail under the swift change of conditions; conventions fall away like dry husks.

Men in distress begin to think about the failure of economic practice and about the necessity of political combinations which are supernational. Only through perils and upheavals can Nations be brought to further developments. May the present upheavals lead to a better world.

Above and beyond this valuation of our time we have this further duty, the care for what is eternal and highest amongst our possessions, that which gives to life its import and which we wish to hand on to our children purer and richer than we received it from our forbears. Toward these purposes, you have affectionately contributed with your blessed services.

Editorial Notes

[a] The English printed version of this speech is based on the first of two drafts that Einstein wrote in pencil [AEA 28-253 and 28-254]. In the second version, Einstein expressed himself much more forcefully against Nazi Germany, Fascism, militarism, and communism.

[b] For Einstein's use of the phrase "good European," see Docs. 3 and 41.

66. Foreword to "The Contribution of the Jews of Germany to German Civilization" (1933)

Cahiers Juifs, Vol. 2, no. 5–6 (septembre-novembre 1933) and *Mein Weltbild* (1934), pp. 161–162.[a]

This text was published in French in *Cahiers juifs*, a journal founded in 1933 by the chief rabbi of Egypt, David Prato, and published six times a year in Alexandria and Paris. The entire issue was devoted to the subject of the contributions of German Jews to German civilization. The German original manuscript [AEA 28-242] and an English translation were published a few months later.

The following pages are dedicated to an appreciation of the achievements of the German Jews. It must be remembered that we are concerned here with a body of people amounting, in numbers, to no more than the population of a moderate-sized town, who have held their own against 100 times as many Germans, in spite of handicaps and prejudices, through the superiority of their ancient cultural traditions. Whatever attitude people may take up toward this little people, nobody who retains a shred of sound judgment in these times of confusion can deny them respect. In these days of the persecution of the German Jews especially, it is time to remind the western world that it owes to the Jewish people (a) its religion and therewith its most valuable moral ideals, and (b), to a large extent, the resurrection of the world of Greek thought. Nor should it be forgotten that it was a translation of the Bible, that is to say, a translation from Hebrew, which brought about the refinement and perfection of the German language.[b] Today the Jews of Germany find their fairest consolation in the thought of all they have produced and achieved for humanity by their efforts in modern times as well; and no oppression however brutal, no campaign of calumny however subtle will blind those who have eyes to see the intellectual and moral qualities inherent in this people.

Editorial Notes

[a] Originally published in *Mein Weltbild* in 1934 by Querido Verlag in Amsterdam and in *The World as I See It*. New York: Covici-Friede 1934, pp. xv–xvi. Translated into English by Alan Harris. In early 1934, Rudolf Kayser, the son-in-law of Einstein, published in German a collection of Einstein's essays, newspaper articles, and statements under the title *Mein Weltbild*. It was translated into English and published a few months later by the Covici-Friede publisher in New York. Elsa Einstein, who was in Paris at the side of her dying daughter Ilse, had requested proofs of the book, which she did not receive in time. She objected to the book's title, since both she and Einstein had made it clear that the book was not by Einstein himself. She proposed the title "Einstein, as he looks at the world." Einstein had written an article with this preliminary German title ("Wie ich in die Welt sehe") already in 1930. It was published as "What I Believe" in *Forum and Century* in November 1930 (see Doc. 53).

[b] Einstein is here referring to Martin Luther's translations of the *New Testament* 1522 and the *Old Testament* 1534.

67. On Germany and Hitler (1935)

O. Nathan and H. Norden, eds. *Einstein on Peace*. New York: Simon & Schuster (1960), pp. 263–264.

> Einstein most likely wrote the text below in 1935 or 1936. It was not published during his lifetime.

To the everlasting shame of Germany, the spectacle unfolding in the heart of Europe is tragic and grotesque, and it reflects no credit on the community of nations which calls itself civilized!

For centuries, the German people have been subject to indoctrination by an unending succession of schoolmasters and drill sergeants. The Germans have been trained in hard work and made to learn many things, but they have also been drilled in slavish submission, military routine and brutality. The postwar democratic constitution of the Weimar Republic fitted the German people about as well as the giant's clothes fitted Tom Thumb. Then came inflation and depression, with everyone living in fear and tension.

Hitler appeared, a man with limited intellectual abilities and unfit for any useful work, bursting with envy and bitterness against all whom nature and fate had favored over him. He was the son of a petty bourgeois family with just enough class arrogance to hate even the working class, which was struggling for greater equality in living standards. But it was the culture and education which had been denied him forever that he hated most of all. In his desperate situation, he found that his confused, hate-filled speeches were met with stormy approval by the oh-so-many whose situation and attitude were similar to his own. He picked up this human flotsam on the streets and in the taverns and organized them around himself. This is the way he launched his political career.

But what made him a "leader" was his bitter hatred of foreign countries and of a defenseless minority, the German Jews. He particularly hated them as representatives of a spirituality that he found uncanny and, not entirely unjustifiably, un-German.

Incessant tirades against these two "enemies" won him the support of the masses, to whom he promised glorious triumphs and a golden age. He shrewdly exploited for his own purposes the centuries-old habitual urge of the masses, who had been militarily drilled to march, to blindly obey orders, to drudgery. Thus he became the *Fuehrer*.

Money flowed plentifully into his coffers, not least from the propertied classes who saw him as a tool for preventing the social and economic liberation of the people, which had its beginning under the Weimar Republic. He played up to the people with the kind of romantic, pseudo-patriotic phrase mongering to which they had become accustomed in the period before the [First] World War, as well as through the hoax of the superiority of an "Aryan" or "Nordic" race invented by the antisemites for their special purposes. His disjointed personality makes it impossible for me to know to what degree he might actually have believed in the nonsense which he kept on dispensing. Those, however, who rallied around him and who came to the surface through the Nazi wave were for the most part hardened cynics, fully aware of the mendacity of their methods.

68. Some Thoughts Concerning Education (1936)

School and Society 44, no. 1141 (1936), pp. 589–592.

> Address at the seventy-second convocation of the State University of New York in Chancellors Hall of the State Education Building at Albany, N.Y., on 15 October 1936, in celebration of the Tercentenary of Higher Education in America.

A day of celebration like this one generally is in the first place dedicated to retrospect, especially to the memory of public figures who have gained special distinction for the development of the country's cultural life. This service of love for our predecessors must indeed not be neglected, especially as such a remembrance of the best of the past is likely to inspire the well-disposed of to-day to a courageous effort. But this should be done by someone who, from his youth, has been connected with the state of New York and is familiar with its past, not by one who, like a gypsy, has wandered about and gathered his experiences in all kinds of countries.

Thus, if I endeavor to do justice to your kind and highly esteemed invitation, there is nothing else left for me but to speak about such questions as, independently of space and time, always have been and will be connected with educational matters. In this attempt, I cannot lay any claim to being an authority, especially as intelligent and well-meaning people of all times have dealt with educational problems and have certainly repeatedly expressed their views clearly about this subject. From what source shall I, as a partial layman in the realm of pedagogy, derive courage to expound to you opinions with no foundations except personal experience and personal conviction?

If it were really a scientific matter, one would probably be tempted to silence by such considerations. However, with the affairs of active human beings it is different. Here knowledge of truth alone does not suffice; on the contrary, this knowledge must continually be renewed by ceaseless effort, if it is not to be lost. It resembles a statue of marble which stands in the desert and is continuously threatened with burial by the shifting sand. The hands of service must ever be at work, in order that the marble continue lastingly to shine in the sun. To these serving hands mine also shall belong.

The school has always been the most important means of transmitting the wealth of tradition from one generation to the next. This applies to-day in an even higher degree than in former time for, through the modern development of economic life, the family as bearer of tradition and education has been weakened. The continued existence and health of human society is therefore in a still higher degree more dependent on the school than formerly.

Sometimes one sees in the school simply the instrument for transferring a certain maximum quantity of knowledge to the growing generation. But that is not so. Knowledge is dead; the school, however, serves the living. It should develop in the young individuals those qualities and capabilities which are of value for the welfare of the commonwealth. But that does not mean that individuality should be destroyed, and the individual become a mere tool of the community, like a bee or an ant. For a community of standardized individuals without personal uniqueness and personal aims would be a poor community without possibilities for development. On the contrary, the aim must be the training of independently acting and thinking individuals, who, however, see in the service of the community their highest mission in life. As far as I can judge, the English school system comes nearest to the realization of this ideal.

But how shall one try to attain this ideal? Should one perhaps try to realize this aim by moralizing? Not at all! Words are and remain an empty sound, and the road to hell has ever been accompanied by lip service to an ideal. But personalities are not formed by what is heard and said, but by work and action.

The most important method of education, accordingly, always has consisted of that in which the pupil was urged to actual performance. This applies as well to the first attempts at writing of the primary boy as to the doctor's thesis on graduation from university, or as to the mere memorizing of a poem, the writing of a composition, the interpretation and translation of a text, the solving of a mathematical problem, or the practice of physical sport.

But behind every achievement stands the motivation which is at the foundation of that achievement, and which in turn is strengthened and nourished by the accomplishment of the undertaking. Here reside the greatest differences, and they are of the greatest importance to the educational value of the school. The same achievement may owe its existence to fear and compulsion, ambitious desire for validation and distinction, or loving interest in the subject and a desire for truth and understanding, and thus to that divine curiosity which every healthy child possesses, but which so often atrophies prematurely. The educational influence which is exercised upon the pupil by the execution of one and the same task may be widely different, depending upon whether fear of hurt, egoistic passion or desire for pleasure and satisfaction are at the bottom of this assignment. And nobody will maintain that the administration of the school and the attitude of the teachers does not have an influence upon the molding of the psychological foundation for pupils.

To me, the worst thing seems to be for a school principally to work with methods of fear, force and artificial authority. Such treatment destroys the sound sentiments, the sincerity, and the self-confidence of the pupil. It produces the submissive subject. It is no wonder that such schools were the rule in Germany and Russia. I know that the schools in this country are free from this worst evil; this also is so in Switzerland and probably in all democratically governed countries. It is comparatively simple to keep the school free from this worst of all evils. Give into the hands of the teacher the fewest possible coercive measures, so that the only source of the pupil's respect for the teacher is the human and intellectual qualities of the latter.

The second-named motive, ambition or, in milder terms, the aiming at recognition and consideration, lies firmly fixed in human nature. With absence of mental stimulus of this kind, human cooperation would be entirely impossible; the desire for the approval of one's fellow man certainly is one of the most important binding forces of society. In this complex of feelings, constructive and destructive forces lie closely together. Desire for approval and recognition is a healthy motive, but the desire to be acknowledged as better, stronger, or more intelligent than a fellow being or fellow pupil easily leads to an excessively egoistic psychological attitude, which may become injurious for the individual and for the community. Therefore, the school and the teacher must guard against employing the easy method of inciting individual ambition in order to induce the pupils to diligent work.

Many have cited Darwin's theory of the struggle for existence and the selectivity connected with it as justification for the encouragement of the spirit of competition. Some have also tried to prove pseudoscientifically the necessity of the anarchic system of economic competition. But this is mistaken, because man owes his strength in the struggle for existence to the fact that he is a socially living animal. As little as a battle between single ants of an ant hill is essential for survival, just so little is this the case with the individual members of a human community.

Therefore, one should guard against preaching success to young people in the customary sense of the aim of life. For a successful person is in most cases he who receives a great deal from his fellow men, usually incomparably more than corresponds to his service to them. The value of a person, however, should be seen in what he gives and not in what he is able to receive.

The most important motive for achievement in the school and in life is the pleasure in work, pleasure in its result, and the knowledge of the value of the result to the community. I see the school's most important task in the awakening and strengthening of these psychological forces in the young person. Such a psychological foundation alone leads to a joyous desire for the highest possessions of mankind, knowledge and artistic creation.

The awakening of these productive psychological powers is certainly less easy than the exercise of compulsion or the awakening of individual ambition but is the more valuable for it. The point is to develop the childlike inclination for play and the childlike desire for knowledge and to guide the child toward socially important fields; it is that education which is based primarily upon the desire for successful activity and understanding. If the school succeeds in working successfully according to such points of view, it will be highly regarded by the rising generation and school assignments will be taken up as a kind of gift. I have known children who preferred school time to vacation.

Such a school demands from the teacher that he be a kind of artist in his province. What can be done to ensure that this spirit prevails in school? For this there is just as little universal remedy as there is for an individual to remain healthy. But there are certain preconditions which can be met. First, teachers should grow up in such schools. Second, the teacher should be given extensive freedom in the selection of the material to be taught and the methods of teaching employed. For it is true also of him that pleasure in the shaping of his work is killed by force and external pressure.

If you have followed attentively my considerations up to this point, you will probably wonder about one thing. I have spoken in detail about the spirit in which, according to my opinion, youth should be instructed. But I have said nothing yet about the choice of subjects for instruction, nor about the method of teaching. Should language predominate, or scientific education?

To this I answer: In my opinion all this is of secondary importance. If a young person has trained his muscles and physical endurance by gymnastics and walking, he will later be capable of every physical work. This is also analogous to the training of the mind and the exercising of the mental and manual skill. Thus, the wit was not wrong who defined education in this way: "Education is that which remains, if one has forgotten everything one has learned in school." For this reason, I am not at all anxious to take sides in the struggle between the followers of the classical philologic-historical education and the education more devoted to natural science.

On the other hand, I want to oppose the idea that the school has to teach directly that special knowledge and those accomplishments which one has to directly apply later in life. The demands of life are much too manifold to allow for the possibility of such a specialized training in school. Apart from that, it seems to me, moreover, objectionable to treat the individual like a dead tool. The school should seek that the young person leave it as a harmonious personality, not as a specialist. This, in my opinion, is true in a certain sense even for technical schools, whose graduates will devote themselves to quite definite professions. The development of a general ability for independent thinking and judgment should always be placed foremost, not the acquisition of specialized knowledge. If one masters the fundamentals of one's subject and has learned to think and work independently, one will better be able to adapt to progress and changes than those whose training principally consists in acquiring detailed knowledge.

Finally, I wish to emphasize once more that what has been said here in somewhat categorical form does not claim to mean more than the personal opinion of a man, which is founded upon *nothing but* his own personal experience, which he has gathered as a student and as a teacher. I thank you for having given me the opportunity to express these opinions before such a distinguished gathering.

69. The Calling of the Jews (1936)

Out of My Later Years. New York: Philosophical Library (1950), p. 268.

An address upon his induction as honorary fellow of the Jewish Academy of Arts and Sciences on 22 March 1936. The previously unpublished German text appeared in English in the collection of essays *Out of My Later Years*, prepared in 1949 by Rudolf Kayser and printed by the Philosophical Library in spring 1950. The initial planned title for the collection was *Science and Life: New Essays by Albert Einstein*. A retranslation of the English edition into German was published in 1952 without Einstein's explicit permission.

This is a time when there seems to be a particular need for men of philosophical persuasion—that is to say, friends of wisdom and truth—to join together. For while it is true that our time has accumulated more knowledge than any earlier age, that love of truth and insight which lent wings to the spirit of the Renaissance has grown cold, giving way to sober specialization rooted in the material spheres of society rather than in the spiritual. But groups such as this one are devoted solely to spiritual aims.

In centuries past, Judaism clung exclusively to its moral and spiritual tradition. Its teachers were its only leaders. But with adaptation to a larger social whole, this spiritual orientation has receded into the background, though even today the Jewish people owe to it their apparently indestructible vigor. If we are to preserve that vigor for the benefit of mankind, we must hold to that spiritual orientation toward life.[a]

The dance about the Golden Calf was not merely a legendary episode in the history of our forefathers—an episode that seems to me in its simplicity more innocent than that total adherence to material and selfish goals threatening Judaism in our own days. At this time, a union of those who rally to the spiritual heritage of our people has supreme justification. This is all the more true for a group[b] that is free of all historical and national narrowness. We Jews should be and remain the carriers and patrons of spiritual values. But we should also always be aware of the fact that these spiritual values are and always have been the common goal of all mankind.

Editorial Notes

[a] In the German original, Einstein added: "that has always been the source of new strength and has led to the development of great personalities."

[b] In the German original, Einstein wrote "community" instead of "group."

70. Why Do They Hate the Jews? (1938)

Collier's 102 (26 November 1938), pp. 9–10, 38.

"*Collier's* asked the most famous victims of antisemitism to explain why the Jewish people are the object of organized persecution in Germany, in Italy, and elsewhere. Why do they inspire bitterness even in democratic countries? Dr. Einstein here presents his views." The magazine carried the cover line "Einstein Comforts the Jews." Einstein donated the original autograph signed German manuscript of this article, titled "Antisemitismus," to a book auction dinner to benefit both the American Jewish Joint Distribution Committee and the Committee for Christian German Refugees at the Plaza Hotel in New York on 8 December 1938.

I should like to begin by telling you an ancient fable, with a few minor changes—a fable that will serve to throw into bold relief the mainsprings of political antisemitism:

The shepherd boy said to the horse: "You are the noblest beast that treads the earth. You deserve to live in untroubled bliss; and indeed your happiness would be complete were it not for the treacherous stag. But the stag has practiced from youth to excel you in fleetness of foot. His faster pace allows him to reach the water holes before you do. He and his tribe drink up the water far and wide, while you and your foal are left to thirst. Stay with me! My wisdom and guidance shall deliver you and your kind from a dismal and ignominious state."

Blinded by envy and hatred of the stag, the horse agreed. He yielded to the shepherd lad's bridle. He lost his freedom and became the shepherd's slave.

The horse in this fable represents a people, and the shepherd lad a class or clique aspiring to absolute rule over the people; the stag, on the other hand, represents the Jews.

I can hear you say: "A most unlikely tale! No creature would be as foolish as the horse in your fable." But let us give it a little more thought. The horse had been suffering the pangs of thirst, and his vanity was often pricked when he saw the nimble stag outrunning him. You, who have known no such pain and vexation, may find it difficult to understand that hatred and blindness should have driven the horse to act with such ill-advised, gullible haste. The horse, however, fell an easy victim to temptation because his earlier tribulations had prepared him for such a blunder. For there is much truth in the saying that it is easy to give just and wise counsel—to others—but hard to act justly and wisely for oneself. I say to you with full conviction: We all have often played the tragic role of the horse and we are in constant danger of yielding to temptation again.

The situation illustrated in this fable happens again and again in the life of individuals and nations. In brief, we may call it the process by which dislike and hatred of a given person or group are diverted to another person or group incapable of effective defense. But why did the role of the stag in the fable so often fall to the Jews? Why did the Jews so often happen to draw the hatred of the masses? Primarily because there are Jews among almost all nations and because they are everywhere too thinly scattered to defend themselves against violent attack.

A few examples from the recent past will prove the point: Toward the end of the nineteenth century the Russian people were chafing under the tyranny of their government. Stupid blunders in foreign policy further strained their temper until it reached the breaking point. In this extremity the rulers of Russia sought to divert unrest by inciting the masses to hatred and violence toward the Jews. These tactics were repeated after the Russian government had drowned the dangerous revolution of 1905 in blood—and this maneuver may well have helped to keep the hated regime in power until near the end of the World War.

When the Germans had lost the World War hatched by their ruling class, immediate attempts were made to blame the Jews, first for instigating the war and then for losing it. In the course of time, success attended these efforts. The hatred engendered against the Jews not only protected the privileged classes, but enabled a small, unscrupulous, and insolent group to place the German people in a state of complete bondage.

The crimes with which the Jews have been charged in the course of history—crimes which were to justify the atrocities perpetrated against them—have changed in rapid succession. They were supposed to have poisoned wells. They were said to have murdered children for ritual purposes. They were falsely charged with a systematic attempt at the economic domination and exploitation of all mankind. Pseudo-scientific books were written to brand them an inferior, dangerous race. They were reputed to foment wars and revolutions for their own selfish purposes. They were presented at once as dangerous innovators and as enemies of true progress. They were charged with falsifying the culture of nations by penetrating the national life under the guise of becoming assimilated. In the same breath they were accused of being so stubbornly inflexible that it was impossible for them to fit into any society.

Almost beyond imagination were the charges brought against them, charges known to their instigators to be untrue all the while, but which time and again influenced the masses. In times of unrest and turmoil the masses are inclined to hatred and cruelty, whereas in times of peace these traits of human nature emerge but stealthily.

Up to this point I have spoken only of violence and oppression against the Jews—not of antisemitism itself as a psychological and social phenomenon existing even in times and circumstances when no special action against the Jews is under way. In this sense, one may speak of latent antisemitism. What is its basis? I believe that in a certain sense one may actually regard it as a normal manifestation in the life of a people.

The members of any group existing in a nation are more closely bound to one another than they are to the remaining population. Hence a nation will never be free of friction while such groups continue to be distinguishable. In my belief, uniformity in a population would not be desirable, even if it were attainable. Common convictions and aims, similar interests, will in every society produce groups that, in a certain sense, act as units. There will always be friction between such groups—the same sort of aversion and rivalry that exists between individuals.

The need for such groupings is perhaps most easily seen in the field of politics, in the formation of political parties. Without parties the political interests of the citizens of any state are bound to languish. There would be no forum for the free exchange of opinions. The individual would be so isolated and unable to assert his convictions. Political convictions, moreover, ripen and grow only through mutual stimulation and criticism offered by individuals of similar disposition and purpose; and politics is no different from any other field of our cultural existence. Thus it is recognized, for example, that in times of intense religious fervor different sects are likely to spring up whose rivalry stimulates religious life in general. It is well known, on the other hand, that centralization—that is, elimination of independent groups—leads to one-sidedness and barrenness in science and art because such centralization checks and even suppresses any rivalry of opinions and research trends.

§1. Just what is a Jew?

The formation of groups has an invigorating effect in all spheres of human striving, perhaps mostly due to the struggle between the convictions and aims represented by the different groups. The Jews too form such a group with a definite character of its own, and antisemitism is nothing but the antagonistic attitude produced in the non-Jews by the Jewish group. This is a normal social reaction. But for the political abuse resulting from it, it might never have been designated by a special name.

What are the characteristics of the Jewish group? What, in the first place, is a Jew? There are no quick answers to this question. The most obvious answer would be the following: A Jew is a person professing the Jewish faith. The superficial character of this answer is easily recognized by means of a simple parallel. Let us ask the question: What is a snail? An answer similar in kind to the one given above might be: A snail is an animal inhabiting a snail shell. This answer is not altogether incorrect; nor, to be sure, is it exhaustive; for the snail shell happens to be but one of the material products of the snail. Similarly, the Jewish faith is but one of the characteristic products of the Jewish community. It is, furthermore, known that a snail can shed its shell without thereby ceasing to be a snail.[a] The Jew who abandons his faith (in the formal sense of the word) is in a similar position. He remains a Jew.

Difficulties of this kind appear whenever one seeks to explain the essential character of a group.

The bond that has united the Jews for thousands of years and that unites them today is, above all, the democratic ideal of social justice, coupled with the ideal of mutual aid and tolerance among all men. Even the most ancient religious scriptures of the Jews are steeped in these social ideals, which have powerfully affected Christianity and Mohammedanism and have had a benign influence upon the social structure of a great part of mankind. The introduction of a weekly day of rest should be remembered here—a profound blessing to all mankind. Personalities such as Moses, Spinoza, and Karl Marx, dissimilar as they may be, all lived and sacrificed themselves for the ideal of social justice; and it was the tradition of their forefathers that led them on this thorny path. The unique accomplishments of the Jews in the field of philanthropy spring from the same source.

The second characteristic trait of Jewish tradition is the high regard in which it holds every form of intellectual aspiration and spiritual effort. I am convinced that this great respect for intellectual striving is solely responsible for the contributions that the Jews have made toward the progress of knowledge, in the broadest sense of the term. In view of their relatively small number and the considerable external obstacles constantly placed in their way on all sides, the extent of those contributions deserves the admiration of all sincere men. I am convinced that this is not due to any special wealth of endowment, but to the fact that the esteem in which intellectual accomplishment is held among the Jews creates an atmosphere particularly favorable to the development of any talents that may exist. At the same time a strong critical spirit prevents blind obeisance to any mortal authority.

I have confined myself here to these two traditional traits, which seem to me the most basic. These standards and ideals find expression in small things as in large. They are transmitted from parents to children; they color conversation and judgment among friends; they fill the religious scriptures; and they give to the community life of the group its characteristic stamp. It is in these distinctive ideals that I see the essence of Jewish nature. That these ideals are but imperfectly realised in the group—in its actual everyday life—is only natural. However, if one seeks to give brief expression to the essential character of a group, the approach must always be by the way of the ideal.

§2. Where oppression is a stimulus

In the foregoing I have conceived of Judaism as a community of tradition. Both friend and foe, on the other hand, have often asserted that the Jews represent a race; that their characteristic behavior is the result of innate qualities transmitted by *heredity* from one generation to the next. This opinion gains weight from the fact that the Jews for thousands of years have predominantly married within their own group. Such a custom may indeed *preserve* a homogeneous race—if it existed originally; it cannot *produce* uniformity of the race—if there was originally a racial intermixture. The Jews, however, are beyond doubt a mixed race, just as are all other groups of our civilization. Sincere anthropologists are agreed on this point; assertions to the contrary all belong to the field of political propaganda and must be rated accordingly.

Perhaps even more than on its own tradition, the Jewish group has thrived on oppression and on the antagonism it has forever met in the world. Here undoubtedly lies one of the main reasons for its continued existence through so many thousands of years.

The Jewish group, which we have briefly characterized in the foregoing, embraces about sixteen million people—less than one per cent of mankind, or about half as many as the population of present-day Poland.[b] Their significance as a political factor is negligible. They are scattered over almost the entire earth and are in no way organized as a whole—which means that they are incapable of concerted action of any kind.

Were anyone to form a picture of the Jews solely from the utterances of their enemies, he would have to reach the conclusion that they represent a world power. At first sight that seems downright absurd; and yet, in my view, there is a certain meaning behind it. The Jews as a group may be powerless, but the *sum* of the achievements of their individual members is everywhere considerable and telling, even though these achievements were made in the face of obstacles. The forces dormant in the individual are mobilized, and the individual himself is stimulated to self-sacrificing effort, by the spirit that is alive in the group.

Hence the hatred of the Jews by those who have reason to shun popular enlightenment. More than anything else in the world, they fear the influence of men of intellectual independence. I see in this the essential cause for the savage hatred of Jews raging in present-day Germany. To the Nazi group the Jews are not merely a means for turning the resentment of the people away from themselves, the oppressors; they see the Jews as a non-assimilable element that cannot be driven into uncritical acceptance of dogma, and that, therefore—as long as it exists at all—threatens their authority because of its insistence on popular enlightenment of the masses.

Proof that this conception goes to the heart of the matter is convincingly furnished by the solemn ceremony of the burning of the books staged by the Nazi regime shortly after its seizure of power. This act, senseless from a political point of view, can only be understood as a spontaneous emotional outburst. For that reason, it seems to me more revealing than many acts of greater purpose and practical importance.

In the field of politics and social science there has grown up a justified distrust of generalizations pushed too far. When thought is too greatly dominated by such generalizations, misinterpretations of specific sequences of cause and effect readily occur, doing injustice to the actual multiplicity of events. Abandonment of generalization, on the other hand, means to relinquish understanding altogether. For that reason I believe one may and must risk generalization, as long as one remains aware of its uncertainty. It is in this spirit that I wish to present in all modesty my conception of antisemitism, considered from a general point of view.

In political life I see two opposed tendencies at work, locked in constant struggle with each other. The first, optimistic, trend proceeds from the belief that the free unfolding of the productive forces of individuals and groups essentially leads to a satisfactory state of society. It recognizes the need for a central power, placed above groups and individuals, but concedes to such power only organizational and regulatory functions. The second, pessimistic, trend assumes that free interplay of individuals and groups leads to the destruction of society; it thus seeks to base society exclusively upon authority, blind obedience, and coercion. In fact, this trend is pessimistic only to a limited extent; for it is optimistic in regard to those who are, and desire to be, the bearers of power and authority. The adherents of this second trend are the enemies of the free groups and of education for independent thought. They are, moreover, the carriers of political antisemitism.

Here in America all pay lip service to the first, optimistic, tendency. Nevertheless, the second group is strongly represented. It appears on the scene everywhere, though for the most part it hides its true nature. Its aim is political and spiritual dominion over the people by a minority, by the circuitous route of control over the means of production. Its proponents have already tried to utilize the weapon of antisemitism as well as of hostility to various other groups. They will repeat the attempt in times to come. So far, all such tendencies have failed because of the people's sound political instinct.

And so it will remain in the future, if we cling to the rule: Beware of flatterers, especially when they come preaching hatred.

Editorial Notes

[a] Horace B. Baker, a Philadelphia specialist in mollusks, drew Einstein's attention to this unfounded belief. Snails apparently cannot survive in the wild without their shells.

[b] The world's population in 1938 was about 2.3 billion, of whom 16.6 million (or 0.7 percent were Jews. It currently stands at about 15.8 million, or 0.2 percent). Einstein erred here in regard to Poland's population, which was about 35 million.

71. Our Debt to Zionism (1938)

New Palestine 28 (29 April), pp. 2–4.

Text of an address delivered at the sixth annual "Third Seder" celebration of the National Labor Committee for Palestine on 17 April at the Hotel Commodore in New York City.

Rarely since the conquest of Jerusalem by Titus has the Jewish community experienced a period of greater oppression than prevails at the present time. In some respects, indeed, our own time is even more troubled, for man's possibilities of emigration are more limited today than they were then.

Yet we shall survive this period too, no matter how much sorrow, no matter how heavy a loss in life it may bring. A community like ours, which is a community purely by reason of tradition, can only be strengthened by pressure from without. For today every Jew feels that to be a Jew means to bear a serious responsibility not only to his own community, but also toward humanity. To be a Jew, after all, means first of all, to acknowledge and follow in practice those fundamentals in humaneness laid down in the Bible—fundamentals without which no sound and happy community of men can exist.

We meet today because of our concern for the development of Palestine. In this hour, one thing, above all, must be emphasized: Judaism owes a great debt of gratitude to Zionism. The Zionist movement has revived among Jews the sense of community. It has performed productive work, surpassing all the expectations anyone could entertain. This productive work in Palestine, to which self-sacrificing Jews throughout the world have contributed, has saved a large number of our brethren from direst need. In particular, it has been possible to lead a not inconsiderable part of our youth toward a life of joyous and creative work.

Now the fateful disease of our time—exaggerated nationalism, borne up by blind hatred—has brought our work in Palestine to a most difficult stage.[a] Fields cultivated by day must have armed protection at night against fanatical Arab outlaws. All economic life suffers from insecurity. The spirit of enterprise languishes, and a certain measure of unemployment (modest when measured by American standards) has made its appearance.

The solidarity and confidence with which our brethren in Palestine face these difficulties deserve our admiration. Voluntary contributions by those still employed keep the unemployed above water. Spirits remain high, in the conviction that reason and calm will ultimately reassert themselves. Everyone knows that the riots are artificially fomented by those directly interested in embarrassing not only ourselves but especially England. Everyone knows that banditry would cease if foreign subsidies were withdrawn.

Our brethren in other countries, however, are in no way behind those in Palestine. They, too, will not lose heart but will resolutely and firmly stand behind the common work. This goes without saying.

Just one more personal word on the question of partition.[b] I should much rather see reasonable agreement with the Arabs based on living together in peace than the creation of a Jewish state. Apart from practical considerations, my awareness of the essential nature of Judaism resists the idea of a Jewish state with borders, an army, and a measure of temporal power, no matter how modest. I am afraid of the inner damage Judaism will sustain—especially from the development of a narrow nationalism within our own ranks, against which we have already had to fight strongly, even without a Jewish state. We are no longer the Jews of the Maccabee period. A return to a nation in the political sense of the word would be equivalent to turning away from the spiritualization of our community, which we owe to the genius of our prophets. If external necessity should after all compel us to assume this burden, let us bear it with tact and patience.

One more word on the present psychological attitude of the world at large, upon which our Jewish destiny also depends. Antisemitism has always been the cheapest means employed by selfish minorities for deceiving the people. A tyranny based on such deception and maintained by terror must inevitably perish from the poison it generates within itself. For the pressure of accumulated injustice strengthens those moral forces in man which lead to a liberation and purification of public life. May our community through its suffering and its work contribute toward the release of those liberating forces.

Editorial Notes

[a] The Great Revolt by Palestinian Arabs in British Mandate Palestine against British colonial rule and rising Jewish immigration lasted from 1936 to 1939.

[b] Einstein here refers to the Palestine Royal Commission (or Peel Commission) report, which concluded that Palestine be partitioned into an Arab state linked to Transjordan (a British protectorate), and a smaller Jewish state.

72. Ten Fateful Years: Living Philosophies, Revised (1938)

C. Fadiman, ed. *I Believe: The Personal Philosophies of Certain Eminent Men and Women of Our Time.* New York: Simon & Schuster (1939), pp. 367–369.[a]

This essay, composed in 1938 and published in 1939, is a follow-up to Einstein's 1930 essay "What I Believe" (see Doc. 53). During the intervening years, Einstein's life had changed profoundly. As a result of Hitler's rise to power, he had emigrated to the United States in 1933. His wife Elsa

72. Ten Fateful Years

> Einstein and his stepdaughter Ilse Einstein had died. *Kristallnacht* and the annexation of Czechoslovakia and Austria had just changed the fate of millions. Europe was on the brink of war.

Reading once again the lines I contributed ten years ago to the volume *Living Philosophies*, I receive two strangely contrasting impressions. What I wrote then still seems essentially true as ever; yet, it all seems curiously remote and strange. How can that be? Has the world changed so profoundly in ten years, or is it merely that I have grown ten years older, and my eyes see everything in a changed, dimmer light? What are ten years in the history of humanity? Must not all those forces that determine the life of man be regarded as constant, compared with such a trifling interval? Is my critical reason so susceptible that the physiological changes in my body during those ten years have been able to influence my concept of life so deeply? It seems clear to me that such considerations cannot throw light upon a change in the emotional approach to the general problems of life. Nor may the reasons for this curious change be sought in my own external circumstances; for I know that these have always played a subordinate part in my thoughts and emotions.

No, something quite different is involved. In these ten years, confidence in the stability, yes, even the vitality, of human society has largely vanished. One senses not only a threat to man's cultural heritage, but also that a lower value is placed upon all that one would like to see defended at all costs.

Conscious man, to be sure, has at all times been keenly aware that life is an adventure, that life must, forever, be wrested from death. In part the dangers were external: one might fall downstairs and break one's neck, lose one's livelihood without fault, be condemned though innocent, or ruined by calamity. Life in human society meant dangers of all sorts, but these dangers were chaotic in nature, subject to chance. Human society, as a whole, seemed stable. Measured by the ideals of taste and morals, it was decidedly imperfect. But, all in all, one felt familiar with it and, apart from the many kinds of accidents, comparatively safe in it. One accepted its intrinsic qualities as a matter of course, as the air one breathed. Even standards of virtue, aspiration, and practical truth were taken for granted as an inviolable heritage, common to all civilized humanity.

The World War has already shaken this feeling of security. The sanctity of life vanished, and the individual was no longer able to do as he pleased, and to go where he liked. The lie was raised to the dignity of a political instrument. The War was, however, widely regarded as an external event, hardly or not at all as the result of man's conscious, planful action. It was thought of as an interruption of man's normal life from the outside, universally considered unfortunate and evil. The feeling of security in regard to human aims and values remained, for the main part, unshaken.

The subsequent development is sharply marked by political events that are not as far-reaching as the less easily grasped socio-psychological background. First a brief, promising step forward characterized by the creation of the League of Nations through the grandiose initiative of Wilson, and the establishment of a system of collective security among nations. Then the formation of the Fascist states, attended by a series of broken pacts and undisguised acts of violence against humanity and weaker nations. The system of collective security collapsed like a house of cards—a collapse the consequence of which cannot be measured even today. It was a manifestation of weakness of character and lack of responsibility on the part of the leaders in the affected countries, and of shortsighted selfishness in the democracies, which prevented any vigorous counterattack.

Things grew even worse than a pessimist of the deepest dye would have dared prophesy. In Europe to the east of the Rhine, free exercise of the intellect exists no longer, the population is terrorized by gangsters who have seized power, and youth is poisoned by systematic lies. The pseudo-success of political adventurers has dazzled the rest of the world; it becomes apparent everywhere that this generation lacks the strength and force which enabled previous generations to win, in painful struggle and at great sacrifice, political and individual freedom of man.

Awareness of this state of affairs overshadows every hour of my present existence, while ten years ago it did not yet occupy my thoughts; it is this that I feel so strongly in re-reading the words written in the past.

And yet I know that, all in all, man changes but little, even though prevailing notions make him appear in a very different light at different times, and even though the current trends like the present bring him unimaginable sorrow. Nothing of all that will remain but a few pitiful pages in the history books, briefly picturing to the youth of future generations the follies of its ancestors.

Editorial Note

[a] Republished with some minor changes as the postscript of Einstein's original *Living Philosophies* essay (Doc. 5) in the 1940 revision of this book, with considerable edits by Einstein himself.

73. Letter to Franklin D. Roosevelt (1939)

Photostat Records Service. Franklin D. Roosevelt Presidential Library and Museum, Hyde Park, N.Y.

Regarded in hindsight as having initiated the U.S. atomic energy project, this letter dated 2 August 1939 was only transmitted to President Franklin D. Roosevelt on 11 October 1939 by Alexander Sachs, an adviser to the president. Several drafts had earlier been composed by Hungarian Jewish émigré scientist Leo Szilard, who also composed a technical addendum. Roosevelt acknowledged receipt of the letter on 19 October 1939 and informed Einstein that he was appointing a committee "to thoroughly investigate the possibilities of your suggestion regarding the element of uranium" [AEA 33-92]. However, the Manhattan Project was only initiated the day after the Pearl Harbor attack, and officially started in August 1942. The letter first became public on 12 August 1945 in the official report of the atomic bomb known as the "Smyth Report," shortly after the bombings of Hiroshima and Nagasaki on 6 and 9 August 1945. Henry DeWolf Smyth was a physicist at Princeton University and member of the National Defense Research Committee's Uranium Section during World War II. The letter was reprinted in the introduction of *A Statement of Purpose* (Princeton, N.J.: Emergency Committee of Atomic Scientists, 1946), pp. 1–2. The committee was founded by Einstein and Szilard in May 1946.

Sir:

Some recent work by E. Fermi and L. Szilard,[a] which has been communicated to me in manuscript, leads me to expect that the element uranium may be turned into a new and important source of energy in the immediate future. Certain aspects of the situation which has arisen seem to call for watchfulness and, if necessary, quick action on the part of the Administration. I believe therefore that it is my duty to bring to your attention the following facts and recommendation.

In the course of the last four months it has been made probable—through the work of Joliot[b] in France as well as Fermi and Szilard in America—that it may become possible to set up a nuclear chain reaction in a large mass of uranium, by which vast amounts of power and large quantities of new radium-like elements would be generated. Now it appears almost certain that this could be achieved in the immediate future.

This new phenomenon would also lead to the construction of bombs, and it is conceivable—though much less certain—that extremely powerful bombs of a new type may thus be constructed. A single bomb of this type, carried by boat and exploded in a port, might very well destroy the whole port together with some of the surrounding territory. However, such bombs might very well prove to be too heavy for transportation by air.

The United States has only very poor ores of uranium in moderate quantities. There is some good ore in Canada and the former Czechoslovakia, while the most important source of uranium is Belgian Congo.

In view of this situation, you may think it desirable to have some permanent contact maintained between the Administration and the group of physicists working on chain reactions in America. One possible way of achieving this might be for you to entrust with this task a person who has your confidence and who could perhaps serve in an unofficial capacity. His task might comprise the following:

a) to approach Government Departments, keep them informed of the further development, and put forward recommendations for Government action, giving particular attention to the problem of securing a supply of uranium ore for the United States;

b) to speed up the experimental work, which is at present being carried on within the limits of the budgets of university laboratories, by providing funds, if such funds be required, through his contacts with private persons who are willing to make contributions for this cause, and perhaps also by obtaining the co-operation of industrial laboratories which have the necessary equipment.

I understand that Germany has actually stopped the sale of uranium from the Czechoslovakian mines which she has taken over. That she should have taken such early action might perhaps be understood on the ground that the son of German Under-Secretary of State, von Weizsäcker, is attached to the Kaiser-Wilhelm-Institut in Berlin where some of the American work on uranium is now being repeated.[c]

Yours very truly,

A. Einstein

Editorial Notes

[a] Nuclear fission was discovered by Otto Hahn, Fritz Strassman, Lise Meitner, and Otto Frisch in December 1938. The physicists Niels Bohr and Léon Rosenfeld first reported on this work to American colleagues in mid-January 1939. In spring 1939, the émigré Italian scientist Enrico Fermi, together with Szilard, Walter H. Zinn, and Herbert L. Anderson, began research at Columbia University on a self-sustaining nuclear chain reaction.

[b] Frédéric Joliot-Curie, Hans von Halban, and Lew Kowarski discovered that neutrons were emitted in the fission of uranium. This was an important step toward a self-sustaining chain reaction.

[c] Carl Friedrich von Weizsäcker was part of Germany's nuclear power program, which was initiated on 17 September 1939.

74. Freedom and Science (1940)

Ruth N. Anshen, ed. *Freedom: Its Meaning*. New York: Harcourt, Brace (1940), pp. 91–93.[a]

This essay appeared in the first installment of Ruth Nanda Anshen's (1900–2003) *Science of Culture* book series, intended as a collaborative project directed at clarifying major concepts from different points of view.

> Franz Boas, Benedetto Croce, J.B.S. Haldane, Thomas Mann, Bertrand Russell, and others contributed to the volume. Part I of this essay was apparently written by Einstein's sister, Maja Einstein Winteler (1881–1951), who was living in Einstein's house in Princeton at the time.

I.

At first glance it seems that freedom and science do not have much relation to one another. In any case, freedom may well exist without science, that is, to the extent that man can live without science, man in whom the impulse of inquiry is innate. But what of science without freedom?

Above all, a man of science requires inward freedom, for he must endeavor to free himself from prejudices and must constantly convince himself anew, when new facts emerge, that what has been established, however authoritatively, is still valid. Intellectual independence is thus a primary necessity for the scientific inquirer. But political liberty is also extraordinarily important for his work. He must be able to utter what seems true to him without concern about, or danger to, his life and livelihood. This is apparent in historical investigations, but it is a vital precondition for all scientific activity, however remote from politics. If certain books are condemned and made inaccessible in so far as their author is not acceptable to the government on account of his political orientation or race, as is largely the case to-day, the inquirer cannot attain an adequate basis on which to build. And how can the building stand if it lacks a secure foundation?

It is self-evident that absolute freedom is an ideal which cannot be realized in our social and political life. But all men of good-will should seek to guard mankind's effort to realize this ideal in ever-increasing measure.

II.

I know that it is a hopeless undertaking to debate about fundamental value judgments. For instance, if someone approves, as a goal, the extirpation of the human race from the earth, one cannot refute such a viewpoint on rational grounds. But if there is agreement on certain goals and values, one can argue rationally about the means by which these objectives may be attained. Let us, then, indicate two goals which may well be agreed upon by nearly all who read these lines.

1. Those instrumental goods which should serve to maintain the life and health of all human beings should be produced by the least possible labor of all.
2. The satisfaction of physical needs is indeed the indispensable precondition of a satisfactory existence, but in itself it is not enough. In order to be content, people must also have the possibility of developing their intellectual and artistic powers to whatever extent accords with their personal characteristics and abilities.

The first of these two goals requires the promotion of all knowledge relating to the laws of nature and the laws of social processes, that is, the promotion of all scientific endeavor. For scientific endeavor is a natural whole, the parts of which mutually support one another in a way which, to be sure, no one can anticipate. However, the progress of science presupposes the possibility of unrestricted communication of all results and judgments—freedom of expression and instruction in all realms of intellectual endeavor. By freedom, I understand social conditions of such a kind that the expression of opinions and assertions about general and particular matters of knowledge will not involve dangers or serious disadvantages for him who expresses them. This freedom of communication is indispensable for the development and extension of scientific knowledge, a consideration of much practical import. In the first instance, it must be guaranteed by law. But laws alone cannot secure freedom of expression; in order that everyone may present their views without penalty, there must be a spirit of tolerance in the entire population. Such an ideal of external liberty can never be fully attained but must be sought unremittingly if scientific thought, and philosophical and creative thinking in general, are to be advanced as far as possible.

If the second goal, that is, the possibility of the spiritual development of all individuals, is to be secured, a second kind of outward freedom is necessary. People should not have to work for the achievement of the necessities of life to such an extent that they have neither time nor strength for personal activities. Without this second kind of outward liberty, freedom of expression is useless. Advances in technology would provide the possibility of this kind of freedom if the problem of a reasonable division of labor were solved.

The development of science and of the creative activities of the spirit in general requires still another kind of freedom, which may be characterized as inward freedom. It is this freedom of the spirit which consists in the independence of thought from the restrictions of authoritarian and social prejudices, as well as from unphilosophical routinizing and habit in general. This inward freedom is an infrequent gift of nature and a worthy objective for the individual. Yet the community can do much to further this achievement, too, at least by not interfering with its development. Thus, schools may interfere with the development of inward freedom through authoritarian influences, and through imposing on young people excessive spiritual burdens; on the other hand, schools may favor such freedom by encouraging independent thought. Only if outward and inner freedom are constantly and consciously pursued is there a possibility of spiritual development and perfection, and thereby of improving the outward and inner life.

Editorial Note

[a] Translated from German by James Gutmann, Professor of Philosophy at Columbia University.

75. The Common Language of Science (1942)

Advancement of Science 2, no. 5 (1942), pp. 109–110.

Einstein recorded this address in English for the Conference on Science and World Order that took place at the Royal Institution, London, during 26–28 September 1941. The conference was organized by the British Association for the Advancement of Science and its Division for the Social and International Relations of Science. The published English text was transcribed from the radio transmission, which was garbled and unclear. The text below contains corrections, based on the original German manuscript [AEA 28-557].

The first step toward language was to link acoustically or otherwise communicable signs to sense impressions. Most likely, all sociable animals have arrived at this primitive kind of communication, at least to a certain degree. A higher development is reached when further signs are introduced and understood which establish relations between those other signs indicating experiences. At this stage, it is already possible to report a somewhat complex series of impressions. We can see that language has come into existence. If language is to lead at all to understanding, there must be rules concerning the relations between signs on the one hand and, on the other hand, there must be a stable correspondence between signs and impressions. In their childhood, individuals connected by the same language grasp these rules and relations mainly by intuition; when these rules enter consciousness, the so-called "grammar" of language is established.

In an early stage, every single word may correspond directly to impressions. At a later stage, this direct mapping is lost in as far as at least some words appear to be connected to experiences only if used in connection with other words—for instance, such words as "is," "or," "thing." Then word-connections are mapped onto groups of experiences, rather than words being mapped onto experiences. When language becomes thus partially independent from the background of impressions, a greater inner coherence and independence is gained.

Only at this higher-level stage of development, where frequent use is made of so-called abstract concepts, does language become an instrument of reasoning in the true sense of the word. But it is also this development which turns language into a dangerous source of error and deception. Everything depends on the degree to which words and word combinations correspond to the world of experiences.

What is it that brings about such an intimate connection between language and thinking? Is there no thinking without the use of language, namely in concepts and concept-relationships to which words need not necessarily be assigned? Has not every one of us struggled for words, although the connection between "things" was already clear?

We might be inclined to attribute to the act of thinking complete independence from language if the individual formed, or were able to form, concepts without verbal communication from others. Yet, the mental shape of an individual growing up under such conditions would most likely be very impoverished. Thus we must conclude that the mental development of the individual and his way of forming concepts depends to a high degree upon language. This makes us realize to what extent a linguistic community means a community of ideas. In this sense, thought is bound to language.

Now, what distinguishes the language of science from language as we ordinarily understand the word? How is it that the language of science is essentially common to all? What science strives for is the greatest possible certainty and clarity of conceptualization—as well as the greatest possible clarity and certainty with regard to the assignment of conceptual formations to sensory data.

As an illustration, let us take the language of Euclidean geometry and algebra. They deal with a small number of independently introduced concepts—namely, symbols such as the integral number, the straight line, the point—as well as fundamental relations that connect these concepts. All statements and all other concepts are built, or defined, on this fundamental basis. The connection between concepts and statements to the world of sensory data is established through acts of counting and measuring, about whose realization there is full clarity, at least within certain limits.

The supranational character of scientific concept formation and scientific language is based on the fact that they have been constructed by the best brains of all countries and all times. They created in solitude, but in effect in common work, the spiritual tools for technical revolutions which have transformed the life of mankind in the last centuries. Their system of concepts has served as a guide in the bewildering chaos of perceptions so that we learned to grasp the general from the particular.

What hopes and fears does the scientific method imply for mankind? I do not think that is the right way to put the question. Whatever this tool in the hand of mankind will produce depends entirely on the nature of the goals alive in this mankind. Once these goals are set, the scientific method furnishes means to realize them. Yet it cannot furnish the very goals. The scientific method itself would not have led anywhere—it would not even have been born at all—without a passionate striving for clear understanding.

Perfection of means and confusion of goals seem, in my opinion, to characterize our age. If we desire sincerely and passionately the safety, the welfare, and the free development of the talents of all people, we will not lack the means to approach such a state. Even if only a small part of mankind strives for such goals, their superiority will prove itself in the long run.

76. Newton's 300th Birthday (1942)

Manchester Guardian (24 December 1942), section 4, pp. 6–7.

Sir Isaac Newton was born on Christmas Day, 1642.

Reason of course is weak when measured against its never-ending task: weak, indeed, compared with the follies and passions of mankind, which, we must admit, almost entirely control our human destinies in great things and small. Yet works of understanding outlast the noisy bustling of generators and spread light and warmth across the centuries. Consoled by this thought, let us turn in these unquiet days to the memory of Isaac Newton, who 300 years ago was given to mankind.

To think of him is to think of his work. For such a man can be understood only by thinking of him as the scene on which the struggle for eternal truth took place. Long before Newton, there had been virile minds which conceived that it ought to be possible by purely logical deductions from simple physical hypotheses to make cogent explanations of phenomena perceptible to the senses. But Newton was the first to succeed in finding a clearly formulated basis from which he could deduce a wide field of phenomena by means of mathematical thinking, logically, quantitatively, and in harmony with experience. Indeed, he might even have hoped that the fundamental basis of his mechanics would come in time to furnish the key to the understanding of all phenomena. So thought his pupils—with more assurance than he himself—and so did his successors up to the end of the eighteenth century.

How did this miracle come to birth in his brain?[a] Forgive me, reader, the illogical question, for if by reason we could deal with the problem of the "How" then there could be no question of a miracle in the proper sense of the word. It is the goal of every activity of intellect to convert a "miracle" into something which it has grasped. If in this case the miracle permits itself to be converted, our admiration for the mind of Newton becomes only the greater thereby.

Galileo, by an ingenious interpretation of the simplest facts of experience, had established the proposition: A body upon which no external force is at work permanently maintains its original velocity (and direction): if it alters its velocity (or the direction of its movement) the change must be referred to an external cause.

To utilize this knowledge quantitatively, the conceptions of velocity and rate of change of velocity—that is acceleration in the case of any given motion of a body conceived as dimensionless (a material point)—must first be interpreted with mathematical exactness. This task led Newton to invent the basis of the differential and integral calculus.

This in itself was a creative achievement of the first order. But for Newton as a physicist, it meant the invention of a new kind of conceptual language that he needed in order to formulate the general laws of motion. For a given body, he had now to put forward the hypothesis that his precisely formulated acceleration both in magnitude and direction was proportional to the force directed upon it. The coefficient of proportionality which characterizes a body with reference to its power of acceleration completely describes the (dimensionless) body with reference to its mechanical quality; thus was discovered the fundamental conception of mass.

All the foregoing might be described—though in the extremely modest manner of speaking—as the exact formulation of something the essence of which had already been recognized by Galileo. But it by no means succeeded in solving the main problem. In other words, the law of motion yields the movement of a body only when the direction and magnitude of the force exerted upon it are known for all times.

Thus the problem reduced itself to another problem: How to find out the operative forces. To a mind any less bold than Newton's it must have seemed hopeless, considering the immeasurable multifariousness of the effects which the bodies of a universe seem to produce upon each other. Moreover, bodies whose motions we perceive are by no means dimensionless points—that is to say, perceptible as material points. How was Newton to deal with such chaos?

If we push a cart moving without friction on a horizontal plane, it follows that the force we exert upon it is given directly. That is the ideal case from which the law of motion is derived. That we are not here dealing with a dimensionless point appears unessential. How does it stand then with a falling body in space? A freely falling body behaves almost as simply as the dimensionless point, if one regards its movement as a whole. It has accelerated downwards. The acceleration, according to Galileo, is independent of its nature and its velocity. The earth, of course, must be decisive for the existence of this acceleration. It seemed, then, that the earth by its mere presence exerted a force upon the body. The earth consists of many parts. The idea seemed inevitable that each of these parts affects the falling body and that all these effects are combined. It seems, then, to be a force which bodies by their very presence exert upon each other through space. These forces seem to be independent of velocities and dependent only upon the relative position and the quantitative property of the various bodies exerting them. This quantitative property might be conditioned by the mass, for the mass seems to characterize the body from the mechanical point of view. This strange effect of things at a distance may be called gravitation.

Now, to gain a precise knowledge of this effect, one has only to find out how strong is the force exerted upon each other by two bodies of a given mass from a given distance. As for its direction, it would probably be no other than the line connecting them. Finally, then, what remains unknown is only the dependence of this force upon the distance between the two bodies. But this one cannot be known *a priori*. Here only experience could be of use.

Such experience, however, was available to Newton. The acceleration of the moon was known from its orbit and could be compared with the acceleration of the freely falling body on the surface of the earth. Furthermore, the movements of the planets about the sun had been determined by Kepler with great exactness and comprehended in simple empirical laws. So it was possible to ascertain how the effects of gravitation coming from the earth and those coming from the sun depended on the factor of distance. Newton found that everything was explainable by a force which was inversely proportional to the square of the distance. And with that, the goal was reached, the science of celestial mechanics was born, confirmed a thousand times over by Newton himself and by those who came after him.

But how about the rest of physics? Gravitation and the law of motion could not explain everything. What determined the equilibrium of the parts of a solid body? How was light to be explained, and how electrical phenomena? By introducing material points and forces of various kinds acting at a distance, everything seemed to be derivable from the law of motion.

That hope has not been fulfilled, and no one any longer believes in the solution of all our problems on this basis. Nevertheless, the thinking of physicists to-day is conditioned to a high degree by Newton's fundamental conceptions. So far, it has not been possible to substitute for the Newtonian unified conception of the universe a similarly unified comprehensive conception. But what we have gained up to now would have been impossible without Newton's clear system.

From the observation of the stars have chiefly come the intellectual tools indispensable to the development of modern technology. For the abuse of this technology in our time creative intellects like Newton's are as little responsible as the stars themselves, whose sight and observation inspired their thoughts. It is necessary to say this because in our time esteem of intellectual values for their own sake is no longer so lively as it was in the centuries of the intellectual renascence.

Editorial Note

[a] The term *annus mirabilis* has been variously translated into English as "year of wonder," "wonderful year," or "miraculous year," and has often been used to describe Newton's achievements in 1665–1666, as well as Einstein's in 1905. Einstein most likely uses "Wunder" as "wonder," rather than "miracle" and here plays on the linguistic connection between "Wunder" and "Bewunderung," which means "admiration." For another discussion of "wonder," see Doc. 92.

77. Remarks on Bertrand Russell's Theory of Knowledge (1944)

Paul A. Schilpp, ed. *The Philosophy of Bertrand Russell.* Evanston, Ill.: Northwestern University Press (1944), pp. 278–291.[a]

Already in 1922, Einstein admired Bertrand Russell's pacifism and wrote a preface to a German edition of Russell's *Political Ideals* (see Doc. 26). In 1940, he agreed to write an essay on Russell's epistemology for Paul A. Schilpp's planned volume on Russell's philosophy, almost at the same time as Russell was denied an appointment as professor at the City College of New York because of his writings against religion and his views on sexuality. The case, which later became a clarion call for academic freedom, reached the Supreme Court in March 1940 (*Kay v. Board of Education*). Russell was prohibited from teaching. Einstein wrote a letter in support of Russell, and also thanked Albert C. Barnes in 1941 for offering Russell a position at the Barnes Foundation.

When the editor asked me to write something about Bertrand Russell, my admiration and respect for that author at once induced me to say yes. I owe innumerable happy hours to the reading of Russell's works, something which I cannot say of any other contemporary scientific writer, with the exception of Thorstein Veblen. Soon, however, I discovered that it is easier to give such a promise than to fulfill it. I had promised to say something about Russell as philosopher and epistemologist. After having in full confidence begun with it, I quickly recognized what a slippery field I had ventured upon, having, due to lack of experience, until now cautiously limited myself to the field of physics. The present difficulties of his science force the physicist to come to grips with philosophical problems to a greater degree than was the case with earlier generations. Although I shall not speak here of those difficulties, it was my concern with them, more than anything else, which led me to the position outlined in this essay.

In the evolution of philosophic thought through the centuries the following question has played a major role: What knowledge is pure thought able to supply independently of sense perception? Is there any such knowledge? If not, what precisely is the relation between our knowledge and the raw-material furnished by sense-impressions? An almost boundless chaos of philosophical opinions corresponds to these questions and to a few others intimately connected with them. Nevertheless, there is visible in this process of relatively fruitless but heroic endeavors a systematic trend of development, namely an increasing skepticism concerning every attempt by means of pure thought to learn something about the "objective world," about the world of "things" in contrast to the world of mere "concepts and ideas." Be it said parenthetically that, just as on the part of a real philosopher, quotation-marks are used here to introduce an illegitimate concept, which the reader is asked to permit for the moment, although the concept is suspect in the eyes of the philosophical police.

During philosophy's childhood it was rather generally believed that it is possible to find everything which can be known by means of mere reflection. It was an illusion which any one can easily understand if, for a moment, one dismisses what one has learned from later philosophy and from natural science; one will not be surprised to find that Plato ascribed a higher reality to "Ideas" than to empirically experienceable things. Even in Spinoza and as late as in Hegel this prejudice was the vitalizing force which seems still to have played the major role. Someone, indeed, might even raise the question whether, without something of this illusion, anything really great can be achieved in the realm of philosophic thought—but we do not wish to ask this question.

This more aristocratic illusion concerning the unlimited penetrative power of thought has as its counterpart the more plebeian illusion of naïve realism, according to which things "are" as they are perceived by us through our senses. This illusion dominates the daily life of humans and of animals; it is also the point of departure in all of the sciences, especially of the natural sciences.

The effort to overcome these two illusions is not independent the one of the other. The overcoming of naïve realism has been relatively simple. In his introduction to his volume, *An Inquiry into Meaning and Truth*,[b] Russell has characterized this process in a marvelously pregnant fashion:

> We all start from "naïve realism," i.e., the doctrine that things are what they seem. We think that grass is green, that stones are hard, and that

snow is cold. But physics assures us that the greenness of grass, the hardness of stones, and the coldness of snow, are not the greenness, hardness, and coldness that we know in our own experience, but something very different. The observer, when he seems to himself to be observing a stone, is really, if physics is to be believed, observing the effects of the stone upon himself. Thus, science seems to be at war with itself: when it most means to be objective, it finds itself plunged into subjectivity against its will. Naïve realism leads to physics, and physics, if true, shows that naïve realism is false. Therefore, naïve realism, if true, is false; therefore, it is false. (pp. 14–15)

Apart from their masterful formulation these lines say something which had never previously occurred to me. For, superficially considered, the mode of thought in Berkeley and Hume seems to stand in contrast to the mode of thought in the natural sciences. However, Russell's just cited remark uncovers a connection: If Berkeley relies upon the fact that we do not directly grasp the "things" of the external world through our senses, but that only events causally connected with the presence of "things" reach our sense-organs, then this is a consideration which gets its persuasive character from our confidence in the physical mode of thought. For, if one doubts the physical mode of thought in even its most general features, there is no necessity to interpolate between the object and the act of vision anything which separates the object from the subject and makes the "existence of the object" problematical.

It was, however, the very same physical mode of thought and its practical successes which have shaken the confidence in the possibility of understanding things and their relations by means of purely speculative thought. Gradually the conviction gained recognition that all knowledge about things is exclusively a working-over of the raw-material furnished by the senses. In this general (and intentionally somewhat vaguely stated) form this sentence is probably today commonly accepted. But this conviction does not rest on the supposition that anyone has actually proved the impossibility of gaining knowledge of reality by means of pure speculation, but rather upon the fact that the empirical (in the above-mentioned sense) procedure alone has shown its capacity to be the source of knowledge. Galileo and Hume first upheld this principle with full clarity and decisiveness.

Hume saw that concepts which we must regard as essential, such as, for example, causal connection, cannot be gained from material given to us by the senses. This insight led him to a skeptical attitude as concerns knowledge of any kind. If one reads Hume's books, one is amazed that many and sometimes even highly esteemed philosophers after him have been able to write so much obscure stuff and even find grateful readers for it. Hume has permanently influenced the development of the best of philosophers who came after him. One senses him in the reading of Russell's philosophical analyses, whose acumen and simplicity of expression have often reminded me of Hume.[c]

Humans have an intense desire for assured knowledge. That is why Hume's clear message seemed crushing: The sensory raw-material, the only source of our knowledge, through habit may lead us to belief and expectation but not to the knowledge and still less to the understanding of law-abiding relations. Then Kant took the stage with an idea which, though certainly untenable in the form in which he put it, signified a step toward the solution of Hume's dilemma: Whatever in knowledge is of empirical origin is never certain (Hume). If, therefore, we have definitely assured knowledge, it must be grounded in reason itself. This is held to be the case, for example, in the propositions of geometry and in the principle of causality. These and certain other types of knowledge are, so to speak, a part of the instrumentality of thinking and therefore do not previously have to be gained from sense data (i.e., they are *a priori* knowledge). Today everyone knows of course that the mentioned concepts contain nothing of the certainty, of the inherent necessity, which Kant had attributed to them. The following, however, appears to me to be correct in Kant's statement of the problem: in thinking we use, with a certain "right," concepts to which there is no access from the materials of sensory experience, if the situation is viewed from the logical point of view.

As a matter of fact, I am convinced that even much more is to be asserted: the concepts which arise in our thought and in our linguistic expressions are all—when viewed logically—the free creations of thought which cannot inductively be gained from sense-experiences. This is not so easily noticed only because we have the habit of combining certain concepts and conceptual relations (propositions) so definitely with certain sense-experiences that we do not become conscious of the gulf—logically unbridgeable—which separates the world of sensory experiences from the world of concepts and propositions.

Thus, for example, the series of integers is obviously an invention of the human mind, a self-created tool which simplifies the ordering of certain sensory experiences. But there is no way in which this concept could be made to grow, as it were, directly out of sense experiences. It is deliberately that I choose here the concept of number, because it belongs to pre-scientific thinking and because, in spite of that fact, its constructive character is still easily recognizable. The more, however, we turn to the most primitive concepts of everyday life, the more difficult it becomes amidst the mass of inveterate habits to recognize the concept as an independent creation of thinking. It was thus that the fateful conception—fateful, that is to say, for an understanding of the here existing conditions—could arise, according to which the concepts originate from experience by way of "abstraction," i.e., through omission of a part of its content. I want to indicate now why this conception appears to me to be so fateful.

As soon as one is at home in Hume's critique, one is easily led to believe that all those concepts and propositions which cannot be deduced from the sensory raw-material are, on account of their "metaphysical" character, to be removed from thinking. For all thought acquires material content only through its relationship with that sensory material. This latter proposition I take to be entirely true, but I hold the prescription for thinking which is grounded on this proposition to be false. For this claim—if only carried through consistently—absolutely excludes thinking of any kind as "metaphysical."

In order that thinking might not degenerate into "metaphysics," or into empty talk, it is only necessary that enough propositions of the conceptual system be firmly enough connected with sensory experiences and that the conceptual system, in view of its task of ordering and surveying sense-experience, should show as much unity and parsimony as possible. Beyond that, however, the "system" is (as regards logic) a free play with symbols according to (logical) arbitrarily given rules of the game. All this applies as much (and in the same manner) to the thinking in daily life as to the more consciously and systematically constructed thought in the sciences.

It will now be clear what is meant if I make the following statement: By his clear critique, Hume did not only advance philosophy in a decisive way but also—though through no fault of his—created a danger for philosophy in that, following his critique, a fateful "fear of metaphysics" arose which has come to be a malady of contemporary empiricist philosophizing; this malady is the counterpart to that earlier philosophizing in the clouds, which thought it could neglect and dispense with what was given by the senses.

No matter how much one may admire the acute analysis which Russell has given us in his latest book on *Meaning and Truth*, it still seems to me that even there the specter of the metaphysical fear has caused some damage. For this fear seems to me, for example, to be the cause for conceiving of the "thing" as a "bundle of qualities," such that the "qualities" are to be taken from the sensory raw-material. Now the fact that two things are said to be one and the same thing, if they coincide in all qualities, forces one to consider the geometrical relations between things as belonging to their qualities. (Otherwise one is forced to look upon the Eiffel Tower in Paris and that in New York as the "same thing.")[1] Over against that, I see no "metaphysical" danger in taking the thing (the object in the sense of physics) as an independent concept into the system together with the proper spatio-temporal structure.

In view of these endeavors, I am particularly pleased to note that, in the last chapter of the book, it finally crops out that one can, after all, not get along without "metaphysics." The only thing to which I take exception there is the bad intellectual conscience which shines through between the lines.[d]

Editorial Notes

[a] An English translation of each page by Paul A. Schilpp is presented opposite the original German text.

[b] Russell delivered the William James Lectures at Harvard University during the fall 1940 academic term. They were published that same year as *An Inquiry into Meaning and Truth* (New York: W. W. Norton & Company, Inc.).

[c] For the significant influence of Hume's thought on Einstein, see also the early essay (Doc. 4) and his *Autobiographical Notes* (Doc. 94).

[d] The book consists of twenty-five chapters, its last being entitled "Language and Metaphysics."

---

[1] Compare Russell's *An Inquiry into Meaning and Truth*, pp. 119–120, chapter on "Proper Names."

78. To the Heroes of the Battle of the Warsaw Ghetto (1944)

Bulletin of the Society of Polish Jews, New York (1944).

Written as a one-year commemoration of the revolt of the Warsaw ghetto Jews, which began on the eve of Passover, 19 April 1943, in response to the continued deportation of Jews to extermination camps. The fighting lasted until 16 May 1943, with 13,000 Jews killed and more than 50,000 deported to the death camps of Majdanek and Treblinka.

They fought and died as members of the Jewish nation, in the struggle against organized bands of German murderers. To us, these sacrifices are a strengthening of the bond between us, the Jews of all the countries. We strive to be one in suffering and in the effort to achieve a better human society, that society which our prophets have so clearly and forcibly set before us as a goal.

The Germans as an entire people are responsible for these mass murders and must be punished as a people if there is justice in the world and if the consciousness of collective responsibility in the nations is not to perish from the earth entirely. Behind the Nazi party stands the German people, who elected Hitler after he had in his book and in his speeches made his shameful intentions clear beyond the possibility of misunderstanding. The Germans are the only people who have not made any serious attempt of counter-action leading to the protection of the innocently persecuted. When they are entirely defeated and begin to lament over their fate, we must not let ourselves be deceived again, but keep in mind that they deliberately used the humanity of others to make preparation for their last and most grievous crime against humanity.

79. On the Atomic Bomb (1945)

Atlantic Monthly 176, no. 5 (November 1945), pp. 43–45.[a]

> The article in *Atlantic Monthly* was preceded by the following editor's note:
>
> Albert Einstein was born in Ulm, Germany, in 1879. He was awarded the Nobel Prize for Physics in 1921. Driven into exile by Hitler's advent to power, Professor Einstein came to this country in 1933, was appointed for life a member of the Institute for Advanced Study at Princeton, and became an American citizen in 1940.
>
> On August 2, 1939, just a month before the outbreak of World War II, Dr. Einstein wrote a letter which made history. The letter was addressed to President Roosevelt, and it starts with the sentence: "Some recent work by E. Fermi and L. Szilard, which has been communicated to me in manuscript, leads me to expect that the element uranium may be turned into a new and important source of energy in the immediate future." Dr. Einstein went on: "This new phenomenon would also lead to the construction of bombs ... extremely powerful bombs. A single bomb of this type, carried by a boat and exploded in a port, might very well destroy the whole port, together with some of the surrounding territory."
>
> It was Einstein's daring formula, E equals mc^2, which led to the concept that atomic energy would some day be unlocked. Here, in these words recorded by Raymond Swing, he explains how mankind must control atomic power.—The Editor

The release of atomic energy has not created a new problem. It has merely made more urgent the necessity of solving an existing one. One could say that it has affected us quantitatively, not qualitatively. As long as there are sovereign nations possessing great power, war is inevitable. That statement is not an attempt to say when war will come, but only that it is sure to come. That fact was true before the atomic bomb was made. What has been changed is the destructiveness of war.

I do not believe that civilization will be wiped out in a war fought with the atomic bomb. Perhaps two thirds of the people of the earth might be killed, but enough men capable of thinking, and enough books, would be left to start again, and civilization could be restored.

I do not believe that the secret of the bomb should be given to the United Nations organization. I do not believe that it should be given to the Soviet Union. Either course would be like the action of a man with capital, who, wishing another man to work with him on some enterprise, should start out by simply giving his prospective partner half of his money. The second man might choose to start a rival enterprise, when what was wanted was his cooperation.

The secret of the bomb should be committed to a World Government, and the United States should immediately announce its readiness to give it to a World Government. This government should be founded by the United States, the Soviet Union and Great Britain—the only three powers with great military strength. The fact that there are only three nations with great military power should make it easier rather than harder to establish such a government.

Since the United States and Great Britain have the secret of the atomic bomb and the Soviet Union does not, they should invite the Soviet Union to prepare and present the first draft of a Constitution for the proposed World Government. That action should help to dispel the distrust with the Russians already feel because the bomb is being kept a secret, chiefly to prevent their having it. Obviously the first draft would not be the final one, but the Russians should be made to feel that the World Government would assure them their security.

It would be wise if this Constitution were to be negotiated by a single American, a single Britisher, and a single Russian. They would have to have advisers, but these advisers should only advise when asked. I believe three men can succeed in writing a workable Constitution acceptable to all three nations. Six or seven men, or more, probably would fail.

After the three great powers have drafted a Constitution and adopted it, the smaller nations should be invited to join the World Government. They should be free to stay out; and though they would be perfectly secure in staying out, I am sure they would wish to join. Naturally they should be entitled to propose changes in the Constitution as drafted by the Big Three. But the Big Three should go ahead and organize the World Government whether the smaller nations join or not.

The World Government would have power over all military matters and need have only one further power: the power to intervene in countries where a minority is oppressing a majority and creating the kind of instability that leads to war. Conditions such as exist in Argentina and Spain should be dealt with. There must be an end to the concept of non-intervention, for to end it is part of keeping the peace.

The establishment of the World Government must not have to wait until the same conditions of freedom are to be found in all three of the great powers. While it is true that in the Soviet Union the minority rules, I do not consider that internal conditions there are of themselves a threat to world peace. One must bear in mind that the people in Russia did not have a long political education, and changes to improve Russian conditions had to be carried through by a minority for the reason that there was no majority capable of doing it. If I had been born a Russian, I believe I could have adjusted myself to this condition.

It is not necessary, in establishing a world organization with a monopoly of military authority, to change the structure of the three great powers. It would be for the three individuals who draft the Constitution to devise ways for the different structures to be fitted together for collaboration.

Do I fear the tyranny of a World Government? Of course I do. But I fear still more the coming of another war of wars. Any government is certain to be evil to some extent. But a World Government is preferable to the far greater evil of wars, particularly with their intensified destructiveness. If a World Government is not established by agreement, I believe it will come in another way and in a much more dangerous form. For war or wars will end in one power's being supreme and dominating the rest of the world by its overwhelming military strength.

Now that we have the atomic secret, we must not lose it, and that is what we should risk doing if we should give it to the United Nations organization or to the Soviet Union. But we must make it clear, as quickly as possible, that we are not keeping the bomb a secret for the sake of our power, but in the hope of establishing peace in a World

Government, and that we will do our utmost to bring this World Government into being.

I appreciate that there are persons who favor a gradual approach to World Government even though they approve of it as the ultimate objective. The trouble about taking little steps, one at a time, in the hope of reaching that ultimate goal is that while they are being taken, we continue to keep the bomb secret without making our reason convincing to those who do not have the secret. That of itself creates fear and suspicion, with the consequence that the relations of rival sovereignties deteriorate dangerously. So, while persons who take only a step at a time may think they are approaching world peace, they actually are contributing, by their slow pace, to the coming of war. We have no time to spend in this way. If war is to be averted, it must be done quickly.

We shall not have the secret very long. I know it is argued that no other country has money enough to spend on the development of the atomic bomb, and this fact assures us the secret for a long time. It is a mistake often made in this country to measure things by the amount of money they cost. But other countries which have the materials and the men can apply them to the work of developing atomic power if they care to do so. For men and materials and the decision to use them, and not money, are all that is needed.

I do not consider myself the father of the release of atomic energy. My part in it was quite indirect. I did not, in fact, foresee that it would be released in my time. I believed only that release was theoretically possible. It became practical through the accidental discovery of chain reactions, and this was not something I could have predicted. It was discovered by Hahn in Berlin, and he himself misinterpreted what he discovered. It was Lise Meitner who provided the correct interpretation and escaped from Germany to place the information in the hands of Niels Bohr.

I do not believe that a great era of atomic science is to be assured by organizing sciences in the way large corporations are organized. One can organize to apply a discovery already made, but not to make one. Only a free individual can make a discovery. There can be a kind of organizing by which scientists are assured their freedom and proper conditions of work. Professors of science in American universities, for instance, should be relieved of some of their teaching so as to have time for more research. Can you imagine an organization of scientists making the discoveries of Charles Darwin?

Nor do I believe that the vast private corporations of the United States are suitable to the needs of these times. If a visitor should come to this country from another planet, would he not find it strange that in this country so much power is given to private corporations without their having commensurate responsibility? I say this to stress that the American government must keep the control of atomic energy, not because socialism is necessarily desirable, but because atomic energy was developed by the government, and it would be unthinkable to turn over this property of the people to any individual or group of individuals. As to socialism, unless it is international to the extent of producing a World Government which controls all military power, it might more easily lead to wars than does capitalism, because it represents a still greater concentration of power.

To give any estimate of when atomic energy can be applied to constructive purposes is impossible. What now is known is only how to use a fairly large quantity of uranium. The use of quantities sufficiently small to operate, say, a car or an airplane is as yet impossible. No doubt it will be achieved, but nobody can say when.

Nor can one predict when materials more common than uranium can be used to supply atomic energy. Presumably all materials used for this purpose will be among the heavier elements of high atomic weight. Those elements are relatively scarce, because of their lesser stability. Most of these materials may already have disappeared by radioactive disintegration. So, though the release of atomic energy can be, and no doubt will be, a great boon to mankind, that may not be for some time.

I myself do not have the gift of explanation by which to persuade large numbers of people of the urgencies of the problems the human race now faces. Hence I should like to commend someone who has this gift of explanation—Emery Reves, whose book, *The Anatomy of Peace*,[b] is intelligent, brief, clear, and, if I may use the abused term, dynamic on the topic of war and the need for World Government.

Since I do not foresee that atomic energy is to be a great boon for a long time, I have to say that for the present it is a menace.[c] Perhaps it is well that it should be. It may intimidate the human race into bringing order into its international affairs, which, without the pressure of fear, it would not do.

Editorial Notes

[a] Einstein's article carried the subtitle "As told to Raymond Swing." In September 1945, Einstein had met with the celebrated ABC broadcaster, who was raising the issue of the dangers of the atomic bomb in his weekly newscasts. The *Atlantic Monthly* November issue was released on October 26. Also excerpted in the *New York Times* (27 October), p. 17:6. The following clarifying statement by Einstein was published in the *New York Times*, 29 October, p. 4:3:

> I am aware that my statement about the necessity of world government has been partly misunderstood and misinterpreted and I am not entirely without blame. Now, to avoid any harm that might be done, I have to say right away that in the present situation nothing is more important than to create an atmosphere of confidence between the great powers so that the great problem of abolition of competitive armament can be solved. It is my belief that to give away the secret under the present anarchic conditions would lead to an accelerated armament race. When I said that the so-called secret of the atomic bomb should not be thrown away in the present state of affairs, it was my intention to express sharply that everything, especially knowledge about the destructive weapons, should be used to bring about centralized organization for security. The major powers, in whom rests the greatest military strength, must set up permanent machinery for control and military security which will make national armament superfluous.

[b] Reves 1945.
[c] The Soviet Union successfully tested an atomic weapon on 29 August 1949. The first power plant to produce small amounts of usable electricity through nuclear reactions began operation on 20 December 1951.

80. On the American Council for Judaism (1945)

Aufbau 11, no. 50 (1945), p. 11.[a]

Established in 1942 following the Central Conference of American Rabbis, the American Council for Judaism was a Reform Jewish organization noted for its active opposition to Zionism. On 4 December 1945, ACJ President Lessing J. Rosenwald met with U.S. President Harry Truman and submitted to him a seven-point proposal to reconsider the "Palestine problem" which denounced Zionist efforts. In response, Einstein made the following comment in the German-Jewish publication *Aufbau*.

am very happy indeed to hear that the platform for which the American Council for Judaism stands is meeting with strong opposition. This organization appears to me to be nothing more than a pitiable attempt to obtain favor and toleration from our enemies by betraying true Jewish ideals and by mimicking those who claim to stand for 100 percent Americanism. I believe this method to be both undignified and ineffective. Our opponents are bound to view it with disdain and even with contempt, and in my opinion justly. He who is untrue to his own cause cannot command the respect of others. Apart from these considerations, the movement in question is a fairly exact copy of the *Zentralverein Deutscher Staatsbürger Jüdischen Glaubens* ("Central Association of German Citizens of Jewish Faith")[b] of unhappy memory, which in the days of our crucial need showed itself utterly impotent and corroded the Jewish group by undermining that inner certitude by which alone our Jewish people could have overcome the trials of this difficult age.

Editorial Notes

[a] This English translation was printed in *Out of My Later Years* in 1956.
[b] The Central-verein was a Jewish-German organization founded in 1893 with the aim to unify German-Jewish people and combat rising antisemitism. It was criticized by Zionists of the German-Jewish community for adopting what was perceived to be an assimilationist stance, as it denied being part of a worldwide Jewish identity and instead advocated for German-Jewish "symbiosis." Einstein had criticized its goals and aims several decades earlier; see Doc. 13.

81. Commemorative Words for Franklin D. Roosevelt (1945)

Aufbau 11, no. 17 (27 April 1945), p. 7.

Einstein's message was read at the commemorative gathering organized by the German-language U.S. weekly journal *Aufbau (Reconstruction)* and its publisher, the New World Club (formerly the German-Jewish

Club), at the Society for Ethical Culture in New York on Sunday, 22 April 1945, at 2:00 pm, ten days after Roosevelt's death. Speakers included Rabbi Stephen S. Wise, Congressman Emmanuel Celler, Manfred George, and Fritz Schlesinger.

We Jews and immigrants are particularly bound and indebted to the dear departed one. It seldom happens that a person who has his heart in the right place also has the political genius and the will power needed in order to have a decisive and lasting influence on the course of history. President Roosevelt recognized the inevitable early on and took care that America would emerge victorious against the menacing danger represented by Germany. Within the limits of political possibilities, he worked successfully for the security of the weak and for a revitalization of the economy. The burden he carried was heavy, but his sense of humor gave him an inner freedom seldom found among those who are constantly faced with the most critical decisions. He was unbelievably bound and determined to attain his final goals, yet amazingly flexible in overcoming the strong resistance any farsighted statesman faces in a democratic country, where even those in the highest office have a limited amount of authority. No matter when this man might have been taken from us, we would have felt we had suffered an irreplaceable loss. It is tragic that he did not live to lend his unique abilities to solving the problem of international security. It is also tragic, especially for us Jews, that he, with his lively sense for justice, was unable to take part in the decisive negotiations which will determine whether our unspeakably hard-tested people will find a refuge, whether the gates of Palestine will be opened for the refugees and persecuted among us. For all people of good will, Roosevelt's death will be felt like that of an old and dear friend. May he have a lasting influence on our thoughts and convictions.

82. The Way Out (1946)

D. Masters and K. Way, eds. *One World or None*. New York: McGraw-Hill (1946), pp. 76–77.

> This article was published as Chapter 14 of the book *One World or None, A Report to the Public on the Full Meaning of the Atomic Bomb*. In November 1945, the envisaged title read: *A Report to the Public on the Full Meaning of the Atomic Bomb ... by the Scientists Who Created It*. The eighty-page booklet contained contributions by many prominent scientists, including Niels Bohr, Arthur H. Compton, J. Robert Oppenheimer, Hans Bethe, Leo Szilard, Harold Urey, Eugene Wigner, and others. It concluded with a summarizing statement from the Federation of American Scientists.

The construction of the atom bomb has brought about the effect that all the people living in cities are threatened, everywhere and constantly, with sudden destruction. There is no doubt that this condition has to be abolished if man is to prove himself worthy, at least to some extent, of the self-chosen name of *Homo sapiens*. However, there still exist widely divergent opinions concerning the degree to which traditional social and political forms, historically developed, will have to be sacrificed in order to achieve the desired security.

After the First World War, we were confronted with a paradoxical situation regarding the solution of international conflicts. An International Court of Justice had been established for a peaceful solution of these conflicts on the basis of international law. Furthermore, a political instrument for securing peace by means of international negotiation in a sort of world parliament had been created in the form of the League of Nations. The nations united in the League had further outlawed as criminal the method of solving conflicts by means of war.

Thus the nations were imbued with an illusion of security that led inevitably to bitter disappointment. For the best court of justice is meaningless unless it is backed by the authority and power to execute its decisions, and exactly the same thing is true of a world parliament. An individual state with sufficient military and economic power can easily resort to violence and voluntarily destroy the entire structure of supranational security built on nothing but words and documents. Moral authority alone is an inadequate means of securing the peace.

The United Nations Organization is now in the process of being tested. It may eventually emerge as the agency of "security without illusion" that we so badly need. But it has not as yet gone beyond the area of moral authority as, in my opinion, it must.

Our situation is rendered more acute by other circumstances, only two of which will be presented here. So long as the individual state, despite its official condemnation of war, has to consider the possibility of engaging in war, it must influence and educate soldiers in the event of war. Therefore it is compelled not only to cultivate a technical military training and type of thinking but also to implant a spirit of national vanity in its people in order to secure their inner readiness for the outbreak of war. Of course, this kind of education counteracts all endeavors to establish moral authority for any supranational security organization.

The danger of war in our time is further heightened by another technical factor. Modern weapons, in particular the atom bomb, have led to a considerable advantage in the means of offense or attack over those of defense. And this could well bring about the result that even responsible statesmen might find themselves compelled to wage a preventive war.

In view of these evident facts, there is, in my opinion, only one way out.

It is necessary that conditions be established that guarantee the individual state the right to solve its conflicts with other states on a legal basis and under international jurisdiction.

It is necessary that the individual state be prevented from making war by a supranational organization supported by a military power that is exclusively under its control.

Only when these two conditions have been fully met can we have some assurance that we shall not vanish into the atmosphere, dissolved into atoms, one of these days.

From the viewpoint of the political mentality prevailing at present, it may seem illusory, even fantastic, to hope for the realization of such conditions within a period of a few years. Yet their realization cannot wait for a gradual historical development to take its course. For, so long as we do not achieve supranational military security, the above-mentioned factors can always and forcibly lead us into war. Even more than the will for power, the fear of sudden attack will prove to be disastrous for us if we do not openly and decisively meet the problem of depriving national spheres of power of their military strength, turning such power over to a supranational authority.

With due consideration for the difficulties involved in this task, I have no doubt about one point. *We shall be able to solve the problem when it will be clearly evident to all that there is no other, no cheaper way out of the present situation.*

Now I feel it my obligation to say something about the individual steps which might lead to a solution of the security problem.

1. Mutual inspection by the leading military powers of methods and installations used for the production of offensive weapons, combined with an interchange of pertinent technical and scientific discoveries, would diminish fear and distrust, at least for the time being. In the breathing spell thus provided, we would have to prepare more thorough measures. For this preliminary step should be taken with conscious awareness that the ultimate goal is the denationalization of military power altogether.

   This first step is necessary to make any successive moves possible. However, we should be wary of believing that its execution would immediately result in security. There still would remain the possibility of an armament race with regard to a possible future war, and there always exists the temptation to resort once more, by "underground" methods, to the military secret, that is, keeping secret the knowledge about methods and means of and actual preparations for warfare. Real security is tied to the denationalization of military power.

2. This denationalization can be prepared through a steadily increasing interchange of military and scientific-technical personnel among the armies of the different nations. The interchange should follow a carefully elaborated plan, aimed at converting the national armies systematically into a supranational military force. A national army, one might say, is the last place where national feeling may be expected to weaken. Even so, the nationalism can be progressively immunized at a rate proportionate at least to the building of the supranational army; and the whole process can be facilitated by integrating it with the recruiting and training of the latter. The process

of interchanging personnel would further lessen the danger of surprise attacks and in itself would lay the psychological foundation for internationalization of military resources.

Simultaneously the strongest military powers could draft the working papers for a supranational security organization and for an arbitration committee, as well as the legal basis for, and the precise stipulation of, obligations, competencies, and restrictions of the latter with respect to the individual nations. They could further decide upon the terms of election for establishing and maintaining these bodies. When an agreement on these points shall have been reached, a guarantee against wars of worldwide dimensions can be assured.
3. The above-named bodies can now begin to function. The vestiges of national armies can then be either disbanded or placed under the high command of the supranational authority.
4. After the cooperation of the nations of highest military importance has been secured, the attempt should be made to incorporate, if possible, all nations into the supranational organization, provided that it is their voluntary decision to join.

This outline may perhaps create the impression that the presently prevailing military powers are to be assigned too dominant a role. I have tried, however, to present the problem with a view to a sufficiently swift realization that will allow us to avoid difficulties greater than those already inherent in the nature of such a task. It may be simpler, of course, to reach preliminary agreement among the strongest military powers than among *all* nations, big and small, for a body of representatives of all nations is a hopelessly clumsy instrument for the speedy achievement of even preliminary results. Even so, the task confronting us requires of all concerned the utmost sagacity and tolerance, which can be achieved only through awareness of the harsh necessity we have to face.

83. Foreword to *Spinoza: Portrait of a Spiritual Hero* (1946)

R. Kayser. *Spinoza: Portrait of a Spiritual Hero*. New York: Philosophical Library (1946), pp. ix–xi.

> Kayser originally published his Spinoza portrait in German in 1932. Einstein supplied a foreword only for the English edition.

Men of insight and sensitivity, assailed by the frightful events of these times, can hardly ward off a feeling of despondency and isolation. The confidence in the sure and constant progress of mankind that inspired people in the nineteenth century has given way to a crippling disillusionment. No one, it may be presumed, can deny the progress made in the realm of knowledge and in the field of technological invention; but we have experienced the disillusioning fact that all these advancements have not essentially alleviated the hardships of human destiny; nor have they ennobled human actions. The habit of casual interpretation of all phenomena, including those in the psychic and social spheres, has deprived the more wide-awake intellectual of the feeling of security and of those consolations which traditional religion, founded on authority, offered to earlier generations. It is a kind of banishment from a paradise of childlike innocence.

Such, in brief, is the description of the distress experienced by a conscious person of our time. Often he escapes from this misery into a fanciedly superior skepticism, or into distractions of all kinds which keep him from coming to his senses. But the effort is vain. We cannot in the long run substitute narcotics for wholesome nourishment.

On the whole, we learn very little about how people wrestle with this situation, unless we happen to be psychiatrists; and even they, as a rule, gain an insight into the travail of only those individuals who may lack the strength of resolving this spiritual conflict by themselves. Except for such instances, we know very little about the way in which our contemporaries settle the problem of their relationship as individuals to the hard facts of given conditions, both human and extra-human, so that they may arrive at that inner peace and security without which no kind of harmonious existence or work is possible. Furthermore, there are only a few individuals whose inner clarity is such as might enable them to impart their own subjective experiences to their fellow men in comprehensible form.

Hence, it is of great importance for the people of our time to become acquainted with the lives and struggles of those eminent personalities who have endured and overcome this same spiritual distress, and whose life story and life work can afford us a penetrating insight into this heroic conflict.

Among such personalities, Baruch Spinoza is one of the most outstanding. That is why it is so gratifying to have the author of this book acquaint us with the life and struggles of such a man. The author views Spinoza not so much with the critical eye of the professional philosopher as with that of the sympathetic historian who has an intuitive comprehension of the motive forces operative within that pure and lonely spirit. Naturally, the reading of this book cannot replace a close study of Spinoza's own works, but it brings Spinoza's personality near to us and thus facilitates the effort of penetrating into his thought.

Although he lived three hundred years before our time, the spiritual situation with which Spinoza had to cope peculiarly resembles our own. The reason for this is that he was utterly convinced of the causal dependence of all phenomena, at a time when the success accompanying the efforts to achieve a knowledge of the causal relationship of natural phenomena was still quite modest. Spinoza's conviction extended not only to inanimate nature but also to human feelings and actions. He had no doubt that our notion of possessing a free will (i.e., independent of causality) was an illusion resulting from our ignorance of the causes operative within us. In the study of this causal relationship he saw a remedy for fear, hate, and bitterness, the only remedy to which a genuinely spiritual man can have recourse. He demonstrated his justification for this conviction not only by the clear, precise formulation of his thoughts, but also by the exemplary fashioning of his own life.

84. A Message to My Adopted Country (1946)

Pageant 1 (January 1946), pp. 36–37.

Published with the introduction: "One who became an American by choice sounds a warning to his fellow-citizens." The text was reprinted in 1950 in Einstein's collection *Out of My Later Years* under the title "The Negro Question."

I am writing as one who has lived among you in America only a little more than ten years. And I am writing seriously and warningly. Many readers may ask: "What right has he to speak out about things which concern us alone, and which no newcomer should touch?"

I do not think such a standpoint is justified. One who has grown up in an environment takes much for granted. On the other hand, one who has come to this country as a mature person may have a keen eye for everything peculiar and characteristic. I believe he should speak out freely on what he sees and feels, for by so doing he may perhaps prove himself useful.

What soon makes the new arrival devoted to this country is the democratic trait among the people. I am not thinking here so much of the democratic political constitution of this country, however highly it must be praised. I am thinking of the relationship between individual people and of the attitude they maintain toward one another.

In the United States everyone feels assured of his worth as an individual. No one humbles himself before another person or class. Even the great difference in wealth, the superior power of a few, cannot undermine this healthy self-confidence and natural respect for the dignity of one's fellow-man.

There is, however, a somber point in the social outlook of Americans. Their sense of equality and human dignity is mainly limited to men of white skins. Even among these there are prejudices of which I as a Jew am clearly conscious; but they are unimportant in comparison with the attitude of the "Whites" toward their fellow-citizens of darker complexion, particularly toward Negroes. The more I feel an American, the more this situation pains me. I can escape the feeling of complicity in it only by speaking out.

Many a sincere person will answer me: "Our attitude toward Negroes is the result of unfavorable experiences which we have had by living side by side with Negroes in this country. They are not our equals in intelligence, sense of responsibility, reliability."

I am firmly convinced that whoever believes this suffers from a fatal misconception. Your ancestors dragged these black people from their homes by force; and in the white man's quest for wealth and an easy life they have been ruthlessly suppressed and exploited, degraded into slavery. The modern prejudice against Negroes is the result of the desire to maintain this unworthy condition.

The ancient Greeks also had slaves. They were not Negroes but white men who had been taken captive in war. There could be no talk of racial differences. And yet Aristotle, one of the great Greek philosophers, declared slaves inferior beings who were justly subdued and deprived of their liberty. It is clear that he was enmeshed in a traditional prejudice from which, despite his extraordinary intellect, he could not free himself.

A large part of our attitude toward things is conditioned by opinions and emotions which we unconsciously absorb as children from our environment. In other words, it is tradition—besides inherited aptitudes and qualities—which makes us what we are. We but rarely reflect how relatively small as compared with the powerful influence of tradition is the influence of our conscious thought upon our conduct and convictions.

It would be foolish to despise tradition. But without growing self-consciousness and increasing intelligence we must begin to control tradition and assume a critical attitude toward it, if human relations are ever to change for the better. We must try to recognize what in our accepted tradition is damaging to our fate and dignity—and shape our lives accordingly.

I believe that whoever tries to think things through honestly will soon recognize how unworthy and even fatal is the traditional bias against Negroes.

What, however, can the man of good will do to combat this deeply rooted prejudice? He must have the courage to set an example by word and deed and must watch lest his children become influenced by this racial bias.

I do not believe there is a way in which this deeply entrenched evil can be quickly healed. But until this goal is reached there is no greater satisfaction for a just and well-meaning person than the knowledge that he has devoted his best energies to the service of the good cause.

That is precisely what I have tried to do in writing this.

85. The Military Mentality (1947)

American Scholar 16, no. 3 (Summer 1947), pp. 353–354.

> Einstein is here replying to an article published in the previous issue of the journal that dealt with the debate over the military or civilian control of scientific projects, especially in the area of nuclear research.

It seems to me that the decisive point in the situation under discussion lies in the fact that the problem before us cannot be viewed as an isolated one. First of all, one may pose the following question: From now on, institutions for learning and research will more and more have to be supported by grants from the state, since, for various reasons, private sources will not suffice. Is it at all reasonable that the distribution of the funds raised for these purposes from the taxpayer should be entrusted to the military? To this question, every prudent person will certainly answer: "No!" For it is evident that the difficult task of the most beneficent distribution should be placed in the hands of people whose training and life's work give proof that they know something about science and scholarship.

If reasonable people nevertheless favor military agencies for the distribution of a major part of the available funds, the reason for this lies in the fact that they subordinate cultural concerns to their general political outlook. We must then focus our attention on these practical political viewpoints, their origins and their implications. In doing so we shall soon recognize that the problem here under discussion is but one of many and can only be fully estimated and properly adjudged when placed in a broader framework.

The tendencies we have mentioned are something new for America. They arose when, under the influence of the two World Wars and the consequent concentration of all forces of a military goal, a predominantly military mentality developed, which with the almost sudden victory became even more accentuated. The characteristic feature of this mentality is that people place the importance of what Bertrand Russell so tellingly terms "naked power" far above all other factors which affect the relations between peoples. The Germans, misled by Bismarck's successes in particular, underwent just such a transformation of their mentality—in consequence of which they were entirely ruined in less than a hundred years.

I must frankly confess that the foreign policy of the United States since the termination of hostilities has reminded me, sometimes irresistibly, of the attitude of Germany under Kaiser Wilhelm II, and I know that, independent of me, this analogy has most painfully occurred to others as well. It is characteristic of the military mentality that non-human factors (atom bombs, strategic bases, weapons of all sorts, the possession of raw materials, etc.) are held essential, while the human being, his desires and thoughts—in short, the psychological factors—are considered as unimportant and secondary. Herein lies a certain resemblance to Marxism, at least insofar as its theoretical side alone is kept in view. The individual is degraded to a mere instrument; he becomes "human material." The normal ends of human aspiration vanish with such a viewpoint. Instead, the military mentality raises "naked power" as a goal in itself—one of the strangest illusions to which men can succumb.

In our time the military mentality is still more dangerous than formerly because the offensive weapons have become much more powerful than the defensive ones. Therefore it leads, by necessity, to preventive war. The general insecurity that goes hand in hand with this results in the sacrifice of the citizen's civil rights to the supposed welfare of the state. Political witch-hunting, controls of all sorts (e.g., control of teaching and research, of the press, and so forth) appear inevitable, and for this reason do not encounter that popular resistance, which, were it not for the military mentality, would provide a protection. A reappraisal of all values gradually takes place insofar as everything that does not clearly serve the utopian ends is regarded and treated as inferior.

I see no other way out of prevailing conditions than a farseeing, honest and courageous policy with the aim of establishing security on supranational foundations. Let us hope that persons will be found, sufficient in number and moral force, to guide the nation on this path so long as a leading role is imposed on her by external circumstances. Then problems such as have been discussed here will cease to exist.

86. In the Shadow of the Atomic Bomb (1947)

Southern Patriot 7 (May 1949), p. 3.

> Einstein sent this essay as a mass-mailed letter on behalf of the Emergency Committee of Atomic Scientists on 6 August 1947, two years before it was printed in the *Southern Patriot* with some paragraphs omitted.

I am writing today to you and other friends who have helped us during the past year. It is a difficult moment at which to write. All about us, we see the wreckage of great hopes which mankind held for the building of peace. The gulf between East and West which men of good will have worked to close is widening daily. Some people believe that no reconciliation is possible and that another World War must decide the issue; we scientists reply that it is no longer possible to decide any issue by such means—an atomic war will bring no real decision but only unprecedented death and devastation on both sides.

Such a time in history breeds defeatism and despair. But there are those among us who believe that man has within him the capacity to meet and overcome even the great tests of our times. What we must not lose, or we lose all, is our willingness to seek the truth and our courage to act upon the truth. If we maintain these, we cannot despair.

We scientists believe upon ample evidence that the time of decision is upon us—that what we do or fail to do within the next few years will determine the fate of our civilization. That is the gist of the enclosed statement of this Committee, which was published on June 30th, 1947. We call for a "higher realism which recognizes that ... our fate is joined with that of our fellow men throughout the world." Great ideas may often be expressed in very simple words. In the shadow of the atomic bomb, it has become apparent that all men are brothers. If we recognize this as truth and act upon this recognition, mankind may go forward to a higher plane of human development. If the angry passions of a nationalistic world engulf us further, we are doomed.

The task of the scientists, as we conceive it, is untiringly to explain these truths, so that the American people will understand all that is at stake. We believe that with such understanding, the American people will choose from among many paths to reach a peaceful solution and that they will move toward such a solution and not toward war. And we believe that, in the long run, security for all nations demands a supranational solution.

We will strive unceasingly to bring this understanding to the American people through all the avenues of public discussion open to us. If we are to have any hope of influencing the Russians, and of convincing them that America wants peace and security for all peoples, we must understand the reasons for their profound distrust of everything foreign, which has become the sickness of a stubborn isolationism. We will work for understanding, not abject appeasement.

This letter goes to you on the second anniversary of the bombing of Hiroshima. On that day, the American people assumed responsibility before the world for the release of the most revolutionary force since the discovery of fire. Each of us, whether as scientists who made the bomb possible, or as citizens of the nation that applied the knowledge, stands accountable for the use we make of this tremendous new force. To our generation has come the possibility of making the most fateful decision in the recorded history of the human race. By an act of the collective will, made now, in this fraction of historic time, we can ensure that this great and painful achievement of man's intellect, instead of turning upon humanity, may be secured for the benefit of future generations.

I believe that mankind, capable of reason, restraint, and courage, will choose this path of peace.

No one can predict the events of the coming year but each of us has it in his power today to act for peace. You have helped us in the past. I ask you to help again.

87. A Plea for International Understanding (1947)

New York Times (12 November 1947), p. 1:7.

> Einstein's remarks were addressed to the members of the Foreign Press Association to the United Nations on 11 November 1947 in accepting the Association's awards as chairman of the Emergency Committee of Atomic Scientists. The award was given "in recognition of his valiant effort to make the world's nations understand the need of outlawing atomic energy as a means of war, and of developing it as an instrument of peace" (*Bulletin of Atomic Scientists*, Vol. 4, Number 1, January 1948).

I am most grateful to the Foreign Press Association for granting me its Award in appreciation of my modest efforts on behalf of a great cause. My happiness is dimmed, however, by a consciousness of the menacing situation in which human society—shrunk into one community with a common fate—finds itself. Everyone is aware of that situation, but only a few act accordingly. Most people go on living their every-day life; half frightened, half indifferent, they behold the ghostly tragi-comedy that is being performed on the international stage. But on that stage, on which the actors under the flood-lights play their ordained parts, our fate of tomorrow, life or death of the nations, is being decided.

It would be different if the problem were not one of the things made by Man himself, such as the atomic bomb. It would be different, for instance, if an epidemic of bubonic plague were threatening the entire world. In such a case conscientious and expert persons would be brought together and they would work out an intelligent plan to combat the plague. After having reached agreement upon the right ways and means, they would submit their plan to the governments. Those would hardly raise serious objections but rather agree speedily on the measures to be taken. They certainly would never think of trying to handle the matter so that their own nation would be spared whereas the next one would be decimated.

But could not our situation be compared to one of a menacing epidemic? People are unable to view this situation in its true light, for their eyes are blinded by passion. General fear and anxiety create hatred and aggressiveness. The adaptation to warlike aims and activities has corrupted the mentality of man; intelligent, objective and humane thinking has hardly any effect and is even persecuted as unpatriotic.

There are, no doubt, in the opposite camps enough people of sound judgment and sense of justice who would be capable and eager to work out together a solution for the factual difficulties. But the efforts of such people are hampered by the fact that it is impossible for them to come together for informal discussions. I am thinking of persons who are accustomed to the objective approach to a problem and who will not be confused by exaggerated nationalism or other passions. This forced separation of the people of both camps is one of the major obstacles to a solution of the burning problem of international security.

As long as contact between the two camps is limited to the official negotiations I can see little prospect of an intelligent agreement, since considerations of national prestige as well as the attempt to talk out of the window for the benefit of the masses make reasonable progress almost impossible. What one party suggests officially is for that reason alone suspected by the other. Also behind all official negotiations stands—though veiled—the threat of naked power. The official method can lead to success only after spade-work of an informal nature has prepared the ground; the conviction that a mutually satisfactory solution can be reached must be gained first; then the actual negotiations can get under way with a fair promise of success.

We scientists believe that what we and our fellow-men do or fail to do within the next few years will determine the fate of our civilization. And we consider it our task to help people realize all that is at stake, and to work, not for appeasement, but for understanding and ultimate agreement between peoples and nations.

It seems to me that the foreign journalists in this country, when they bestowed their Award upon a scientist, must have been led by similar considerations as those which I have expressed here. May they succeed in contributing to the general acceptance of that point of view.

88. Quantum Mechanics and Reality (1948)

Dialectica 2, nos. 3–4 (1948), pp. 320–324.

> This essay was written in German in April 1948 for a special issue of the journal *Dialectica*, edited by Wolfgang Pauli, and containing contributions by Niels Bohr, Louis de Broglie, Werner Heisenberg, and others, and published in Switzerland. An English translation appeared in M. Born, *The Natural Philosophy of Cause and Chance* (1949). Einstein here essentially argues that quantum mechanics, as it stood then, was an incomplete description of physical states. He had expressed these views since 1927. For a more technical paper, see Doc. 32 in Volume 1 of this series.

In the following, I will show why I do not consider the method of quantum mechanics satisfactory in principle in a brief and elementary fashion. On the other hand, I will note that I by no means deny that this theory represents a meaningful and, in a certain sense, even definitive advance in physical knowledge. I conceive that this theory will perhaps be contained in a later one, like the path optics in the wave optics. The relationship will remain; the foundations, however, will be deepened by a comprehensive rebuilding.

I

I consider a free particle, described (in the sense of quantum mechanics) at a certain time by a spatially restricted ψ-function. According to this description, the particle possesses neither a sharply defined momentum nor a sharply defined position.

In which sense shall I imagine that this representation describes a real, individual state of affairs? Two possible points of view seem to me possible and obvious, and we will weigh one against the other:

(a) The (free) particle really has a definite position and momentum, even if they cannot be ascertained simultaneously by measurement in the same individual case. According to this point of view, the ψ-function represents an *incomplete* description of the real state of affairs. This point of view is not the one physicists accept. Its acceptance would lead to an attempt to obtain an incomplete description of the real state of affairs alongside the complete one, and to discover physical laws for such a description. The theoretical framework of quantum mechanics would then be exploded.

(b) In reality, the particle has neither a definite momentum nor a definite position; the description by ψ-function is, in principle, a complete description. The sharply defined position of the particle, obtained by measuring the position, cannot be interpreted as the position of the particle prior to the measurement. The sharp localization that appears as a result of the measurement is brought about only as a result of the unavoidable (but not unimportant) operation of measurement. The result of the measurement depends not only on the real particle situation, but also on the nature of the measuring mechanism, which is incompletely known in principle. An analogous situation arises when the momentum or any other observable relating to the particle is measured. This is presumably the interpretation preferred by physicists at present, and one must admit that it alone does justice naturally to the empirical state of affairs expressed in Heisenberg's principle within the framework of quantum mechanics.

According to this point of view, two ψ-functions, which differ in more than trivialities, always describe two different real situations (for example, the particle with a well-defined position and one with a well-defined momentum).

The above is also valid, *mutatis mutandis*, to describe systems consisting of several particles. Here, too, we assume (in the sense of interpretation Ib) that the ψ-function completely describes a real state of affairs and that two (essentially) different ψ-functions describe two different real states of affairs, even if they could lead to identical results when a complete measurement is made. If the results of the measurement tally, it is put down to the partly unknown influence of the measurement arrangements.

II

If one asks what, irrespective of quantum mechanics, is characteristic of the world of ideas of physics, one is first of all struck by the following: The concepts of physics relate to a real outside world—i.e., ideas are established relating to things such as bodies, fields, etc. claiming a "real existence" independent of the perceiving subject. On the other hand, they have been brought into as secure a relationship as possible with the sense-data. Another characteristic of these physical objects is that they are thought of as arranged in a space-time continuum. An essential aspect of this arrangement of things in physics is that these things claim, at a certain time, an existence independent of one another, provided these objects "are situated in different parts of space." Without this assumption of the independence of the existence (the "being-thus") of objects far apart from one another in space, which stems from everyday thinking in the first place, physical thinking in the familiar sense would not be possible. It is also hard to see any way of formulating and testing the laws of physics unless one makes a clear distinction of this kind. Field theory has carried this principle to the extreme by localizing the elementary objects on which it is based and which exist independently of each other, as well as the elementary laws that have been postulated for it, in the infinitely small (four-dimensional) elements of space.

The following idea characterizes the relative independence of objects far apart in space (A and B): External influence on A has no direct influence on B; this is known as the "principle of contiguity," which is used consistently only in the field theory. If this axiom were completely abolished, the idea of the existence of (quasi-)enclosed systems, and thereby the postulation of laws empirically verifiable in the accepted sense, would become impossible.

III

I now assert that the interpretation of quantum mechanics (according to Ib) is not consistent with principle II. Let us consider a physical system S_{12}, which consists of two subsystems S_1 and S_2. These two subsystems may have been in a state of mutual physical interaction at an earlier time. We are, however, considering them at a time after this interaction has ended. Let the entire system be completely described, in the quantum mechanical sense, by a ψ-function ψ_{12} of the coordinates $q_1, \ldots$ and $q_2, \ldots$ of the two subsystems (ψ_{12} cannot be represented as a product of the form $\psi_1 \psi_2$, only as a sum of such products). At time t, let the two subsystems be separated from each other in space, in such a way that ψ_{12} only differs from 0 when $q_1, \ldots$ belong to a limited part R_1 of space, and $q_2, \ldots$ belong to a part R_2 separated from R_1.

The ψ-functions of the single subsystems S_1 and S_2 are then already unknown—that is, they do not exist at all. The methods of quantum mechanics, however, allow us to determine ψ_2 of S_2 from ψ_{12} if a complete measurement of subsystem S_1, in the sense of quantum mechanics, is also available. Instead of the original ψ_{12} of S_{12}, one thus obtains the ψ-function ψ_2 of the subsystem S_2.

But the kind of complete measurement undertaken on the part of system S_1 in the quantum theoretical sense—that is, which observable we are measuring—is crucial for this determination. For example, if S_1 consists of a single particle, then we have the choice of measuring either its position or its momentum components.

The resulting ψ_2 depends on this choice, so that different kinds of (statistical) predictions regarding measurements to be carried out later on S_2 are obtained according to the choice of measurement carried out on S_1. From the point of view of the interpretations of Ib, this means a different real situation is being created in regard to S_2, according to the choice of complete measurement of S_1, which can be described variously by ψ_2, ψ_2', ψ_2'', etc.

Seen from the perspective of quantum mechanics alone, this does not present any difficulty; for, according to the choice of measurement to be carried out on S_1, a different real situation is created, and the necessity of having to attach two or more different ψ-functions $\psi_2, \psi_2', \ldots$ to the same system S_1 cannot arise.

It is a different matter, however, when one tries to adhere to the principles of quantum mechanics and principle II—i.e., the independent existence of the real state of affairs existing in two separate parts of space R_1 and R_2. In our example, the complete measurement on S_1 represents a physical operation that only affects part R_1 of space.

Such an operation, however, can have no direct influence on the physical reality in a remote part R_2 of space. It follows that every statement about S_2 we arrive at as a result of a complete measurement of S_1 has to be valid for system S_2, even if no measurement whatsoever is carried out on S_1. This would mean all statements that can be deduced from the settlement of ψ_2 or ψ_2' must simultaneously be valid for S_2. This is, of course, impossible if ψ_2, ψ_2', etc. should represent different real states of affairs for S_2—that is, one comes into conflict with the Ib interpretation of the ψ-function.

There seems, to me, no doubt that those physicists who regard the descriptive methods of quantum mechanics as definitive in principle would react to this line of thought in the following way: They would drop the requirement II for the independent existence of the physical reality present in different parts of space; they would be justified in pointing out that the quantum theory nowhere makes explicit use of this requirement.

I admit this, but would point out that, when I consider the physical phenomena known to me, especially those so successfully encompassed by quantum mechanics, I still cannot find any fact anywhere that would make it appear likely that requirement II will have to be abandoned.

I am therefore inclined to believe that the description of quantum mechanics in the sense of Ia has to be regarded as an incomplete and indirect description of reality, to be replaced at some later date by a more complete and direct one.

At all events, one should beware, in my opinion, of committing oneself too dogmatically to the present theory in searching for a unified basis (i.e., a continuous field theory) for the whole of physics.

89. A Reply to the Soviet Scientists (1948)

Bulletin of the Atomic Scientists 4 (1948), pp. 35–37.

This essay was in response to a letter criticizing Einstein's support of World Government, published in the New Times, a Moscow English-language newspaper, on 26 November 1947, with the title, "About Certain Fallacies of Professor Albert Einstein." The letter was signed by distinguished and well-known members of the Academy of Sciences of the USSR: Sergei Vavilov, A. N. Frumkin, Abram F. Ioffe, and N. N. Semyonov. The letter, together with Einstein's reply below, were published sequentially in the February 1948 issue of the Bulletin of Atomic Scientists.

Four of my Russian colleagues have published a benevolent attack upon me in an open letter carried by the *New Times*. I appreciate the effort they have made and I appreciate even more the fact that they have expressed their point of view so candidly and straightforwardly. To act intelligently in human affairs is only possible if an attempt is made to understand the thoughts, motives and apprehensions of one's opponent so fully that one can see the world through his eyes. All well-meaning people should try to contribute as much as possible to improving such mutual understanding. It is in this spirit that I should like to ask my Russian colleagues and any other reader to accept the following answer to their letter. It is the reply of a man who anxiously tries to find a feasible solution without having the illusion that he himself knows "the truth" or "the right path" to follow. If in the following I shall express my views somewhat dogmatically, I do it only for the sake of clarity and simplicity.

Although your letter, in the main, is clothed in an attack upon the non-socialistic foreign countries, particularly the United States, I believe that behind the aggressive front there lies a defensive mental attitude which is nothing else but the trend toward an almost unlimited isolationism. The escape into isolationism is not difficult to understand if one realizes what Russia has suffered at the hands of foreign countries during the last three decades—the German invasions with planned mass murder of the civilian population, foreign interventions during the civil war, the systematic campaign of calumnies in the western press, the support of Hitler as an alleged tool to fight Russia. However understandable this desire for isolation may be, it remains no less disastrous to Russia and to all other nations; I shall say more about it later on.

Socialism is a framework not a solution

The chief object of your attack against me concerns my support of "world government."[a] I should like to discuss this important problem only after having said a few words about the antagonism between socialism and capitalism; for your attitude on the significance of this antagonism seems to dominate completely your views on international problems. If the socioeconomic problem is considered objectively, it appears as follows: technological development has led to increasing centralization of the economic mechanism. It is this development which is also responsible for the fact that economic power in all widely industrialized countries has become concentrated in the hands of relatively few. These people, in capitalist countries, do not need to account for their actions to the public as a whole; they must do so in socialist countries in which they are civil servants similar to those who exercise political power.

I share your view that a socialist economy possesses advantages which definitely counterbalance its disadvantages whenever the management lives up, at least to some extent, to adequate standards. No doubt, the day will come when all nations (as far as such nations still exist) will be grateful to Russia for having demonstrated, for the first time, by vigorous action the practical possibility of a planned economy in spite of exceedingly great difficulties. I also believe that capitalism, or, we should say, the system of free enterprise, will prove unable to check unemployment, which will become increasingly chronic because of technological progress, and unable to maintain a healthy balance between production and the purchasing power of the people.

On the other hand we should not make the mistake of blaming capitalism for all existing social and political evils, and of assuming that the very establishment of socialism would be able to cure all the social and political ills of humanity. The danger of such a belief lies, first, in the fact that it encourages fanatical intolerance on the part of all the "faithfuls" by making a possible social method into a type of church which brands all those who do not belong to it as traitors or as nasty evildoers. Once this stage has been reached, the ability to understand the convictions and actions of the "unfaithfuls" vanishes completely. You know, I am sure, from history how much unnecessary suffering such rigid beliefs have inflicted upon mankind.

Anarchy: Economic and political

Any government is in itself an evil insofar as it carries within it the tendency to deteriorate into tyranny. However, except for a very small number of anarchists, every one of us is convinced that civilized society cannot exist without a government. In a healthy nation there is a kind of dynamic balance between the will of the people and the government which prevents its degeneration into tyranny. It is obvious that the danger of such deterioration is more acute in a country in which the government has authority not only over the armed forces but also over all the channels of education and information as well as over the economic existence of every single citizen. I say this merely to indicate that socialism as such cannot be considered the solution to all social problems but merely as a framework within which such a solution is possible.

What has surprised me most in your general attitude, expressed in your letter, is the following aspect: You are such passionate opponents of anarchy in the economic sphere, and yet equally passionate advocates of anarchy, e.g., unlimited sovereignty, in the sphere of international politics. The proposition to curtail the sovereignty of individual states appears to you in itself reprehensible, as a kind of violation of a natural right. In addition, you try to prove that behind the idea of curtailing sovereignty the United States is hiding her intention of economic domination and exploitation of the rest of the world without going to war. You attempt to justify this indictment by analyzing in your fashion the individual actions of this government since the end of the last war. You attempt to show that the Assembly of the United Nations is a mere puppet show controlled by the United States and hence the American capitalists.

Such arguments impress me as a kind of mythology; they are not convincing. They make obvious, however, the deep estrangement among the intellectuals of our two countries, which is the result of a regrettable and artificial mutual isolation. If a free personal exchange of views should be made possible and should be encouraged, the intellectuals, possibly more than anyone else, could help to create an atmosphere of mutual understanding between the two nations and their problems. Such an atmosphere is a necessary prerequisite for the fruitful development of political cooperation. However, since for the time being we depend upon the cumbersome method of "open letters," I shall want to indicate briefly my reaction to your arguments.

Economic oligarchy not entirely dominant

Nobody would want to deny that the influence of the economic oligarchy upon all branches of our public life is very powerful. This influence, however, should not be overestimated. Franklin Delano Roosevelt was elected president in spite of desperate opposition by these very powerful groups and was reelected three times; and this took place at a time when decisions of great consequence had to be made.

Concerning the policies of the American Government since the end of the war, I am neither willing, nor able, nor entitled to justify or explain them. It cannot be denied, however, that the suggestions of the American Government with regard to atomic weapons represented at least an attempt toward the creation of a supranational security organization.[b] If they were not acceptable, they could at least have served as a basis of discussion for a real solution of the problems of international security. It is, indeed, the attitude of the Soviet Government that was partly negative and partly dilatory which has made it so difficult for well-meaning people in this country to use their political influence as they would have wanted, and to oppose the "war mongers."[c] With regard to the influence of the United States upon the United Nations Assembly, I wish to say that, in my opinion, it stems not only from the economic and military power of the United States but also from the efforts of the United States and the United Nations to lead toward a genuine solution of the security problem.

Concerning the controversial veto power, I believe that the efforts to eliminate it or to make it ineffective have their primary cause less in specific intentions of the United States than in the manner in which the veto privilege has been abused.

U.S. forced to emphasize exports

Let me come now to your suggestion that the policy of the United States seeks to obtain economic domination and exploitation of other nations. It is a precarious undertaking to say anything reliable about aims and intentions. Let us rather examine the objective factors involved. The United States is fortunate in producing all the important industrial products and foods in her own country, in sufficient quantities. The country also possesses almost all important raw materials. Because of her tenacious belief in "free enterprise" she cannot succeed in keeping the purchasing power of the people in balance with the productive capacity of the country. For these very same reasons there is a constant danger that unemployment will reach threatening dimensions.

Because of these circumstances the United States is compelled to emphasize her export trade. Without it, she could not permanently keep her total productive machinery fully utilized. These conditions would not be harmful if the exports were balanced by imports of about the same value. Exploitation of foreign nations would then consist in the fact that the labor value of imports would considerably exceed that of exports. However, every effort is being made to avoid this since almost every import would make a part of the productive machinery idle.

Danger lies not in gold but in loans

This is why foreign countries are not able to pay for the export commodities of the United States, payment which, in the long run, would indeed be possible only through imports by the latter. This explains why a large portion of all the gold has come to the United States. On the whole, this gold cannot be utilized except for the purchase of foreign commodities, which because of the reasons already stated, is not practicable. There it lies, this gold, carefully protected against theft, a monument to governmental wisdom and to economic science! The reasons which I have just indicated make it difficult for me to take the alleged exploitation of the world by the United States very seriously.

However, the situation just described has a serious political facet. The United States, for the reasons indicated, is compelled to ship part of its production to foreign countries. These exports are financed through loans which the United States is granting foreign countries.[d] It is, indeed, difficult to imagine how these loans will ever be repaid. For all practical purposes, therefore, these loans must be considered gifts which may be used as weapons in the arena of power politics. In view of the existing conditions and in view of the general characteristics of human beings, this, I frankly admit, represents a real danger. Is it not true, however, that we have stumbled into a state of international affairs which tends to make every invention of our minds and every material good into a weapon and, consequently, into a danger for mankind?

This question brings us to the most important matter, in comparison to which everything else appears insignificant indeed. We all know that power politics, sooner or later, necessarily leads to war, and that war, under present circumstances, would mean a mass destruction of human beings and material goods the dimension of which are much, much greater than anything that has ever before happened in history.

Danger of annihilation supersedes all others

It is really unavoidable that, because of our passions and our inherited customs, we should be condemned to annihilate each other so thoroughly that nothing would be left over which would deserve to be conserved? Is it not true that all the controversies and differences of opinion which we have touched upon in our strange exchange of letters are insignificant pettinesses compared to the danger in which we all find ourselves? Should we not do everything in our power to eliminate the danger which threatens all nations alike?

If we hold fast to the concept and practice of unlimited sovereignty of nations it only means that each country reserves the right for itself of pursuing its objectives through warlike means. Under the circumstances, every nation must be prepared for that possibility; this means it must try with all its might to be superior to anyone else. This objective will dominate more and more our entire public life and will poison our youth long before the catastrophe is itself actually upon us. We must not tolerate this, however, as long as we still retain a tiny bit of calm reasoning and human feelings.

This alone is on my mind in supporting the idea of "World Government," without any regard to what other people may have in mind when working for the same objective. I advocate world government because I am convinced that there is no other possible way of eliminating the most terrible danger in which man has ever found himself. The objective of avoiding total destruction must have priority over any other objective.

I am sure you are convinced that this letter is written with all the seriousness and honesty at my command; I trust you will accept it in the same spirit.

Editorial Notes

[a] For Einstein's detailed exposition on this point, see Document 79.

[b] Most likely a reference to the Acheson-Lilienthal report and the Baruch Plan of June 1946, to the effect that, under the authority of the United Nations Atomic Energy Commission, the United States would agree to decommission all of its atomic weapons on the condition that no country would produce atomic weapons. It also proposed other measures of verification and control, international exchanges, and cooperation on nuclear power.

[c] The Soviet Union, at the time not yet in possession of an atomic bomb, rejected the plan with the argument that it would maintain U.S. monopoly of nuclear technology, and that the United Nations was a tool of the U.S. government.

[d] Einstein is referring here to the Marshall Plan, or European Recovery Program, that provided loans and grants and helped imports deemed necessary for reconstruction.

90. Religion and Science: Irreconcilable? (1948)

Christian Register 127 (June 1948), pp. 19–20.

Einstein composed this essay as a letter of greetings to the American Unitarian Association on the occasion of its 123rd anniversary gathering in Boston, where the letter was read aloud by Reverend Jacob Trapp of the Unitarian church of Summit, N.J. Einstein had given a similar address in 1940, unpublished at the time, to the First Conference on Science, Philosophy, and Religion and Their Relation to the Democratic Way of Life, founded by seventy-nine leading American intellectuals. See also Doc. 54 on the same subject.

Does there truly exist an insuperable contradiction between religion and science? Can religion be superseded by science? The answers to these questions have, for centuries, given rise to considerable dispute and, indeed, bitter fighting. Yet, in my own mind there can be no doubt that in both cases a dispassionate consideration can only lead to a negative answer. What complicates the solution, however, is the fact that while most people readily agree on what is meant by "science," they are likely to differ on the meaning of "religion."

As to science, we may well define it for our purpose as "methodical thinking directed toward finding regulative connections between our sensual experiences." Science, in the immediate, produces knowledge and, indirectly, means of action. It leads to methodical action if definite goals are set up in advance. For the function of setting up goals and passing statements of value transcends its domain. While it is true that science, to the extent of its grasp of causative connections, may reach important conclusions as to the compatibility and incompatibility of goals and evaluations, the independent and fundamental definitions regarding goals and values remain beyond science's reach.

As regards religion, on the other hand, one is generally agreed that it deals with goals and evaluations and, in general, with the emotional foundation of human thinking and acting, as far as these are not predetermined by the inalterable hereditary disposition of the human species. Religion is concerned with man's attitude toward nature at large, with the establishing of ideals for the individual and communal life, and with mutual human relationship. These ideals religion attempts to attain by exerting an educational influence on tradition and through the development and promulgation of certain easily accessible thoughts and narratives (epics and myths) which are apt to influence evaluation and action along the lines of the accepted ideals.

It is this mythical, or rather this symbolic content of the religious traditions which is likely to come into conflict with science. This occurs whenever this religious stock of ideas contains dogmatically fixed statements on subjects which belong in the domain of science. Thus, it is of vital importance for the preservation of true religion that such conflicts be avoided when they arise from subjects which, in fact, are not really essential for the pursuance of the religious aims.

When we consider the various existing religions as to their essential substance, that is, divested of their myths, they do not seem to me to differ as basically from each other as the proponents of the "relativistic" or conventional theory wish us to believe. And this is by no means surprising. For the moral attitudes of a people that is supported by religion need always aim at preserving and promoting the sanity and vitality of the community and its individuals, since otherwise this community is bound to perish. A people that were to honor falsehood, defamation, fraud and murder would be unable, indeed, to subsist for very long.

When confronted with a specific case, however, it is no easy task to determine clearly what is desirable and what should be eschewed, just as we find it difficult to decide what exactly it is that makes good painting or good music. It is something that may be felt intuitively more easily than rationally comprehended. Likewise, the great moral teachers of humanity were, in a way, artistic geniuses in the art of living. In addition to the most elementary precepts directly motivated by the preservation of life and the sparing of unnecessary suffering, there are others to which, although they are apparently not quite commensurable to the basic precepts, we nevertheless attach considerable importance. Should truth, for instance, be sought unconditionally even where its attainment and its accessibility to all would entail heavy sacrifices in toil and happiness? There are many such questions which, from a rational vantage point, cannot easily be answered or cannot be answered at all. Yet, I do not think that the so-called "relativistic" viewpoint is correct, not even when dealing with the more subtle moral decisions.

When considering the actual living conditions of present day civilized humanity from the standpoint of even the most elementary religious commands, one is bound to experience a feeling of deep and painful disappointment at what one sees. For while religion prescribes brotherly love in the relations among the individuals and groups, the actual spectacle more resembles a battlefield than an orchestra. Everywhere, in economic as well as in the political life, the guiding principle is one of ruthless striving for success at the expense of one's fellow men. This competitive spirit prevails even in school and, destroying all feelings of human fraternity and cooperation, conceives of achievement not as derived from the love for productive and thoughtful work, but as springing from personal ambition and fear of rejection.

There are pessimists who hold that such a state of affairs is necessarily inherent in human nature; it is those who propound such views that are the enemies of true religion, for they imply thereby that religious teachings are utopian ideals and unsuited to afford guidance in human affairs. The study of the social patterns in certain so-called primitive cultures, however, seems to have made it sufficiently evident that such a defeatist view is wholly unwarranted. Whoever is concerned with this problem, a crucial one in the study of religion as such, is advised to read the description of the Pueblo Indians in Ruth Benedict's book, *Patterns of Culture*. Under the hardest living conditions, this tribe has apparently accomplished the difficult task of delivering its people from the scourge of competitive spirit and to foster in it a temperate, cooperative conduct of life, free of external pressure and without any curtailment of happiness.

The interpretation of religion, as here advanced, implies a dependence of science on the religious attitude, a relation which, in our predominantly materialistic age, is only too easily overlooked. While it is true that scientific results are entirely independent from religious or moral considerations, those individuals to whom we owe the great creative achievements of science were all of them imbued with the truly religious conviction that this universe of ours is something perfect and susceptible to the rational striving for knowledge. If this conviction had not been a strongly emotional one and if those searching for knowledge had not been inspired by Spinoza's *Amor Dei Intellectualis*,[a] they would hardly have been capable of that untiring devotion which alone enables man to attain his greatest achievements.

Editorial Note

[a] For Spinoza and his concept, see Doc. 4 and its endnote [a] and Doc. 50.

91. Why Socialism? (1949)

Monthly Review 1, no. 1 (1949), pp. 9–15.

This essay featured in the first volume of the *Monthly Review: An Independent Socialist Magazine*, edited by Leo Huberman and Paul M. Sweezy, published in May 1949.

Is it advisable for one who is not an expert on economic and social issues to express views on the subject of socialism? I believe for a number of reasons that it is.

Let us first consider the question from the point of view of scientific knowledge. It might appear that there are no essential methodological differences between astronomy and economics: scientists in both fields attempt to discover laws of general acceptability for a circumscribed group of phenomena in order to make the interconnection of these phenomena as clearly understandable as possible. But in reality such methodological differences do exist. The discovery of general laws in the field of economics is made difficult by the circumstance that observed economic phenomena are often affected by many factors, which are very hard to evaluate separately. In addition, the experience which has accumulated since the beginning of the so-called civilized period of human history has—as is well known—been largely influenced and limited by causes which are by no means exclusively economic in nature. For example, most of the major states of history owed their existence to conquest. The conquering peoples established themselves, legally and economically, as the privileged class of the conquered country. They seized for themselves a monopoly of the land ownership and appointed a priesthood from among their own ranks. The priests, in control of education, made the class division of society into a permanent institution and created a system of values by which the people were thenceforth, to a large extent unconsciously, guided in their social behavior.

But historic tradition is, so to speak, of yesterday; nowhere have we really overcome what Thorstein Veblen called "the predatory phase" of human development. The observable economic facts belong to that phase, and even such laws as we can derive from them are not applicable to other phases. Since the real purpose of socialism is precisely to overcome and advance beyond the predatory phase of human development, economic science in its present state can throw little light on the socialist society of the future.

Second, socialism is directed toward a social-ethical end. Science, however, cannot create ends and, even less, instill them in human beings; science, at most, can supply the means by which to attain certain ends. But the ends themselves are conceived by personalities with lofty ethical ideals and—if these ends are not stillborn, but vital and vigorous—are adopted and carried forward by those many human beings who, half unconsciously, determine the slow evolution of society.

For these reasons, we should be on our guard not to overestimate science and scientific methods when it is a question of human problems, and we should not assume that experts are the only ones who have a right to express themselves on questions affecting the organization of society.

Innumerable voices have been asserting for some time now that human society is passing through a crisis, that its stability has been gravely shattered. It is characteristic of such a situation that individuals feel indifferent or even hostile toward the group, small or large, to which they belong. In order to illustrate my meaning, let me record here a personal experience. I recently discussed with an intelligent and well-disposed man the threat of another war, which in my opinion would seriously endanger the existence of mankind, and I remarked that only a supra-national organization would offer protection from that danger. Thereupon my visitor, very calmly and coolly, said to me: "Why are you so deeply opposed to the disappearance of the human race?"

I am sure that as little as a century ago, no one would have so lightly made a statement of this kind. It is the statement of a man who has striven in vain to attain an equilibrium within himself, and has more or less lost hope of succeeding. It is the expression of a painful solitude and isolation from which so many people are suffering in these days. What is the cause? Is there a way out?

It is easy to raise such questions, but difficult to answer them with any degree of assurance. I must try, however, as best I can, although I am very conscious of the fact that our feelings and strivings are often contradictory and obscure and that they cannot be expressed in easy and simple formulas.

Man is, at one and the same time, a solitary being and a social being. As a solitary being, he attempts to protect his own existence and that of those who are closest to him, to satisfy his personal desires, and to develop his innate abilities. As a social being, he seeks to gain the recognition and affection of his fellow human beings, to share in their pleasures, to comfort them in their sorrows, and to improve their conditions of life. Only the existence of these varied, frequently conflicting, strivings accounts for the special character of a man, and their specific combination determines the extent to which an individual can achieve an inner equilibrium and can contribute to the well-being of society. It is quite possible that the relative strength of these two drives is, in the main, fixed by inheritance. But the personality that finally emerges is largely formed by the environment in which a man happens to find himself during his development, by the structure of the society in which he grows up, by the tradition of that society, and by its appraisal of particular types of behavior. The abstract concept "society" means to the individual human being the sum total of his direct and indirect relations to his contemporaries and to all the people of earlier generations. The individual is able to think, feel, strive, and work by himself, but he depends so much upon society—in his physical, intellectual, and emotional existence—that it is impossible to think of him, or to understand him, outside the framework of society. It is "society" which provides man with food, clothing, a home, the tools of work, language, the forms of thought, and most of the content of thought; his life is made possible through the labor and the accomplishments of the many millions past and present who are all hidden behind the small word "society."

It is evident, therefore, that the dependence of the individual upon society is a fact of nature which cannot be abolished—just as in the case of ants and bees. However, while the whole life process of ants and bees is fixed down to the smallest detail by rigid, hereditary instincts, the social pattern and interrelationships of human beings are very variable and susceptible to change. Memory, the capacity to make new combinations, the gift of oral communication have made possible developments among human beings which are not dictated by biological necessities. Such developments manifest themselves in traditions, institutions, and organizations; in literature; in scientific and engineering accomplishments; in works of art. This explains how it happens that, in a certain sense, man can influence his life through his own conduct, and that in this process conscious thinking and wanting can play a part.

Man acquires at birth, through heredity, a biological constitution which we must consider fixed and unalterable, including the natural urges which are characteristic of the human species. In addition, during his lifetime, he acquires a cultural constitution which he adopts from society through communication and through many other types of influences. It is this cultural constitution which, with the passage of time, is subject to change and which determines to a very large extent the relationship between the individual and society. Modern anthropology has taught us, through comparative investigation of so-called primitive cultures, that the social behavior of human beings may differ greatly, depending upon prevailing cultural patterns and the types of organization which predominate in society. It is on this that those who are striving to improve the lot of man may ground their hopes: human beings are *not* condemned, because of their biological constitution, to annihilate each other or to be at the mercy of a cruel, self-inflicted fate.

If we ask ourselves how the structure of society and the cultural attitude of man should be changed in order to make human life as satisfying as possible, we should constantly be conscious of the fact that there are certain conditions which we are unable to modify. As mentioned before, the biological nature of man is, for all practical purposes, not subject to change. Furthermore, technological and demographic developments of the last few centuries have created conditions which are here to stay. In relatively densely settled populations with the goods which are indispensable to their continued existence, an extreme division of labor and a highly-centralized productive apparatus are absolutely necessary. The time—which, looking back, seems so idyllic—is gone forever when individuals or relatively small groups could be completely self-sufficient. It is only a slight exaggeration to say that mankind constitutes even now a planetary community of production and consumption.

I have now reached the point where I may indicate briefly what to me constitutes the essence of the crisis of our time. It concerns the relationship of the individual to society. The individual has become more conscious than ever of his dependence upon society. But he does not experience this dependence as a positive asset, as an organic tie, as a protective force, but rather as a threat to his natural rights, or even to his economic existence. Moreover, his position in society is such that the egotistical drives of his make-up are constantly being accentuated, while his social drives, which are by nature weaker, progressively deteriorate. All human beings, whatever their position in society, are suffering from this process of deterioration. Unknowingly prisoners of their own egotism, they feel insecure, lonely, and deprived of the naive, simple, and unsophisticated enjoyment of life. Man can find meaning in life, short and perilous as it is, only through devoting himself to society.

The economic anarchy of capitalist society as it exists today is, in my opinion, the real source of the evil. We see before us a huge community of producers, the members of which are unceasingly striving to deprive each other of the fruits of their collective labor—not by force, but on the whole in faithful compliance with legally established rules. In this respect, it is important to realize that the means of production—that is to say, the entire productive capacity that is needed for producing consumer goods as well as additional capital goods—may legally be, and for the most part are, the private property of individuals.

For the sake of simplicity, in the discussion that follows I shall call "workers" all those who do not share in the ownership of the means of production—although this does not quite correspond to the customary use of the term. The owner of the means of production is in a position to purchase the labor power of the worker. By using the means of production, the worker produces new goods, which become the property of the capitalist. The essential point about this process is the relation between what the worker produces and what he is paid, both measured in terms of real value. Insofar as the labor contract is "free," what the worker receives is determined not by the real value of the goods he produces, but by his minimum needs and the capitalists' requirements for labor power in relation to the number of workers competing for jobs. It is important to understand that even in theory the payment of the worker is not determined by the value of his product.

Private capital tends to become concentrated in few hands, partly because of competition among the capitalists, and partly because technological development and the increasing division of labor encourage the formation of larger units of production at the expense of the smaller ones. The result of these developments is an oligarchy of private capital, the enormous power of which cannot be effectively checked even by a democratically organized political society. This is true since the members of legislative bodies are selected by political parties, largely financed or otherwise influenced by private capitalists who, for all practical purposes, separate the electorate from the legislature. The consequence is that the representatives of the people do not, in fact, sufficiently protect the interests of the underprivileged sections of the population. Moreover, under existing conditions, private capitalists inevitably control, directly or indirectly, the main sources of information (press, radio, education). It is thus extremely difficult, and indeed, in most cases quite impossible, for the individual citizen to come to objective conclusions and to make intelligent use of his political rights.

The situation prevailing in an economy based on the private ownership of capital is thus characterized by two main principles: first, means of production (capital) are privately owned and the owners dispose of them as they see fit; second, the labor contract is free. Of course, there is no such thing as a *pure* capitalist society in this sense. In particular, it should be noted that the workers, through long and bitter political struggles, have succeeded in securing a somewhat improved form of the "free labor contract" for certain categories of workers. But taken as a whole, the present day economy does not differ much from "pure" capitalism.

Production is carried on for profit, not for use. There is no provision that all those able and willing to work will always be in a position to find employment; an "army of unemployed" almost always exists. The worker is constantly in fear of losing his job. Since unemployed and poorly paid workers do not provide a profitable market, the production of consumers' goods is restricted, and great hardship is the consequence. Technological progress frequently results in more unemployment rather than in an easing of the burden of work for all. The profit motive, in conjunction with competition among capitalists, is responsible for an instability in the accumulation and utilization of capital which leads to increasingly severe depressions. Unlimited competition leads to a huge waste of labor, and to that crippling of the social consciousness of individuals which I mentioned before.

This crippling of individuals I consider the worst evil of capitalism. Our whole educational system suffers from this evil. An exaggerated competitive attitude is inculcated into the student, who is trained to worship acquisitive success as a preparation for his future career.

I am convinced there is only *one* way to eliminate these grave evils, namely through the establishment of a socialist economy, accompanied by an educational system which would be oriented toward social goals. In such an economy, the means of production are owned by society itself and are utilized in a planned fashion. A planned economy, which adjusts production to the needs of the community, would distribute the work to be done among all those able to work and would guarantee a livelihood to every man, woman, and child. The education of the individual, in addition to promoting his own innate abilities, would attempt to develop in him a sense of responsibility for his fellow men in place of the glorification of power and success in our present society.

Nevertheless, it is necessary to remember that a planned economy is not yet socialism. A planned economy as such may be accompanied by the complete enslavement of the individual. The achievement of socialism requires the solution of some extremely difficult socio-political problems: how is it possible, in view of the far-reaching centralization of political and economic power, to prevent bureaucracy from becoming all-powerful and overweening? How can the rights of the individual be protected and therewith a democratic counterweight to the power of bureaucracy be assured?

Clarity about the aims and problems of socialism is of greatest significance in our age of transition. Since, under present circumstances, free and unhindered discussion of these problems has come under a powerful taboo, I consider the foundation of this magazine to be an important public service.

92. Autobiographical Notes (1949)

Paul A. Schilpp, ed. *Albert Einstein: Philospher-Scientist*. Evanston, Ill.: Northwestern University Press (1949), pp. 2–95.

The original German text was published together with an English translation by Paul Arthur Schilpp on facing pages. The volume also contained essays by many contemporaries of Einstein, and a response by Einstein to some of their comments. Einstein had completed this autobiographical essay in 1946, but the slow rate at which other contributions arrived delayed the volume's publication until 1949. The volume was clearly intended as a monument to the totality of Einstein's scientific thought. Einstein's essay is unconventional, in the sense that it addresses only superficially his outward life and focuses almost exclusively on the development of his scientific ideas and their connection to the methods and assumptions of physical theories. Einstein himself summarizes

> the goal here: "For the essential in the being of a man of my type lies precisely in *what* he thinks and *how* he thinks, not in what he does or suffers."[a]

Here I sit in order to write, at the age of 67, something like my own obituary. I am doing this not merely because Dr. Schilpp has persuaded me to do it; but because I do, in fact, believe that it is a good thing to show those who are striving alongside of us, how one's own striving and searching appears to one in retrospect. After some reflection, I felt how insufficient any such attempt is bound to be. For, however brief and limited one's working life may be, and however predominant may be the ways of error, the exposition of that which is worthy of communication does nonetheless not come easy—today's person of 67 is by no means the same as was the one of 50, or 30, or of 20. Every reminiscence is colored by today's being what it is, and therefore by a deceptive point of view. This consideration could very well deter. Nevertheless, much can be lifted out of one's own experience which is not open to another consciousness.

Even when I was a fairly precocious young man,[b] the nothingness of the hopes and strivings which chases most men restlessly through life came to my consciousness with considerable vitality. Moreover, I soon discovered the cruelty of that chase, which in those years was much more carefully covered up by hypocrisy and glittering words than is the case today. By the mere existence of his stomach, everyone was condemned to participate in that chase. Moreover, it was possible to satisfy the stomach by such participation, but not man in so far as he is a thinking and feeling being. As the first way out, there was religion, which is implanted into every child by way of the traditional education machine. Thus I came—despite the fact that I was the son of entirely irreligious (Jewish) parents—to a deep religiosity, which, however, found an abrupt ending at the age of 12. Through the reading of popular scientific books, I soon reached the conviction that much in the stories of the Bible could not be true. The consequence was a positively fanatic [orgy of] freethinking[c] coupled with the impression that youth is intentionally being deceived by the state through lies; it was a crushing impression. Suspicion against every kind of authority grew out of this experience, a skeptical attitude toward the convictions that were alive in any specific social environment—an attitude which has never again left me, even though later on, because of a better insight into the causal connections, it lost some of its original poignancy.

It is quite clear to me that the religious paradise of youth, which was thus lost, was a first attempt to free myself from the chains of the "merely personal," from an existence which is dominated by wishes, hopes, and primitive feelings. Out yonder, there was this huge world, which exists independently of us human beings and which stands before us like a great, eternal riddle, at least partially accessible to our inspection and thinking. The contemplation of this world beckoned like a liberation, and I soon noticed that many a man whom I had learned to esteem and to admire had found inner freedom and security in devoted occupation with it. The mental grasp of this extra-personal world within the frame of the given possibilities swam as highest aim, half consciously and half unconsciously, before my mind's eye. Similarly motivated men of the present and of the past, as well as the insights which they had achieved, were the friends who could not be lost. The road to this paradise was not as comfortable and alluring as the road to the religious paradise, but it has proved itself as trustworthy, and I have never regretted having chosen it.

What I have here said is true only within a certain sense, just as a drawing consisting of a few strokes can do justice to a complicated object, full of perplexing details, only in a very limited sense. If an individual enjoys well-ordered thoughts, it is quite possible that this side of his nature may grow more pronounced at the cost of other sides, and thus may determine his mentality in increasing degree. In this case it is well possible that such an individual in retrospect sees a uniformly systematic development, whereas the actual experience takes place in kaleidoscopic particular situations. The manifoldness of the external situations and the narrowness of the momentary content of consciousness bring about a sort of atomizing of the life of every human being. In a man of my type, the turning-point of the development lies in the fact that gradually the major interest disengages itself to a far-reaching degree from the momentary and the merely personal and turns toward the striving for a mental grasp of things. Looked at from this point of view, the above schematic remarks contain as much truth as can be uttered in such brevity.

What, precisely, is "thinking"? When, at the reception of sense-impressions, memory-pictures emerge, this is not yet "thinking." And when such pictures form a series, each member of which calls forth another, this too is not yet "thinking." When, however, a certain picture turns up in many such series, then—precisely through such return—it becomes an ordering element for such series, in that it connects series which in themselves are unconnected. Such an element becomes an instrument, a concept. I think that the transition from free association or "dreaming" to thinking is characterized by the more or less dominating role which the "concept" plays in it. It is by no means necessary that a concept must be connected with a sensorially cognizable and reproducible sign (word); but when this is the case, thinking becomes by means of that fact communicable.

With what right—the reader will ask—does this man operate so carelessly and primitively with ideas in such a problematic realm without making even the least effort to prove anything? My defense: all our thinking is of this nature of a free play with concepts; the justification for this play lies in the measure of survey over the experience of the senses which we are able to achieve with its aid. The concept of "truth" cannot yet be applied to such a structure; to my thinking, this concept can come in question only when a far-reaching agreement (*convention*) concerning the elements and rules of the game is already at hand.

For me, it is not dubious that our thinking goes on for the most part without use of signs (words), and beyond that, to a considerable degree, unconsciously. For how, otherwise, should it happen that sometimes we "wonder" quite spontaneously about some experience? This "wondering" seems to occur when an experience comes into conflict with a world of concepts which is already sufficiently fixed in us. Whenever such a conflict is experienced hard and intensively, it reacts back upon our thought world in a decisive way. The development of this thought world is in a certain sense a continuous flight from "wonder."

A wonder of such nature I experienced as a child of 4 or 5 years when my father showed me a compass. That this needle behaved in such a determined way did not at all fit into the nature of events, which could find a place in the unconscious world of concepts (effect connected with direct "touch"). I can still remember—or at least believe I can remember—that this experience made a deep and lasting impression upon me. Something deeply hidden had to be behind things. What man sees before him from infancy causes no reaction of this kind; he is not surprised over the falling of bodies, concerning wind and rain, nor concerning the moon or about the fact that the moon does not fall down, nor concerning the differences between living and non-living matter.

At the age of 12 I experienced a second wonder of a totally different nature: in a little book dealing with Euclidean plane geometry, which came into my hands at the beginning of a school year. Here were assertions, as for example, the intersection of the three altitudes of a triangle in one point, which—though by no means evident—could nevertheless be proved with such certainty that any doubt appeared to be out of the question. This lucidity and certainty made an indescribable impression upon me. That the axiom had to be accepted unproved did not disturb me. In any case, it was quite sufficient for me if I could peg proofs upon propositions, the validity of which did not seem to me to be dubious. For example, I remember that an uncle told me the Pythagorean theorem before the holy geometry booklet had come into my hands. After much effort, I succeeded in "proving" this theorem on the basis of the similarity of triangles; in doing so, it seemed to me "evident" that the relations of the sides of the right-angled triangles would have to be completely determined by one of the acute angles. Only something which did not in similar fashion seem to be "evident" appeared to me to be in need of any proof at all. Also, the objects with which geometry deals seemed to be of no different type than the objects of sensory perception, "which can be seen and touched." This primitive idea, which probably also lies at the bottom of the well-known Kantian problematic concerning the possibility of "synthetic judgments *a priori*," rests obviously upon the fact that the relation of geometrical concepts to objects of direct experience (rigid rod, finite interval, etc.) was unconsciously present.

If thus it appeared that it was possible to get certain knowledge of the objects of experience by means of pure thinking, this "wonder" rested upon an error. Nevertheless, for anyone who experiences it for the first time, it is marvelous enough that man is capable at all to reach such a degree of certainty and purity in pure thinking as the Greeks showed us for the first time to be possible in geometry.

Now that I have allowed myself to be carried away sufficiently to interrupt my scantily begun obituary, I shall not hesitate to state here in a few sentences my epistemological credo, although in what precedes something has already incidentally been said about this. This credo actually evolved only much later and very slowly, and does not correspond with the point of view I held in younger years.

I see on the one side the totality of sense-experiences, and, on the other, the totality of the concepts and propositions which are laid down in books. The relations between the concepts and propositions among themselves and each other are of a logical nature, and the business of logical thinking is strictly limited to the achievement of the connection between concepts and propositions among each other according to firmly laid down rules, which are the concern of logic. The concepts and propositions get "meaning," that is to say, "content," only through their connection with sense-experiences. The connection of the latter with the former is purely intuitive, not itself of a logical nature. The degree of certainty with which this connection, that is, intuitive combination, can be undertaken, and nothing else, differentiates empty fantasy from scientific "truth." The system of concepts is a creation of man together with the rules of syntax, which constitute the structure of the conceptual systems. Although the conceptual systems are logically entirely arbitrary, they are bound by the aim to permit the most nearly possible certain (intuitive) and complete coordination with the totality of sense-experiences; secondly they aim at greatest possible scarcity of their logically independent elements (basic concepts and axioms), i.e., undefined concepts and underived [postulated] propositions.

A proposition is correct if, within a logical system, it is deduced according to the accepted logical rules. A system has truth-content according to the certainty and completeness of its coordination-possibility to the totality of experience. A correct proposition borrows its "truth" from the truth-content of the system to which it belongs.

A remark on the historical development. Hume saw clearly that certain concepts, as, for example, that of causality, cannot be deduced from the material of experience by logical methods. Kant, thoroughly convinced of the indispensability of certain concepts, took them—just as they are selected—to be the necessary premises of every kind of thinking and differentiated them from concepts of empirical origin. I am convinced, however, that this differentiation is erroneous, i.e., that it does not do justice to the problem in a natural way. All concepts, even those which are closest to experience, are from the point of view of logic freely chosen conventions, just as is the case with the concept of causality, with which this problematic concerned itself in the first instance.

And now back to the obituary. At the age of 12–16, I familiarized myself with the elements of mathematics together with the principles of differential and integral calculus. In doing so I had the good fortune of hitting up books which were not too particular in their logical rigor, but which made up for this by permitting the main thoughts to stand out clearly and synoptically. This occupation was, on the whole, truly fascinating; climaxes were reached whose impression could easily compete with that of elementary geometry—the basic idea of analytical geometry, the infinite series, the concepts of differential and integral. I also had the good fortune of getting to know the essential results and methods of the entire field of the natural sciences in an excellent popular exposition, which limited itself almost throughout to qualitative aspects (Bernstein's *People's Books on Natural Science*, a work of 5 or 6 volumes), a work which I read with breathless attention. I had also already studied some theoretical physics when, at the age of 17, I entered the Polytechnic Institute of Zürich as a student of mathematics and physics.

There I had excellent teachers (for example, Hurwitz, Minkowski), so that I really could have gotten a sound mathematical education. However, I worked most of the time in the physical laboratory, fascinated by the direct contact with experience. The balance of the time I used in the main in order to study at home the works of Kirchhoff, Helmholtz, Hertz, etc. The fact that I neglected mathematics to a certain extent had its cause not merely in my stronger interest in the natural sciences than in mathematics but also in the following strange experience. I saw that mathematics was split up into numerous specialties, each of which could easily absorb the short lifetime granted to us. Consequently, I saw myself in the position of Buridan's ass, which was unable to decide upon any specific bundle of hay. This was obviously due to the fact that my intuition was not strong enough in the field of mathematics in order to differentiate clearly the fundamentally important, that which is really basic, from the rest of the more or less dispensable erudition. Beyond this, however, my interest in the knowledge of nature was also unqualifiedly stronger; it was not clear to me as a student that the approach to a more profound knowledge of the basic principles of physics is tied up with the most intricate mathematical methods. This dawned upon me only gradually after years of independent scientific work. True enough, physics also was divided into separate fields, each of which was capable of devouring a short lifetime of work without having satisfied the hunger for deeper knowledge. The mass of insufficiently connected experimental data was overwhelming here also. In this field, however, I soon learned to scent out that which was able to lead to fundamentals and to turn aside from everything else, from the multitude of things which clutter up the mind and divert it from the essential. The hitch in this was, of course, the fact that one had to cram all this stuff into one's mind for the examinations, whether one liked it or not. This coercion had such a deterring effect [upon me] that, after I had passed the final examination, I found the consideration of any scientific problems distasteful to me for an entire year. In justice, I must add, moreover, that in Switzerland we had to suffer far less under such coercion, which smothers every truly scientific impulse, than is the case in many another locality. There were altogether only two examinations; aside from these, one could just about do as one pleased. This was especially the case if one had a friend, as did I, who attended the lectures regularly and who worked over their content conscientiously.[d] This gave one freedom in the choice of pursuits until a few months before the examination, a freedom which I enjoyed to a great extent and have gladly taken into the bargain the bad conscience connected with it as by far the lesser evil. It is, in fact, nothing short of a miracle that the modern methods of instruction have not yet entirely strangled the holy curiosity of inquiry; for this delicate little plant, aside from stimulation, stands mainly in need of freedom; without this it goes to wrack and ruin without fail. It is a very grave mistake to think that the enjoyment of seeing and searching can be promoted by means of coercion and a sense of duty. To the contrary, I believe that it would be possible to rob even a healthy beast of prey of its voraciousness, if it were possible, with the aid of a whip, to force the beast to devour continuously, even when not hungry, especially if the food, handed out under such coercion, were to be selected accordingly.—

Now to the field of physics as it presented itself at that time. In spite of all the fruitfulness in particulars, dogmatic rigidity prevailed in matters of principles: In the beginning (if there was such a thing) God created Newton's laws of motion together with the necessary masses and forces. This is all; everything beyond this follows from the development of appropriate mathematical methods by means of deduction. What the nineteenth century achieved on the strength of this basis, especially through the application of the partial differential equations, was bound to arouse the admiration of every receptive person. Newton was probably the first to reveal, in his theory of sound-transmission, the efficacy of partial differential equations. Euler had already created the foundation of hydrodynamics. But the more precise development of the mechanics of discrete masses, as the basis of all physics, was the achievement of the 19th century. What made the greatest impression upon the student, however, was less the technical construction of mechanics or the solution of complicated problems than the achievements of mechanics in areas which apparently had nothing to do with mechanics: the mechanical theory of light, which conceived of light as the wave-motion of a quasi-rigid elastic ether, and above all the kinetic theory of gases:— the independence of the specific heat of monatomic gases of the atomic weight, the derivation of the equation of state of a gas and its relation to the specific heat, the kinetic theory of the dissociation of gases, and above all the quantitative connection of viscosity, heat-conduction and diffusion of gases, which also furnished the absolute magnitude of the atom. These results supported at the same time mechanics as the foundation of physics and of the atomic hypothesis, which latter was already firmly anchored in chemistry. However, in chemistry only the ratios of the atomic masses played any role, not their absolute magnitudes, so that atomic theory could be viewed more as a visualizing symbol than as knowledge concerning the factual construction of matter. Apart from this, it was also of profound interest that the statistical theory of classical mechanics was able to deduce the basic laws of thermodynamics, something which was in essence already accomplished by Boltzmann.

We must not be surprised, therefore, that, so to speak, all physicists of the last century saw in classical mechanics a firm and final foundation for all physics, yes, indeed, for all natural science, and that they never grew tired in their attempts to base Maxwell's theory of electromagnetism, which, in the meantime, was slowly beginning to win out, upon mechanics as well. Even Maxwell and H. Hertz, who in retrospect appear as those who demolished the faith in mechanics as the final basis of all physical thinking, in their conscious thinking adhered throughout to mechanics as the secured basis of physics. It was Ernst Mach who, in his *History of Mechanics*, shook this dogmatic faith; this book exercised a profound influence upon me in this regard while I was a student. I see Mach's greatness in his incorruptible skepticism and independence; in my younger years, however, Mach's epistemological position also influenced me very greatly, a position which today appears to me to be essentially untenable. For he did not place in the correct light the essentially constructive and speculative nature of thought and more especially of scientific thought; in consequence of which he condemned theory on precisely those points where its constructive-speculative character unconcealably comes to light, as for example in the kinetic atomic theory.

Before I enter upon a critique of mechanics as the foundation of physics, something of a broadly general nature will first have to be said concerning the points of view according to which it is possible to criticize physical theories at all. The first point of view is obvious: the theory must not contradict empirical facts. However evident this demand may in the first place appear, its application turns out to be quite delicate. For it is often, perhaps even always, possible to adhere to a general theoretical foundation by securing the adaptation of the theory to the facts by means of artificial additional assumptions. In any case, however, this first point of view is concerned with the confirmation of the theoretical foundation by the available empirical facts.

The second point of view is not concerned with the relation to the material of observation, but with the premises of the theory itself, with what may briefly but vaguely be characterized as the "naturalness" or "logical simplicity" of the premises (of the basic concepts and of the relations between these which are taken as a basis). This point of view, an exact formulation of which meets with great difficulties, has played an important role in the selection and evaluation of theories since time immemorial. The problem here is not simply one of a kind of enumeration of the logically independent premises (if anything like this were at all unequivocally possible), but that of a kind of reciprocal weighing of incommensurable qualities. Furthermore, among theories of equally "simple" foundation that one is to be taken as superior which most sharply delimits the qualities of systems in the abstract (i.e., contains the most definite claims). Of the "realm" of theories I need not speak here, inasmuch as we are confining ourselves to such theories whose object is the totality of all physical appearances. The second point of view may briefly be characterized as concerning itself with the "inner perfection" of the theory, whereas the first point of view refers to the "external confirmation." The following I reckon as also belonging to the "inner perfection" of a theory: We prize a theory more highly if, from the logical standpoint, it is not the result of an arbitrary choice among theories which, among themselves, are of equal value and analogously constructed.

The meager precision of the assertions contained in the last two paragraphs I shall not attempt to excuse by lack of sufficient printing space at my disposal, but confess herewith that I am not, without more ado [immediately], and perhaps not at all, capable to replace these hints by more precise definitions. I believe, however, that a sharper formulation would be possible. In any case, it turns out that among the "augurs" there usually is agreement in judging the "inner perfection" of the theories and even more so concerning the "degree" of "external confirmation."

And now to the critique of mechanics as the basis of physics.

From the first point of view (confirmation by experiment) the incorporation of wave-optics into the mechanical picture of the world was bound to arouse serious misgivings. If light was to be interpreted as undulatory motion in an elastic body (ether), this had to be a medium which permeates everything; because of the transversality of the light-waves in the main similar to a solid body, yet incompressible, so that longitudinal waves did not exist. This ether had to lead a ghostly existence alongside the rest of matter, inasmuch as it seemed to offer no resistance whatever to the motion of "ponderable" bodies. In order to explain the refraction-indices of transparent bodies as well as the processes of emission and absorption of radiation, one would have had to assume complicated reciprocal actions between the two types of matter, something which was not even seriously tried, let alone achieved.

Furthermore, the electromagnetic forces necessitated the introduction of electric masses, which, although they had no noticeable inertia, yet interacted with each other, and whose interaction was, moreover, in contrast to the force of gravitation, of a polar type.

The factor which finally succeeded, after long hesitation, to bring the physicists slowly around to give up the faith in the possibility that all of physics could be founded upon Newton's mechanics, was the electrodynamics of Faraday and Maxwell. For this theory and its confirmation by Hertz's experiments showed that there are electromagnetic phenomena which by their very nature are detached from every ponderable matter—namely the waves in empty space which consist of electromagnetic "fields." If mechanics was to be maintained as the foundation of physics, Maxwell's equations had to be interpreted mechanically. This was zealously but fruitlessly attempted, while the equations were proving themselves fruitful in mounting degree. One got used to operating with these fields as independent substances without finding it necessary to give one's self an account of their mechanical nature; thus mechanics as the basis of physics was being abandoned, almost unnoticeably, because its adaptability to the facts presented itself finally as hopeless. Since then, there exist two types of conceptual elements, on the one hand, material points with forces at a distance between them, and, on the other hand, the continuous field. It presents an intermediate state in physics without a uniform basis for the entirety, which—although unsatisfactory—is far from having been superseded.

Now for a few remarks to the critique of mechanics as the foundation of physics from the second, the "interior," point of view. In today's state of science, i.e., after the departure from the mechanical foundation, such critique has only an interest in method left. But such a critique is well suited to show the type of argumentation which, in the choice of theories in the future will have to play an all the greater role the more the basic concepts and axioms distance themselves from what is directly observable, so that the confrontation of the implications of theory by the facts becomes constantly more difficult and more drawn out. First in line to be mentioned is Mach's argument, which, however, had already been clearly recognized by Newton (bucket experiment). From the standpoint of purely geometrical description, all "rigid" coordinate systems are among themselves logically equivalent. The equations of mechanics (for example, this is already true of the law of inertia) claim validity only when referred to a specific class of such systems, i.e., the "inertial systems." In this the coordinate system as bodily object is without any significance. It is necessary, therefore, in order to justify the necessity of the specific choice, to look for something which lies outside of the objects (masses, distances) with which the theory is concerned. For this reason, "absolute space" as originally determinative was quite explicitly introduced by Newton as the omnipresent active participant in all mechanical events; by "absolute" he obviously means uninfluenced by the masses and by their motion. What makes this state of affairs appear particularly offensive is the fact that there are supposed to be infinitely many inertial systems, relative to each other in uniform translation, which are supposed to be distinguished among all other rigid systems.

Mach conjectures that in a truly rational theory inertia would have to depend upon the interaction of the masses, precisely as was true for Newton's other forces, a conception which for a long time I considered as in principle the correct one. It presupposes implicitly, however, that the basic theory should be of the general type of Newton's mechanics: masses and their interaction as the original concepts. The attempt at such a solution does not fit into a consistent field theory, as will be immediately recognized.

How sound, however, Mach's critique is in essence can be seen particularly clearly from the following analogy. Let us imagine people construct a mechanics, who know only a very small part of the earth's surface and who also cannot see any stars. They will be inclined to ascribe special physical attributes to the vertical dimension of space (direction of the acceleration of falling bodies) and, on the ground of such a conceptual basis, will offer reasons that the earth is in most places horizontal. They might not permit themselves to be influenced by the argument that, as concerns the geometrical properties, space is isotropic, and that it is therefore supposed to be unsatisfactory to postulate basic physical laws, according to which there is supposed to be a preferential direction; they will probably be inclined (analogously to Newton) to assert the absoluteness of the vertical, as proved by experience as something with which one simply would have to come to terms. The preference given to the vertical over all other spatial directions is precisely analogous to the preference given to inertial systems over other rigid coordination systems.

Now to [a consideration of] other arguments which also concern themselves with the inner simplicity, i.e., naturalness, of mechanics. If one puts up with the concepts of space (including geometry) and time without critical doubts, then there exists no reason to object to the idea of action-at-a-distance, even though such a concept is unsuited to the ideas which one forms on the basis of the raw experience of daily life.

However, there is another consideration which causes mechanics, taken as the basis of physics, to appear as primitive. Essentially, there exist two laws

(1) the law of motion
(2) the expression for force or potential energy

The law of motion is precise, although empty, as long as the expression for the forces is not given. In postulating the latter, however, there exists great latitude for arbitrary [choice], especially if one omits the demand, which is not very natural in any case, that the forces depend only on the coordinates (and, for example, not on their differential quotients with respect to time). Within the framework of theory alone, it is entirely arbitrary that the forces of gravitation (and electricity), which come from one point are governed by the potential function $(1/r)$. Additional remark: it has long been known that this function is the central-symmetrical solution of the simplest (rotation-invariant) differential equation $\delta\varphi = 0$; it would therefore have been a suggestive idea to regard this as a sign that this function is to be regarded as determined by a law of space, a procedure by which the arbitrariness in the choice of the law of energy would have been removed. This is really the first insight which suggests a turning away from the theory of distant forces, a development which—prepared by Faraday, Maxwell and Hertz—really begins only later on under the external pressure of experimental data.

I would also like to mention, as one internal asymmetry of this theory, that the inert mass occurring in the law of motion also appears in the expression for the gravitational force, but not in the expression for the other forces. Finally, I would like to point to the fact that the division of energy into two essentially different parts, kinetic and potential energy, must be felt as unnatural; H. Hertz felt this as so disturbing that, in his very last work, he attempted to free mechanics from the concept of potential energy (i.e., from the concept of force).—

Enough of this. Newton, forgive me; you found the only way which, in your age, was just about possible for a man of highest thought and creative power. The concepts, which you created, are even today still guiding our thinking in physics, although we now know that they will have to be replaced by others farther removed from the sphere of immediate experience, if we aim at a profounder understanding of relationships.

"Is this supposed to be an obituary?" the astonished reader will likely ask. I would like to reply: essentially yes. For the essential in the being of a man of my type lies precisely in *what* he thinks and *how* he thinks, not in what he does or suffers. Consequently, the obituary can limit itself in the main to the communicating of thoughts which have played a considerable role in my endeavors.—A theory is the more impressive the greater the simplicity of its premises is, the more different kinds of things it relates, and the more extended is its area of applicability. Therefore the deep impression which classical thermodynamics made upon me. It is the only physical theory of universal content concerning which I am convinced that, within the framework of the applicability of its basic concepts, it will never be overthrown (for the special attention of those who are skeptics on principle).

The most fascinating subject at the time that I was a student was Maxwell's theory. What made this theory appear revolutionary was the transition from forces at a distance to fields as fundamental variables. The incorporation of optics into the theory of electromagnetism, with its relation of the speed of light to the electric and magnetic absolute system of units as well as the relation of the refraction coefficient to the dielectric constant, the qualitative relation between the reflection coefficient and the metallic conductivity of the body—it was like a revelation. Aside from the transition to field-theory, i.e., the expression of the elementary laws through differential equations, Maxwell needed only one single hypothetical step—the introduction of the electrical displacement current in the vacuum and in the dielectrics and its magnetic effect, an innovation which was almost prescribed by the formal properties of the differential equations. In this connection, I cannot suppress the remark that the pair Faraday-Maxwell has a most remarkable inner similarity with the pair Galileo-Newton—the former of each pair grasping the relations intuitively, and the second one formulating those relations exactly and applying them quantitatively.

What rendered the insight into the essence of electromagnetic theory so much more difficult at that time was the following peculiar situation. Electric or magnetic "field intensities" and "displacements" were treated as equally elementary variables, empty space as a special instance of a dielectric body. *Matter* appeared as the bearer of the field, not *space*. By this it was implied that the carrier of the field could have velocity, and this was naturally to apply to the "vacuum" (ether) also. Hertz's electrodynamics of moving bodies rests entirely upon this fundamental attitude.

It was the great merit of H. A. Lorentz that he brought about a change here in a convincing fashion. In principle, a field exists, according to him, only in empty space. Matter—considered as atoms—is the only seat of electric charges; between the material particles there is empty space, the seat of the electromagnetic field, which is created by the position and velocity of the point charges which are located on the material particles. Dielectricity, conductivity, etc., are determined exclusively by the type of mechanical tie connecting the particles, of which the bodies consist. The particle-charges create the field, which, on the other hand, exerts forces upon the charges of the particles, thus determining the motion of the latter according to Newton's law of motion. If one compares this with Newton's system, the change consists in this: action at a distance is replaced by the field, which thus also describes the radiation. Gravitation is usually not taken into account because of its relative smallness; its consideration, however, was always possible by means of the enrichment of the structure of the field, i.e., expansion of Maxwell's law of the field. The physicist of the present generation regards the point of view achieved by Lorentz as the only possible one; at that time, however, it was a surprising and audacious step, without which the later development would not have been possible.

If one views this phase of the development of theory critically, one is struck by the dualism which lies in the fact that the material point in Newton's sense and the field as continuum are used as elementary concepts side by side. Kinetic energy and field-energy appear as essentially different things. This appears all the more unsatisfactory inasmuch as, according to Maxwell's theory, the magnetic field of a moving electric charge represents inertia. Why not then total inertia? Then only field-energy would be left, and the particle would be merely an area of special density of field-energy. In that case, one could hope to deduce the concept of the mass-point together with the equations of the motion of the particles from the field equations—the disturbing dualism would have been removed.

H. A. Lorentz knew this very well. However, Maxwell's equations did not permit the derivations of the equilibrium of the electricity which constitutes a particle. Only other, nonlinear field equations could possibly accomplish such a thing. But no method existed by which this kind of field equations could be discovered without deteriorating into adventurous arbitrariness. In any case, one could believe that it would be possible by and by to find a new and secure foundation for all of physics upon the path which had been so successfully begun by Faraday and Maxwell.

Accordingly, the revolution begun by the introduction of the field was by no means finished. Then it happened that, around the turn of the century, independently of what we have just been discussing, a second fundamental crisis set in, the seriousness of which was suddenly recognized due to Max Planck's investigations into heat radiation (1900). The history of this event is all the more remarkable because, at least in its first phase, it was not in any way influenced by any surprising discoveries of an experimental nature.

On thermodynamic grounds, Kirchhoff had concluded that the energy density and the spectral composition of radiation in a *Hohlraum*, surrounded by impenetrable walls of the temperature T, would be independent of the nature of the walls.[e] That is to say, the nonchromatic density of radiation ρ is a universal function of the frequency ν and of the absolute temperature T. Thus arose the interesting problem of determining this function $\rho(\nu, T)$. What could theoretically be ascertained about this function? According to Maxwell's theory the radiation had to exert a pressure on the walls, determined by the total energy density. From this, Boltzmann concluded, by means of pure thermodynamics, that the entire energy density of the radiation ($\int \rho d\nu$) is proportional to T^4. In this way, he found a theoretical justification of a law which had previously been discovered empirically by Stefan, i.e., in this way he connected this empirical law with the basis of Maxwell's theory. Thereafter, by way of an ingenious thermodynamic consideration, which also made use of Maxwell's theory, W. Wien found that the universal function ρ of the two variables ν and T would have to be of the form

$$\rho \approx \nu^3 f\left(\frac{\nu}{T}\right),$$

Whereby $f(v/T)$ is a universal function of one variable v/T only. It was clear that the theoretical determination of this universal function f was of fundamental importance—this was precisely the task which confronted Planck. Careful measurements had led to a very precise empirical determination of the function f. Relying on those empirical measurements, he succeeded in the first place in finding a statement which rendered the measurements very well indeed:

$$\rho = \frac{8\pi h v^3}{c^3} \frac{1}{exp(hv/kT) - 1}$$

whereby h and k are two universal constants, the first of which led to quantum theory. Because of the denominator, this formula looks a bit queer. Was it possible to derive it theoretically? Planck actually did find a derivation, the imperfections of which remained at first hidden, which latter fact was most fortunate for the development of physics. If this formula was correct, it permitted, with the aid of Maxwell's theory, the calculation of the average energy E of a quasi-monochromatic oscillator within the field of radiation:

$$E = \frac{hv}{exp(hv/kT) - 1}$$

Planck preferred to attempt calculating this latter magnitude theoretically. In this effort, thermodynamics, for the time being, proved no longer helpful, and neither did Maxwell's theory. The following circumstance was unusually encouraging in this formula. For high temperatures (with a fixed v) it yielded the expression

$$E = kT.$$

This is the same expression as the kinetic theory of gases yields for the average energy of a mass-point which is capable of oscillating elastically in one dimension. For in kinetic gas theory one gets

$$E = (R/N)T,$$

whereby R means the constant of the equation of state of a gas and N the number of molecules per mol, from which constant one can compute the absolute size of the atom. Putting these two expressions equal to

each other, one gets
$$N = R/k.$$

The one constant of Planck's formula consequently furnishes exactly the correct size of the atom. The numerical value agreed satisfactorily with the determinations of N by means of kinetic gas theory, even though these latter were not very accurate.

This was a great success, which Planck clearly recognized. But the matter has a serious drawback, which Planck fortunately overlooked at first. For the same considerations demand, in fact, that the relation $E = kT$ would also have to be valid for low temperatures. In that case, however, it would be all over with Planck's formula and with the constant h. From the existing theory, therefore, the correct conclusion would have been: the average kinetic energy of the oscillator is either given incorrectly by the theory of gases, which would imply a refutation of [statistical] mechanics, or else the average energy of the oscillator follows incorrectly from Maxwell's theory, which would imply a refutation of the latter. Under such circumstances, it is most probable that both theories are correct only at the limits, but are otherwise false; this is indeed the situation, as we shall see in what follows. If Planck had drawn this conclusion, he probably would not have made his great discovery, because pure reflection would have been deprived of its foundation.

Now back to Planck's reasoning. On the basis of the kinetic theory of gases, Boltzmann had discovered that, aside from a constant factor, entropy is equivalent to the logarithm of the "probability" of the state under consideration. Through this insight, he recognized the nature of courses of events which, in the sense of thermodynamics, are "irreversible." Seen from the molecular-mechanical point of view, however, all courses of events are reversible. If one calls a molecular-theoretically defined state a microscopically described one—or, more briefly, microstate—and a state described in terms of thermodynamics a macro-state, then an immensely large number (Z) of states belong to a macroscopic condition. Z is then a measure of the probability of a chosen macro-state. This idea appears to be of outstanding importance also because of the fact that its usefulness is not limited to microscopic description on the basis of mechanics. Planck recognized this and applied the Boltzmann principle to a system which consists of very many resonators of the same frequency ν. The macroscopic situation is given through the total energy of the oscillation of all resonators, a micro-condition through determination of the (instantaneous) energy of each individual resonator. In order then to be able to express the number of the microstates belonging to a macro-state by means of a finite number, he [Planck] divided the total energy into a large but finite number of identical energy-elements ε and asked: in how many ways can these energy-elements be divided among the resonators. The logarithm of this number, then, furnishes the entropy and thus (via thermodynamics) the temperature of the system. Planck got his radiation-formula if he chose his energy-elements ε of the magnitude $\varepsilon = h\nu$. The decisive element in doing this lies in the fact that the result depends on taking for ε a definite finite value, i.e., that one does not go to the limit $\varepsilon = 0$. This form of reasoning does not make obvious the fact that it contradicts the mechanical and electrodynamic basis, upon which the derivation otherwise depends. Actually, however, the derivation presupposes implicitly that energy can be absorbed and emitted by the individual resonator only in "quanta" of magnitude $h\nu$, i.e., that the energy of a mechanical structure capable of oscillations as well as the energy of radiation can be transferred only in such quanta—in contradiction to the laws of mechanics and electrodynamics. The contradiction with dynamics was here fundamental, whereas the contradiction with electrodynamics could be less fundamental. For the expression for the density of radiation-energy, although it is *compatible* with Maxwell's equations, is not a necessary consequence of these equations. That this expression furnishes important average-values is shown by the fact that the Stefan-Boltzmann law and Wien's law, which are based on it, are in agreement with experience.

All of this was quite clear to me shortly after the appearance of Planck's fundamental work; so that, without having a substitute for classical mechanics, I could nevertheless see to what kind of consequences this law of temperature-radiation leads for the photo-electric effect and for other related phenomena of the transformation of radiation-energy, as well as for the specific heat of (especially) solid bodies. All my attempts, however, to adapt the theoretical foundation of physics to this [new type of] knowledge failed completely. It was as if the ground had been pulled out from under one, with no firm foundation to be seen anywhere, upon which one could have built. That this insecure and contradictory foundation was sufficient to enable a man of Bohr's unique instinct and tact[f] to discover the major laws of the spectral lines and of the electron-shells of the atoms together with their significance for chemistry appeared to me like a miracle—and appears to me a miracle even today. This is the highest form of musicality in the sphere of thought.

My own interest in those years was less concerned with the detailed consequences of Planck's results, however important these might be. My major question was: What general conclusions can be drawn from the radiation-formula concerning the structure of radiation and even more generally concerning the electromagnetic foundation of physics? Before I take this up, I must briefly mention a number of investigations which relate to the Brownian motion and related objects (fluctuation-phenomena) and which in essence rest upon classical molecular mechanics. Not acquainted with the earlier investigations of Boltzmann and Gibbs, which had appeared earlier and actually exhausted the subject, I developed the statistical mechanics and the molecular-kinetic theory of thermodynamics which was based on the former. My major aim in this was to find facts which would guarantee as much as possible the existence of atoms of definite finite size. In the midst of this I discovered that, according to atomistic theory, there would have to be a movement of suspended microscopic particles open to observation, without knowing that observations concerning the Brownian motion were already long familiar. The simplest derivation rested upon the following consideration. If the molecular-kinetic theory is essentially correct, a suspension of visible particles must possess the same kind of osmotic pressure fulfilling the laws of gases as a solution of molecules. This osmotic pressure depends upon the actual magnitude of the molecules, i.e., upon the number of molecules in a gram-equivalent. If the density of the suspension is inhomogeneous, the osmotic pressure is inhomogeneous, too, and gives rise to a compensating diffusion, which can be calculated from the well-known mobility of the particles. This diffusion can, on the other hand, also be considered as the result of the random displacement—unknown in magnitude originally—of the suspended particles due to thermal agitation. By comparing the amounts obtained for the diffusion current from both types of reasoning one reaches quantitatively the statistical law for those displacements, i.e., the law of the Brownian motion. The agreement of these considerations with experience together with

Planck's determination of the true molecular size from the law of radiation (for high temperatures) convinced the skeptics, who were quite numerous at that time (Ostwald, Mach) of the reality of atoms. The antipathy of these scholars toward atomic theory can indubitably be traced back to their positivistic philosophical attitude. This is an interesting example of the fact that even scholars of audacious spirit and fine instinct can be obstructed in the interpretation of facts by philosophical prejudices. The prejudice—which has by no means died out in the meantime—consists in the faith that facts by themselves can and should yield scientific knowledge without free conceptual construction. Such a misconception is possible only because one does not easily become aware of the free choice of such concepts, which, through verification and long usage, appear to be immediately connected with the empirical material.[g]

The success of the theory of the Brownian motion showed again conclusively that classical mechanics always offered trustworthy results whenever it was applied to motions in which the higher time derivatives of velocity are negligibly small. Upon this recognition, a relatively direct method can be based, which permits us to learn something concerning the constitution of radiation from Planck's formula. One may conclude, in fact, that, in a space filled with radiation, a (vertically to its plane) freely moving, quasi monochromatically reflecting mirror would have to go through a kind of Brownian movement, the average kinetic energy of which equals $\frac{1}{2}(R/N)T$ ($R =$ constant of the gas-equation for one gram-molecule, N equals the number of the molecules per mol, $T =$ absolute temperature). If radiation were not subject to local fluctuations, the mirror would gradually come to rest, because, due to its motion, it reflects more radiation on its front than on its reverse side. However, the mirror must experience certain random fluctuations of the pressure exerted upon it due to the fact that the wave-packets, constituting the radiation, interfere with one another. These can be computed from Maxwell's theory. This calculation, then, shows that these pressure variations (especially in the case of small radiation-densities) are by no means sufficient to impart to the mirror the average kinetic energy $\frac{1}{2}(R/N)T$. In order to get this result, one has to assume rather that there exists a second type of pressure variations, which cannot be derived from Maxwell's theory, which corresponds to the assumption that radiation energy consists of indivisible point-like localized quanta of the energy $h\nu$ (and of momentum $(h\nu/c)$, ($c =$ velocity of light)), which are reflected undivided. This way of looking at the problem showed in a drastic and direct way that a type of immediate reality has to be ascribed to Planck's quanta, that radiation must, therefore, possess a kind of molecular structure in energy, which of course contradicts Maxwell's theory. Considerations concerning radiation which are based directly on Boltzmann's entropy-probability-relation (probability taken equal to statistical temporal frequency) also lead to the same result. This double nature of radiation (and of material corpuscles) is a major property of reality, which has been interpreted by quantum-mechanics in an ingenious and amazingly successful fashion. This interpretation, which is looked upon as essentially final by almost all contemporary physicists, appears to me as only a temporary way out; a few remarks to this [point] will follow later.—

Reflections of this type made it clear to me as long ago as shortly after 1900, i.e., shortly after Planck's trailblazing work, that neither mechanics nor thermodynamics could (except in limiting cases) claim exact validity. By and by, I despaired of the possibility of discovering the true laws by means of constructive efforts based on known facts. The longer and the more despairingly I tried, the more I came to the conviction that only the discovery of a universal formal principle could lead us to assured results. The example I saw before me was thermodynamics. The general principle was there given in the theorem: the laws of nature are such that it is impossible to construct a *perpetuum mobile* (of the first and second kind). How, then, could such a universal principle be found? After ten years of reflection, such a principle resulted from a paradox upon which I had already hit at the age of sixteen: If I pursue a beam of light with the velocity c (velocity of light in a vacuum), I should observe such a beam of light as a spatially oscillatory electromagnetic field at rest. However, there seems to be no such thing, whether on the basis of experience or according to Maxwell's equations. From the very beginning, it appeared to me intuitively clear that, judged from the standpoint of such an observer, everything would have to happen according to the same laws as for an observer who, relative to the earth, was at rest. For how, otherwise, should the first observer know, i.e., be able to determine, that he is in a state of fast uniform motion?

One sees that in this paradox the germ of the special relativity theory is already contained. Today everyone knows, of course, that all attempts to clarify this paradox satisfactorily were condemned to failure as long as the axiom of the absolute character of time, namely, of simultaneity, was anchored unrecognized in the unconscious. Clearly, to recognize this axiom and its arbitrary character really implies already the solution of the problem. The type of critical reasoning which was required for the discovery of this central point was decisively furthered, in my case, especially by the reading of David Hume's and Ernst Mach's philosophical writings.

One had to understand clearly what the spatial coordinates and the temporal duration of events meant in physics. The physical interpretation of the spatial coordinates presupposed a fixed body of reference, which, moreover, had to be in a more or less definite state of motion (inertial system). In a given inertial system the coordinates meant the results of certain measurements with rigid (stationary) rods. (One should always be conscious of the fact that the presupposition of the existence in principle of rigid rods is a presupposition suggested by approximate experience, but which is, in principle, arbitrary.) With such an interpretation of the spatial coordinates, the question of the validity of Euclidean geometry becomes a problem of physics.

If, then, one tries to interpret the time of an event analogously, one needs a means for the measurement of the difference in time (in itself determined periodic process realized by a system of sufficiently small spatial extension). A clock at rest relative to the system of inertia defines a local time. The local times of all space points taken together are the "time," which belongs to the selected system of inertia, if a means is given to "set" these clocks relative to each other. One sees that *a priori* it is not at all necessary that the "times" thus defined in different inertial systems agree with one another. One would have noticed this long ago, if, for the practical experience of everyday life, light did not appear (because of the high value of c) as the means for the statement of absolute simultaneity.

The presupposition of the existence (in principle) of (ideal, namely, perfect) measuring rods and clocks is not independent of each other; since a light signal, which is reflected back and forth between the ends of a rigid rod, constitutes an ideal clock, provided that the postulate of the constancy of the light-velocity in vacuum does not lead to contradictions.

The above paradox may then be formulated as follows. According to the rules of connection, used in classical physics, of the spatial coordinates and of the time of events in the transition from one inertial system to another, the two assumptions of

(1) the constancy of the light velocity
(2) the independence of the laws (thus specially also of the law of the constancy of the light velocity) of the choice of the inertial system (principle of special relativity) are mutually incompatible (even though both taken separately are based on experience).

The insight which is fundamental for the special theory of relativity is this: The assumptions (1) and (2) are compatible if relations of a new type ("Lorentz-transformation") are postulated for the conversion of coordinates and the times of events. With the given physical interpretation of coordinates and time, this is by no means merely a conventional step, but implies certain hypotheses concerning the actual behavior of moving measuring-rods and clocks, which can be experimentally validated or disproved.

The universal principle of the special theory of relativity is contained in the postulate: The laws of physics are invariant with respect to the Lorentz-transformations (for the transition from one inertial system to any other arbitrarily chosen system of inertia). This is a restricting principle for natural laws, comparable to the restricting principle of the non-existence of the *perpetuum mobile* which underlies thermodynamics.

First a remark concerning the relation of the theory to "four-dimensional space." It is a wide-spread error that the special theory of relativity is supposed to have, to a certain extent, first discovered, or at any rate, newly introduced, the four-dimensionality of the physical continuum. This, of course, is not the case. Classical mechanics, too, is based on the four-dimensional continuum of space and time. But in the four-dimensional continuum of classical physics the subspaces with constant time value have an absolute reality, independent of the choice of the reference system. Because of this [fact], the four-dimensional continuum falls naturally into a three-dimensional and a one-dimensional (time), so that the four-dimensional point of view does not force itself upon one as *necessary*. The special theory of relativity, on the other hand, creates a formal dependence between the way in which the spatial coordinates, on the one hand, and the temporal coordinates, on the other, have to enter into the natural laws.

Minkowski's important contribution to the theory lies in the following: Before Minkowski's investigation it was necessary to carry out a Lorentz-transformation on a law in order to test its invariance under such transformations; he, on the other hand, succeeded in introducing a formalism such that the mathematical form of the law itself guarantees its invariance under Lorentz-transformations. By creating a four-dimensional tensor-calculus, he achieved the same thing for the four-dimensional space which the ordinary vector-calculus achieves for the three spatial dimensions. He also showed that the Lorentz-transformation (apart from a different algebraic sign due to the special character of time) is nothing but a rotation of the coordinate system in the four-dimensional space.

First, a remark concerning the theory as it is characterized above. One is struck [by the fact] that the theory (except for the four-dimensional space) introduces two kinds of physical things, i.e., (1) measuring rods and clocks, (2) all other things, e.g., the electromagnetic field, the material point, etc. This, in a certain sense, is inconsistent; strictly speaking, measuring rods and clocks would have to be represented as solutions of the basic equations (objects consisting of moving atomic configurations), not, as it were, as theoretically self-sufficient entities. However, the procedure justifies itself because it was clear from the very beginning that the postulates of the theory are not strong enough to deduce from them sufficiently complete equations for physical events sufficiently free from arbitrariness, in order to base upon such a foundation a theory of measuring rods and clocks. If one did not wish to forego a physical interpretation of the coordinates in general (something which, in itself, would be possible), it was better to permit such inconsistency—with the obligation, however, of eliminating it at a later stage of the theory. But one must not legalize the mentioned sin so far as to imagine that intervals are physical entities of a special type, intrinsically different from other physical variables ("reducing physics to geometry," etc.).

We now shall inquire into the insights of definite nature which physics owes to the special theory of relativity.

(1) There is no such thing as simultaneity of distant events; consequently, there is also no such thing as immediate action at a distance in the sense of Newtonian mechanics. Although the introduction of actions at a distance, which propagate with the speed of light, remains thinkable, according to this theory, it appears unnatural; for in such a theory there could be no such thing as a reasonable statement of the principle of conservation of energy. It therefore appears unavoidable that physical reality must be described in terms of continuous functions in space. The material point, therefore, can hardly be conceived any more as the basic concept of the theory.

(2) The principles of the conservation of momentum and of the conservation of energy are fused into one single principle. The inert mass of a closed system is identical with its energy, thus eliminating mass as an independent concept.

Remark. The speed of light c is one of the quantities which occurs as "universal constant" in physical equations. If, however, one introduces as unit of time, instead of the second, the time in which light travels 1 cm, c no longer occurs in the equations. In this sense, one could say that the constant c is only an apparently universal constant.

It is obvious and generally accepted that one could eliminate two more universal constants from physics by introducing, instead of the gram and the centimeter, properly chosen "natural" units (for example, mass and radius of the electron).

If one considers this done, then only "dimension-less" constants could occur in the basic equations of physics. Concerning such I would like to state a theorem which at present cannot be based upon anything more than upon a faith in the simplicity, i.e., intelligibility, of nature: there are no *arbitrary* constants of this kind; that is to say, nature is so constituted that it is possible logically to lay down such strongly determined laws that within these laws only rationally completely determined constants occur (not constants, therefore, whose numerical value could be changed without destroying the theory).—

The special theory of relativity owes its origin to Maxwell's equations of the electromagnetic field. Inversely, the latter can be grasped formally in satisfactory fashion only by way of the special theory of relativity. Maxwell's equations are the simplest Lorentz-invariant field equations which can be postulated for an antisymmetric tensor derived from a vector field. This in itself would be satisfactory, if we did not know from quantum phenomena that Maxwell's theory does not do justice to the energetic properties of radiation. But how Maxwell's theory would have to be modified in a natural fashion, for this even the special theory of relativity offers no adequate foothold. Also, to Mach's question: "how does it come about that inertial systems are physically distinguished above all other coordinate systems?" This theory offers no answer.

That the special theory of relativity is only the first step of a necessary development became completely clear to me only in my efforts to represent gravitation in the framework of this theory. In classical mechanics, interpreted in terms of the field, the potential of gravitation appears as a *scalar* field (the simplest theoretical possibility of a field with a single component). Such a scalar theory of the gravitational field can easily be made invariant under the group of Lorentz-transformations. The following program appears natural, therefore: The total physical field consists of a scalar field (gravitation) and a vector field (electromagnetic field); later insights may eventually make necessary the introduction of still more complicated types of fields; but to begin with, one did not need to bother about this.

The possibility of the realization of this program was, however, dubious from the very first, because the theory had to combine the following things:

(1) From the general considerations of special relativity theory it was clear that the *inert* mass of a physical system increases with the total energy (therefore, e.g., with the kinetic energy).
(2) From very accurate experiments (especially from the torsion balance experiments of Eötvös) it was empirically known with very high accuracy that the gravitational mass of a body is exactly equal to its inert mass.

It followed from (1) and (2) that the *weight* of a system depends in a precisely known manner on its total energy. If the theory did not accomplish this or could not do it naturally, it was to be rejected. The condition is most naturally expressed as follows: the acceleration of a system falling freely in a given gravitational field is independent of the nature of the falling system (specially therefore also of its energy content).

It then appeared that, in the framework of the program sketched, this elementary state of affairs could not at all or at any rate not in any natural fashion, be represented in a satisfactory way. This convinced me that, within the frame of the special theory of relativity, there is no room for a satisfactory theory of gravitation.

Now it came to me: The fact of the equality of inert and heavy mass, i.e., the fact of the independence of the gravitational acceleration of the nature of the falling substance, may be expressed as follows: In a gravitational field (of small spatial extension) things behave as they do in a space free of gravitation, if one introduces in it, in place of an "inertial system," a reference system which is accelerated relative to an inertial system.

If then one conceives of the behavior of a body, in reference to the latter reference system, as caused by a "real" (not merely apparent) gravitational field, it is possible to regard this reference system as an "inertial system" with as much justification as the original reference system.

So, if one regards as possible, gravitational fields of arbitrary extension which are not initially restricted by spatial limitations, the concept of the "inertial system" becomes completely empty. The concept, "acceleration relative to space," then loses every meaning and with it the principle of inertia together with the entire paradox of Mach.

The fact of the equality of inert and heavy mass thus leads quite naturally to the recognition that the basic demand of the special theory of relativity (invariance of the laws under Lorentz-transformations) is too narrow, i.e., that an invariance of the laws must be postulated also relative to *non-linear* transformations of the coordinates in the four-dimensional continuum.

This happened in 1908. Why were another seven years required for the construction of the general theory of relativity? The main reason lies in the fact that it is not so easy to free oneself from the idea that coordinates must have an immediate metrical meaning. The transformation took place in approximately the following fashion.

We start with an empty, field-free space, as it occurs—related to an inertial system—in the sense of the special theory of relativity, as the simplest of all imaginable physical situations. If we now think of a non-inertial system introduced by assuming that the new system is uniformly accelerated against the inertial system (in a three-dimensional description) in one direction (conveniently defined), then there exists with reference to this system a static parallel gravitational field. The reference system may thereby be chosen as rigid, of Euclidean type, in three-dimensional metric relations. But the time, in which the field appears as static, is *not* measured by *equally constituted* stationary clocks. From this special example, one can already recognize that the immediate metric significance of the coordinates is lost if one admits non-linear transformations of coordinates at all. To do the latter is, however, *obligatory* if one wants to do justice to the equality of gravitational and inert mass by means of the basis of the theory, and if one wants to overcome Mach's paradox as concerns the inertial systems.

If, then, one must give up the attempt to give the coordinates an immediate metric meaning (differences of coordinates = measurable lengths, or times), one will not be able to avoid treating as equivalent all coordinate systems, which can be created by the continuous transformations of the coordinates.

The general theory of relativity, accordingly, proceeds from the following principle. Natural laws are to be expressed by equations which are covariant under the group of continuous coordinate transformations. This group replaces the group of the Lorentz-transformations of the special theory of relativity, which forms a subgroup of the former.

This demand by itself is, of course, not sufficient to serve as point of departure for the derivation of the basic concepts of physics. In the first instance, one may even contest [the idea] that the demand by itself contains a real restriction for the physical laws; for it will always be possible thus to reformulate a law, postulated at first only for certain coordinate systems, such that the new formulation becomes formally universally covariant. Beyond this it is clear from the beginning that an infinitely large number of field-laws can be formulated which have this property of covariance. The eminent heuristic significance of the general principles of relativity lies in the fact that it leads us to the search for those systems of equations which are *in their general covariant* formulation the *simplest ones possible*; among these we shall have to look for the field equations of physical space. Fields which can be transformed into each other by such transformations describe the same real situation.

The major question for anyone doing research in this field is this: Of which mathematical type are the variables (functions of the coordinates) which permit the expression of the physical properties of space ("structure")? Only after that: Which equations are satisfied by those variables?

The answer to these questions is today by no means certain. The path chosen by the first formulation of the general theory of relativity can be characterized as follows. Even though we do not know by what type of field-variables (structure) physical space is to be characterized, we do know with certainty a special case: Such a space is characterized by the fact that for a properly chosen coordinate system the expression

$$ds^2 = dx_1^2 + dx_2^2 + dx_3^2 - dx_4^2 \tag{1}$$

belonging to two neighboring points, represents a measurable quantity (square of distance), and thus has a real physical meaning. Referred to an arbitrary system, this quantity is expressed as follows:

$$ds^2 = g_{ik} dx_i dx_k \tag{2}$$

whereby the indices run from 1 to 4. The g_{ik} form a (real) symmetrical tensor. If, after carrying out a transformation on field (1), the first derivatives of the g_{ik} with respect to the coordinates do not vanish, there exists a gravitational field with reference to this system of coordinates in the sense of the above consideration, a gravitational field, moreover, of a very special type. Thanks to Riemann's investigation of n-dimensional metrical spaces, this special field can be invariably characterized:

(1) Riemann's curvature-tensor R_{iklm}, formed from the coefficients of the metric (2) vanishes.
(2) The orbit of a mass-point in reference to the inertial system (relative to which (1) is valid) is a straight line, therefore an extremal (geodesic). The latter, however, is already a characterization of the law of motion based on (2).

The *universal* law of physical space must now be a generalization of the law just characterized. I now assume that there are two steps of generalization:

(a) pure gravitational field
(b) general field (in which quantities corresponding somehow to the electromagnetic field occur, too).

The instance (a) was characterized by the fact that the field can still be represented by a Riemann-metric (2), i.e., by a symmetric tensor, whereby, however, there is no representation in the form (1) (except in infinitesimal regions). This means that in the case (a) the Riemann-tensor does not vanish. It is clear, however, that in this case a field-law must be valid, which is a generalization (loosening) of this law. If this law also is to be of the second order of differentiation and linear in the second derivatives, then only the equation, to be obtained by a single contraction

$$0 = R_{kl} = g^{im} R_{iklm}$$

came under consideration as field-equation in the case of (a). It appears natural, moreover, to assume that also in the case of (a) the geodetic line is still to be taken as representing the law of motion of the material point.

It seemed hopeless to me at that time to venture the attempt of representing the total field (b) and to ascertain field-laws for it. I preferred, therefore, to set up a preliminary formal frame for the representation of the entire physical reality; this was necessary in order to be able to investigate, at least preliminarily, the usefulness of the basic idea of general relativity. This was done as follows.

In Newton's theory one can write the field-law of gravitation thus:

$$\Delta \varphi = 0$$

(φ = gravitation-potential) at points, where the density of matter, ρ, vanishes. In general, one may write (Poisson equation)

$$\Delta \varphi = 4\pi k \rho \qquad (\rho = \text{mass density}).$$

In the case of the relativistic theory of the gravitational field R_{ik} takes the place of $\Delta \varphi$. On the right side we shall then have to place a tensor also in place of ρ. Since we know from the special theory of relativity that the (inert) mass equals energy, we shall have to put on the right side the tensor of energy-density—more precisely the entire energy-density, insofar as it does not belong to the pure gravitational field. In this way, one gets the field-equations

$$R_{ik} - \tfrac{1}{2}g_{ik}R = -kT_{ik}$$

The second member on the left side is added because of formal reasons; for the left side is written in such a way that its divergence disappears identically in the sense of the absolute differential calculus. The right side is a formal condensation of all things whose comprehension in the sense of a field-theory is still problematic. Not for a moment, of course, did I doubt that this formulation was merely a makeshift in order to give the general principle of relativity a preliminary closed expression. For it was essentially not anything *more* than a theory of the gravitational field, which was somewhat artificially isolated from a total field of as yet unknown structure.

If anything in the theory as sketched—apart from the demand of the invariance of the equations under the group of the continuous coordinate-transformations—can possibly make the claim to final significance, then it is the theory of the limiting case of the pure gravitational field and its relation to the metric structure of space. For this reason, in what immediately follows, we shall speak only of the equations of the pure gravitational field.

The peculiarity of these equations lies, on the one hand, in their complicated construction, especially their non-linear character as regards the field-variables and their derivatives, and, on the other hand, in the almost compelling necessity with which the transformation-group determines this complicated field-law. If one had stopped with the special theory of relativity, i.e., with the invariance under the Lorentz-group, then the field-law $R_{ik} = 0$ would remain invariant also within the frame of this narrower group. But, from the point of view of the narrower group there would at first exist no reason for representing gravitation by so complicated a structure as is represented by the symmetric tensor g_{ik}. If, nonetheless, one would find sufficient reasons for it, there would then arise an immense number of field-laws out of quantities g_{ik}, all of which are covariant under Lorentz-transformations (not, however, under the general group). However, even if, of all the conceivable Lorentz-invariant laws, one had accidentally guessed precisely the law which belongs to the wider group, one would still not yet be at the stage of insight achieved by the general principle of relativity. For, from the standpoint of the Lorentz-group two solutions would incorrectly have to be viewed as physically different from each other, if they can be transformed into each other by a non-linear transformation of coordinates, i.e., if they are, from the point of view of the wider field, only different representations of the same field.

One more general remark concerning field-structure and the group. It is clear that, in general, one will judge a theory to be the more nearly perfect the simpler a "structure" it postulates and the broader the group is concerning which the field-equations are invariant. One sees now that these two demands get in each other's way. For example: according to the special theory of relativity (Lorentz-Group) one can set up a covariant law for the simplest structure imaginable (a scalar field), whereas in the general theory of relativity (wider group of the continuous transformations of coordinates) there is an invariant field-law only for the more complicated structure of the symmetric tensor. We have already given *physical* reasons for the fact that in physics invariance under the wider group has to be demanded:[1] from a purely mathematical standpoint I can see no necessity for sacrificing the simpler structure to the generality of the group.

The group of the general relativity is the first one which demands that the simplest invariant law be no longer linear or homogeneous in the field-variables and in their differential quotients. This is of fundamental importance for the following reason. If the field-law is linear (and homogeneous), then the sum of two solutions is again a solution; as, for example: in Maxwell's field-equations for the vacuum. In such a theory it is impossible to deduce from the field equations alone an interaction between bodies, which can be described separately by means of solutions of the system. For this reason, all theories up to now required, in addition to the field equations, special equations for the motion of material bodies under the influence of the fields. In the relativistic theory of gravitation, it is true, the law of motion (geodetic line) was originally postulated independently in addition to the field-law equations. Afterward, however, it became apparent that the law of motion need not (and must not) be assumed independently, but that it is already implicitly contained within the law of the gravitational field.

---

[1] To remain with the narrower group and at the same time to base the relativity theory of gravitation upon the more complicated (tensor-) structure implies a naïve inconsequence. Sin remains sin, even if it is committed by otherwise ever so respectable men.

The essence of this genuinely complicated situation can be visualized as follows: A single material point at rest will be represented by a gravitational field which is everywhere finite and regular, except at the position where the material point is located: there the field has a singularity. If, however, one computes by means of the integration of the field-equations the field which belongs to two material points at rest, then this field has, in addition to the singularities at the positions of the material points, a line consisting of singular points, which connects the two points. However, it is possible to stipulate a motion of the material points in such a way that the gravitational field which is determined by them does not become singular anywhere at all except at the material points. These are precisely those motions which are described in first approximation by Newton's laws. One may say, therefore: The masses move in such fashion that the solution of the field-equation is nowhere singular except in the mass points. This attribute of the gravitational equations is intimately connected with their non-linearity, and this is a consequence of the wider group of transformations.

Now it would of course be possible to object: If singularities are permitted at the positions of the material points, what justification is there for forbidding the occurrence of singularities in the rest of space? This objection would be justified if the equations of gravitation were to be considered as equations of the total field. [Since this is not the case], however, one will have to say that the field of a material particle may the less be viewed as a *pure gravitational field* the closer one comes to the position of the particle. If one had the field-equation of the total field, one would be compelled to demand that the particles themselves would *everywhere* be describable as singularity-free solutions of the completed field-equations. Only then would the general theory of relativity be a *complete* theory.

Before I enter upon the question of the completion of the general theory of relativity, I must take a stand with reference to the most successful physical theory of our period, to wit, the statistical quantum theory which, about twenty-five years ago, took on a consistent logical form (Schrödinger, Heisenberg, Dirac, Born). This is the only theory at present which permits a unitary grasp of experiences concerning the quantum character of micromechanical events. This theory, on the one hand, and the theory of relativity on the other, are both considered correct in a certain sense, although their combination has resisted all efforts up to now. This is probably the reason why, among contemporary theoretical physicists, there exist entirely differing opinions concerning the question as to how the theoretical foundation of the physics of the future will appear. Will it be a field theory; will it be in essence a statistical theory? I shall briefly indicate my own thoughts on this point.

Physics is an attempt conceptually to grasp reality as it is thought independently of its being observed. In this sense one speaks of "physical reality." In pre-quantum physics, there was no doubt as to how this was to be understood. In Newton's theory reality was determined by a material point in space and time; in Maxwell's' theory, by the field in space and time. In quantum mechanics, it is not so easily seen. If one asks: does a ψ-function of the quantum theory represent a real factual situation in the same sense in which this is the case of a material system of points or of an electromagnetic field, one hesitates to reply with a simple "yes" or "no"; why? What the ψ-function (at a definite time) asserts, is this: What is the probability for finding a definite physical magnitude q (or p) in a definitely given interval if I measure it at time t? The probability is here to be viewed as an empirically determinable, and therefore certainly as a "real" quantity which I may determine if I create the same ψ-function very often and perform a q-measurement each time. But what about the single measured value of q? Did the respective individual system have this q-value even before the measurement? To this question, there is no definite answer within the framework of the [existing] theory, since the measurement is a process which implies a finite disturbance of the system from the outside; it would therefore be thinkable that the system obtains a definite numerical value for q (or p), i.e., the measured numerical value, only through the measurement itself. For the further discussion I shall assume two physicists, A and B, who represent different conceptions with reference to the real situation as described by the ψ-function.

A. The individual system (before the measurement) has a definite value of q (i.e., p) for all variables of the system, and more specifically, that value which is determined by a measurement of this variable. Proceeding from this conception, he will state: The ψ-function is no exhaustive description of the real situation of the system but an incomplete description; it expresses only what we know on the basis of former measurements concerning the system.

B. The individual system (before the measurement) has no definite value of q (i.e., p). The value of the measurement only arises in cooperation with the unique probability which is given to it in view of the ψ-function only through the act of measurement itself. Proceeding from this conception, he will (or, at least, he may) state: the ψ-function is an exhaustive description of the real situation of the system.

We now present to these two physicists the following instance: There is to be a system which at the time t of our observation consists of two partial systems S_1 and S_2, which at this time are spatially separated and (in the sense of the classical physics) are without significant reciprocity. The total system is to be completely described through a known ψ-function ψ_{12} in the sense of quantum mechanics. All quantum theoreticians now agree upon the following: If I make a complete measurement of S_1, I get from the results of the measurement and from ψ_{12} an entirely definite ψ-function ψ_2 of the system S_2. The character of ψ_2 then depends upon what kind of measurement I undertake on S_1.

Now, it appears to me that one may speak of the real factual situation of the partial system S_2. Of this real factual situation, we know to begin with, before the measurement of S_1, even less than we know of a system described by the ψ-function. But on one supposition we should, in my opinion, absolutely hold fast: the real factual situation of the system S_2 is independent of what is done with the system S_1, which is spatially separated from the former. According to the type of measurement which I make of S_1, I get, however, a very different ψ_2 for the second partial system ($\psi_2, \psi_2^{\,1}, \ldots$). Now, however, the real situation of S_2 must be independent of what happens to S_1. For the same real situation of S_2 it is possible therefore to find, according to one's choice, different types of ψ-function. (One can escape from this conclusion only by either assuming that the measurement of S_1 ((telepathically)) changes the real situation of S_2 or by denying independent real situations as such to things which are spatially separated from each other. Both alternatives appear to me entirely unacceptable.)

If now the physicists, A and B, accept this consideration as valid, then B will have to give up his position that the ψ-function constitutes a complete description of a real factual situation. For in this case it would be impossible that two different types of ψ-functions could be coordinated with the identical factual situation of S_2.

The statistical character of the present theory would then have to be a necessary consequence of the incompleteness of the description of the systems in quantum mechanics, and there would no longer exist any ground for the supposition that a future basis of physics must be based upon statistics.—

It is my opinion that the contemporary quantum theory by means of certain definitely laid down basic concepts, which on the whole have been taken over from classical mechanics, constitutes an optimum formulation of the connections. I believe, however, that this theory offers no useful point of departure for future development. This is the point at which my expectation departs most widely from that of contemporary physicists. They are convinced that it is impossible to account for the essential aspects of quantum phenomena (apparently discontinuous and temporally not determined changes of the situation of a system, and at the same time corpuscular and undulatory qualities of the elementary bodies of energy) by means of a theory which describes the real state of things [objects] by continuous functions of space for which differential equations are valid. They are also of the opinion that in this way one cannot understand the atomic structure of matter and of radiation. They rather expect that systems of differential equations, which could come under consideration for such a theory, in any case would have no solutions which would be regular (free from singularity) everywhere in four-dimensional space. Above everything else, however, they believe that the apparently discontinuous character of elementary events can be described only by means of an essentially statistical theory, in which the discontinuous changes of the systems are taken into account by way of the continuous changes of the probabilities of the possible states.

All of these remarks seem to me to be quite impressive. However, the question which is really determinative appears to me to be as follows: What can be attempted with some hope of success in view of the present situation of physical theory? At this point, it is the experiences with the theory of gravitation which determine my expectations. These equations give, from my point of view, more warrant for the expectation to assert something *precise* than all other equations of physics. One may, for example, call on Maxwell's equations of empty space by way of comparison. These are formulations which coincide with the experiences of infinitely weak electromagnetic fields. This empirical origin already determines their linear form; it has, however, already been emphasized above that the true laws cannot be linear. Such linear laws fulfill the super-position-principle for their solutions, but contain no assertions concerning the interaction of elementary bodies. The true laws cannot be linear, nor can they be derived from such. I have learned something else from the theory of gravitation: No ever so inclusive collection of empirical facts can ever lead to the setting up of such complicated equations. A theory can be tested by experience, but there is no way from experience to the setting up of a theory. Equations of such complexity as are the equations of the gravitational field can be found only through the discovery of a logically simple mathematical condition which determines the equations completely or [at least] almost completely. Once one has those sufficiently strong formal conditions, one requires only little knowledge of facts for the setting up of a theory; in the case of the equations of gravitation it is the four-dimensionality and the symmetric tensor as expression for the structure of space which, together with the invariance concerning the continuous transformation-group, determine the equations almost completely.

Our problem is that of finding the field equations for the total field. The desired structure must be a generalization of the symmetric tensor. The group must not be any narrower than that of the continuous transformations of coordinates. If one introduces a richer structure, then the group will no longer determine the equations as strongly as in the case of the symmetrical tensor as structure. Therefore, it would be most beautiful if one were to succeed in expanding the group once more, analogous to the step which led from special relativity to general relativity. More specifically, I have attempted to draw upon the group of the complex transformations of the coordinates. All such endeavors were unsuccessful. I also gave up an open or concealed raising of the number of dimensions of space, an endeavor which was originally undertaken by Kaluza and which, with its projective variant, even today has its adherents. We shall limit ourselves to the four-dimensional space and to the group of the continuous real transformations of coordinates. After many years of fruitless searching, I consider the solution sketched in what follows as the logically most satisfactory.

In place of the symmetrical g_{ik} ($g_{ik} = g_{ki}$), the non-symmetrical tensor g_{ik} is introduced. This magnitude is constituted by a symmetric part s_{ik} and by a real or purely imaginary antisymmetric a_{ik}, thus:

$$g_{ik} = s_{ik} + a_{ik}.$$

Viewed from the standpoint of the group, the combination of s and a is arbitrary, because the tensors s and a individually have tensor-character. It turns out, however, that these g_{ik} (viewed as a whole) play a quite analogous role in the construction of the new theory as the symmetric g_{ik} in the theory of the pure gravitational field.

This generalization of the space structure seems natural also from the standpoint of our physical knowledge, because we know that the electromagnetic field has to do with an antisymmetric tensor.

For the theory of gravitation, it is furthermore essential that from the symmetric g_{ik} it is possible to form the scalar density $\sqrt{|g_{ik}|}$ as well as the contravariant tensor g^{ik} according to the definition

$$g_{ik}g^{il} = \delta_k^l \qquad (\delta_k^l = \text{Kronecker-Tensor}).$$

These concepts can be defined in precisely corresponding manner for the non-symmetric g_{ik}, also for tensor-densities.

In the theory of gravitation, it is further essential that for a given symmetrical g_{ik}-field a field Γ^l_{ik} can be defined, which is symmetric in the lower indices and which, considered geometrically, governs the parallel displacement of a vector. Analogously for the non-symmetric g_{ik} a non-symmetric Γ^l_{ik} can be defined, according to the formula

$$g_{ik,l} - g_{sk}\Gamma^s_{il} - g_{is}\Gamma^s_{kl} = 0, \tag{A}$$

which coincides with the respective relation of the symmetrical g, only that it is, of course, necessary to pay attention here to the position of the lower indices in the g and Γ.

Just as in the theory of a symmetrical g_{ik}, it is possible to form a curvature R_{iklm} out of the Γ and a contracted curvature R_{kl}. Finally, with the use of a variation principle, together with (A), it is possible to find compatible field-equations:

$$\underset{\sim}{\mathfrak{g}}^{is}{}_{,s} = \frac{1}{2}(g^{ik} - g^{ki})\sqrt{-|g_{ik}|} \tag{B_1}$$

$$\underset{\sim}{\Gamma}^s_{is} = 0 \quad \left(\underset{\sim}{\Gamma}^s_{is} = \tfrac{1}{2}(\Gamma^s_{is} - \Gamma^s_{si})\right) \tag{B_2}$$

$$R_{\underline{kl}} = 0 \qquad (C_1)$$

$$R_{\underline{kl},m} + R_{\underline{lm},k} + R_{\underline{mk},l} \qquad (C_2)$$

Each of the two equations (B$_1$), (B$_2$) is a consequence of the other if (A) is satisfied. $R_{\underline{kl}}$ means the symmetric, $R_{\underset{\sim}{kl}}$ the antisymmetric part of R_{ik}.

If the antisymmetric part of g_{ik} vanishes, these formulas reduce to (A) and (C$_1$)—the case of the pure gravitational field.

I believe that these equations constitute the most natural generalization of the equations of gravitation.[2] The proof of their physical usefulness is a tremendously difficult task, inasmuch as mere approximations will not suffice. The question is: What are the everywhere regular solutions of these equations?—

This exposition has fulfilled its purpose if it shows the reader how the efforts of a life hang together and why they have led to expectations of a definite form.

Editorial Notes

[a] For a republication and detailed analysis of this text, see Gutfreund and Renn 2020.
[b] Einstein had initially written "precocious child" instead of "precocious young man."
[c] In his manuscript, Einstein wrote: "Die Folge war eine geradezu fanatische Freigeisterei." The Englsih text in brackets was introduced by Schilpp. The noun "Freigeisterei" has no exact analogue in English. It is to be read as "free-spiritness" or "free-thinking". The concept is closely associated with Friedrich Nietzsche's *Human All Too Human* 1878 and his later work.
[d] The friend was Marcel Grossmann (1878–1936), with whom Einstein later collaborated on the general theory of relativity. See Doc. 96.
[e] The quantum theory of (Hohlraum) "black-body" radiation.
[f] Einstein used the word "Feingefühl," which in the present context most likely means "acuteness" or "sensitivity."
[g] See also Doc. 77 for Einstein's discussion of philosophical prejudices.

---

[2] The theory here proposed, according to my view, represents a fair probability of being found valid, if the way to an exhaustive description of physical reality on the basis of the continuum turns out to be possible at all.

93. Statement to the Society for Social Responsibility in Science (1950)

Science 112 (1950), pp. 760–761.

"When Albert Einstein joined the Society for Social Responsibility in Science during the past summer, he made a public statement for the society to use as it pleased. The SSRS feels that Dr. Einstein's statement deserves the thoughtful attention of as wide as possible a group of colleagues. His statement follows:"

—William F. Hewitt, Jr.

Dear Fellow-Scientists:

The problem of how man should act, if his government prescribes actions or society expects an attitude which his own conscience considers wrong, is indeed an old one. It is easy to say that the individual cannot be held responsible for acts carried out under irresistible compulsion, because the individual is fully dependent upon the society in which he is living and therefore must accept its rules. But the very formulation of this idea makes it obvious to what extent such a concept contradicts our sense of justice.

External compulsion can, to a certain extent, reduce but never cancel the responsibility of the individual. In the Nuremberg trials, this idea was considered to be self-evident.[a] Whatever is morally important in our institutions, laws, and mores can be traced back to interpretation of the sense of justice of countless individuals. Institutions are, in a moral sense, impotent unless they are supported by the sense of responsibility of living individuals. An effort to arouse and strengthen this sense of responsibility of the individual is an important service to mankind.

In our times, scientists and engineers carry particular moral responsibility, because the development of military means of mass destruction is within their sphere of activity. I feel, therefore, that the formation of the Society for Social Responsibility in Science satisfies a true need. This society, through discussion of the inherent problems, will make it easier for the individual to clarify his mind and arrive at a clear position as to his own stand; moreover, mutual help is essential for those who face difficulties because they follow their conscience.

Editorial Note

[a] Between November 1945 and October 1946, the International Military Tribunal prosecuted 22 prominent representatives of Nazi Germany in Nuremberg. Further trials were held subsequently in different Allied countries. One of the most significant outcomes of the proceedings was the finding that, even when acting under duress, individuals can be held responsible for violating international law.

94. Letter Declining the Presidency of Israel (1952)

Ministry of Defense, Israel Defense Forces Archive.

Chaim Weizmann, the first president of the new state of Israel, died on 9 November 1952. One week later, Abba Eban, Israel's ambassador to the United States, was tasked by Prime Minister David Ben-Gurion with offering the presidency to Einstein.

I am deeply moved by the offer from our State of Israel, and at once saddened and ashamed that I cannot accept it. All my life I have dealt with objective matters; hence I lack both the natural aptitude and the experience to deal properly with people and to exercise official functions. For these reasons alone, I should be unsuited to fulfill the duties of that high office, even if advancing age was not making increasing inroads on my strength.

I am the more distressed over these circumstances because my relationship to the Jewish people has become my strongest human bond, ever since I became fully aware of our precarious situation among the nations of the world.

Now that we have lost the man [Chaim Weizmann] who for so many years, against such great and tragic odds, bore the heavy burden of leading us toward political independence, I hope with all my heart that a successor may be found whose experience and personality will enable him to accept the formidable and responsible task.

95. Elementary Considerations on the Interpretation of the Foundations of Quantum Mechanics (1953)

Scientific Papers Presented to Max Born. Edinburgh: Oliver & Boyd (1953), pp. 33–40.

A realist foundational reflection on the difficulties of Max Born's statistical interpretation of the quantum mechanical wave function. Einstein had expressed his reservations in detail in 1946 (see Doc. 92) and had continued to exchange personal correspondence on the topic with Born in the intervening years.

The peculiar character of the present situation is as follows: there are no doubts about the mathematical formalism of the theory, but indeed there are doubts about the physical interpretation of its assertions. What is the relation between the ψ-function and the concrete, unique matter of fact, i.e., the individual state of a single system? Or: what does the ψ-function assert about the (individual) "real-state" of the system?

To begin with, it is doubtful whether the very question has any meaning at all. For one can take the following standpoint: "real" is only the individual result of observation, but not something that exists objectively in space and time independently of the act of observing. If one adopts this purely positivist standpoint, one obviously no longer needs to worry about how "real-state" is to be conceived of in the framework of quantum theory. The exercise then seems like a fencing match against a ghost.

But, when carried out consistently, this tidy positivist point of view has an irreparable weakness: it leads one to declare any linguistically expressible proposition to be meaningless. Does one have the right to declare a description of a single observation result as meaningful, or as true or false? Could such a description not be based on lies, or on experiences which we may interpret as memories of a dream or a hallucination? Does the distinction between waking and dream experiences have any objective meaning at all? In the end, all that remains as "real" are the experiences of an I without any possibility of saying anything about them; for in a tidy positivist analysis, without exception, the terms used in the assertions are rendered meaningless.

In truth, the independent concepts and conceptual systems used in our statements are human creations, self-created tools whose justification and value are based exclusively on the fact that they can be assigned expediently to the experiences. In other words, these tools are justified insofar as they can "explain" the experiences.[1]

The justification of concepts and systems of concepts is only to be judged from the point of view of expediency. This also pertains to the concepts of "physical reality," or "reality of the outer world," "real-state of a system."

A priori, there is no justification to postulate or prohibit them as necessary; what decides is only the evidence.

Beneath these word-symbols lies a program that was absolutely decisive in the development of physical thinking up to the establishment of quantum theory: everything is to be traced back to conceptual objects from the spatio-temporal realm and to law-like relationships that are to apply to these objects. Nothing appears in this description that refers to an empirical knowledge concerning these objects. The moon is ascribed a spatial position (with reference to some coordinate system) at any given point in time, regardless of whether there exist any perceptions about this position. This type of description is meant when one speaks of the physical description of a "real external world," whatever the choice of the elementary building blocks (material points, field, etc.) underlying such a description.

The justification of this program was not seriously doubted by physicists as long as it seemed that everything featured in such a

---

[1] The linguistic kinship between the terms "wahr" (true) and "bewähren" (succeed) is based on an affinity; it is just that care has to be taken that this insight not be misunderstood in a utilitarian sense.

description could in principle be determined empirically in each individual case. It was only with Heisenberg that the fact that this is an illusion in the domain of quantum phenomena was convincingly demonstrated to physicists.

The concept of "physical reality" was now perceived as problematic and the question arose as to what theoretical physics is actually trying to describe (through quantum mechanics) and what the laws it established referred to. This question was answered in quite different ways.

To get closer to an answer, we consider what quantum mechanics asserts about macro-systems, i.e., about such objects that we perceive as "directly observable." About such objects, we know that they, and the laws valid for them, can be represented by classical physics with significant if not unlimited precision. We do not doubt that for such objects there is a real spatial configuration (position) at any point in time as well as a velocity (or a momentum), i.e., a *real situation*—all this to the extent determined by the approximation required by the quantum structure.

We ask: does quantum mechanics imply (to the expected approximation) the real description for macro-bodies provided by classical mechanics? Or—if this question cannot readily be answered by "yes"— in what sense is this the case? Let us consider this using a concrete example.

The special example

The system consists of a sphere of about 1 mm diameter, which moves back and forth between two parallel walls (about 1 m apart) along the x-axis of a coordinate system. The collisions are assumed to be ideally elastic. In this idealized macro-system we imagine the walls to be replaced by "sharp" potential energy expressions in which only the coordinates of the material points constituting the sphere are included. By cunning and trickery, it is arranged in such a way that these reflection processes do not produce a coupling between the center-of-mass coordinate x of the sphere and its "inner" coordinates (including the angular coordinates). In this way, we achieve that for our purposes the position of the sphere (apart from its radius) can be described by x alone.

In terms of quantum mechanics, we are dealing with a process of precisely defined energy. The De Broglie-wave (ψ-function) is then harmonic in its time coordinate. Also, it differs from 0 only between $x = -l/2$ and $x = +l/2$. At the endpoints of the trajectory, the continuous connection to the vanishing of the ψ-function outside the trajectory is achieved by the postulate that $\psi = 0$ for $x = \pm\frac{l}{2}$.

The ψ-function is then a standing wave, which may be represented within the trajectory by the superposition of two harmonic waves traveling in opposite directions:

$$\psi = \tfrac{1}{2}Ae^{i(at-bx)} + \tfrac{1}{2}Ae^{i(at+bx)} \qquad (1)$$

or

$$\psi = Ae^{iat}\cos(bx). \qquad (2)$$

It can be seen from (1a) that the factor A must be chosen to be the same in both terms in order to satisfy the boundary conditions at the ends of the rod. A can be chosen to be real without loss of generality. b is determined by the Schrödinger equation by a and by the mass m.[a] The factor A is normalized in the well-known manner.

In order for a comparison of the example with the corresponding classical problem to be fruitful, we still need to establish that the De Broglie wavelength $\frac{2\pi}{b}$ be small compared to l.

We now take Born's probability interpretation as a basis for interpreting the ψ-function in the usual way:

$$W = \int \psi\overline{\psi}\,dx = A^2 \int \cos^2(bx)\,dx.$$

This is the probability of finding the center-of-mass coordinate x of the sphere in a given interval Δx. Apart from an undulatory "fine structure," whose physical reality is certain, it is simply const. Δx.

What about the probability of the momentum values, or the velocities of the sphere? These probabilities are obtained by Fourier decomposition of ψ. If (1) were valid from $-\infty$ to $+\infty$, (2) would already give the desired Fourier decomposition. There would then be two equally sharp values of momentum, pointing in opposite directions with equal probability. But since both wave trains are limited, they contribute per term a continuous Fourier decomposition whose spectral range is getting smaller as the number of De Broglie wave lengths contained in the range l increases. This comes down to the fact that there exist only two almost sharp values of equal momenta, pointing in opposite directions, which values, by the way, coincide with those of the classical case with equal probability.

Apart from small deviations arising from the quantum structure, these two statistical results are the same as those given for the "time ensemble" of systems in classical theory. In this respect, the theory is completely satisfactory.

But now we ask ourselves: can this theory provide a real-description of an individual case? This question must be answered with "No." For this decision it is essential that we are dealing with a "macro system." Because with a macro-system we are certain it is always in a "real state," which is correctly described to some approximation by classical mechanics. The individual macro system of the kind we are considering therefore always has an almost completely determined center-of-mass coordinate—at least averaged over a short time interval—and an almost completely determined momentum (determined also with respect to its direction). Neither of these two indications can be gathered from the ψ-function (1). From this function we can obtain (by means of Born's interpretation) only such information which refers to a *statistical ensemble* of the type of systems consideredk here.

With the macro system at hand, not every ψ-function that solves the Schrödinger equation has an approximate corresponding real-description in the sense of classical mechanics. This becomes particularly clear when we look at a ψ-function that arises from the superposition of two solutions of the type (1), whose frequencies (or energies), however, are considerably different. This is so because no corresponding real-case of classical mechanics corresponds to such a superposition (but only a statistical ensemble of such real cases in the sense of Born's interpretation).

By way of generalization, we conclude: quantum mechanics describes the overall behavior of systems, not the individual system. The description by the ψ-function is in this sense an incomplete description of the individual system, not a description of its real state.

Remark: One could object to this conclusion as follows. The case of the most extreme sharpness of frequency of the ψ-function considered by us is a limiting case, for which perhaps the requirement of similarity with a problem in classical mechanics could exceptionally fail. If one allows for a finite but small range of frequencies, one can achieve, by an appropriate choice of amplitudes and phases of the superposed ψ-function, that the resulting ψ-function becomes almost sharp with respect to position and momentum. Could one not try to restrict the admissible ψ-functions to be allowed from this point of view and thus achieve that the permitted ψ-functions could be interpreted as a representation of the individual systems?

Such a possibility must be denied for the very reason that the spatial sharpness of such a representation cannot be achieved for all times.—

The circumstance that the Schrödinger equation together with Born's interpretation does not lead to a description of the real-states of individual systems naturally stimulates the search for a theory that is free from such restriction.

So far there have been two efforts in this direction, which have in common the adherence to the Schrödinger equation and the abandonment of Born's interpretation. The first attempt goes back to De Broglie and was pursued with much ingenuity by Bohm.

Just as Schrödinger's original investigation derives the wave equation in analogy to classical mechanics (linearization of Jacobi's equation of analytical mechanics), the equation of motion of the quantized individual system—based on a solution ψ of the Schrödinger equation—shall also be derived by analogy.

The rule is the following. ψ is to be put into the form

$$\psi = Re^{iS}.$$

As a result, the (real) functions R and S are obtained from ψ, depending on the coordinates. The derivative of S with respect to the coordinates should then give us the momenta or velocities of the system, respectively, as functions of time, if the coordinates of the individual system in question are given for a certain time value.

A look at (1a) shows us that in our case $\frac{\partial S}{\partial x}$ vanishes and with it the velocity vanishes as well. This objection against this theoretical attempt, raised by Pauli already a quarter of a century ago, is particularly serious in the case of our example. The vanishing of the velocity contradicts the well-founded requirement that, in the case of a macro-system, the motion must agree with the motion resulting from classical mechanics.

The second attempt to arrive at a real-description of an individual system based on the Schrödinger equation has recently been made by Schrödinger himself. His idea is briefly this. The ψ-function itself represents reality and does not require Born's statistical interpretation. The atomistic entities about which the ψ-function was previously supposed to make assertions, do not exist at all. Applied to our macro-system, this means: the macro-body does not exist as such at all; in any case, nothing like the position of its center of mass exists—not even approximately. Here, too, the requirement that the quantum theoretical description of a macro-system approximately corresponds to the corresponding description according to classical mechanics is violated.

The result of our analysis is this. The only previously acceptable interpretation of the Schrödinger equation is the statistical interpretation given by Born. However, this interpretation does not provide a real description of the individual system but only statistical assertions about the system ensembles.

In my opinion, it is in principle unsatisfactory to base physics on such a theoretical approach, especially since one cannot renounce the objective describability of the individual *macro*-systems (description of the "real state") without the physical worldview dissolving to a certain extent into a fog. At the end of the day, the conception that physics must strive for a real description of the individual system becomes unavoidable. Nature as a whole can only be thought of as an individual (uniquely existing) system and not as a "system ensemble."

Editorial Note

[a] The preceding sentence is corrupted in the published original, and has been completed here on the basis of the original manuscript [AEA 1-172].

96. Recollections–Souvenirs (1955)

Schweizerische Hochschulzeitung 28, Sonderheft 100 Jahre ETH (1955), pp. 145–153.

Einstein wrote this essay in late March 1955 for a special issue dedicated to the hundredth anniversary of the Zurich Polytechnic, eventually renamed Eidgenössische Technische Hochschule, which he had attended from 1896 to 1900. Einstein died three weeks later, on 18 April 1955, in Princeton, N.J., at age seventy-six.

96. Recollections

The editors of this celebratory volume have most kindly urged me to write a contribution. At first, I did not know how to begin and reacted with embarrassing silence. But I gave in when I realized it would not be possible to get off lightly and somewhat graciously. Since I did not feel capable of saying anything terribly worthwhile of an objective nature about the Federal Polytechnic, the only way out was to recount something of my personal experience insofar as it relates, to some extent, to the Polytechnic. To this end, it was first necessary to overcome an inner resistance regarding the professional psychology of the exact scientist. Though he is no freer of vanity than any other member of the species that describes itself euphemistically as "Homo sapiens," it nevertheless goes against his grain to publish something about himself. His education and scientific striving limit him to objective things, those that can be conceptualized.

Here I am, intentionally sinning against this beneficial and liberating situation. But I am not sinning without a plan, and not intemperately. After all, it may be of some interest to the objectively minded reader to learn what has led someone on his path and forced him to develop in a particular manner. This sin also gives me the welcome opportunity to remember certain personalities to whom I am very much indebted.

In 1895, I came as a sixteen-year-old from Italy to Zurich after having spent a year with my parents, without going to school and without having any teachers. My goal had been admittance to the Polytechnic, yet I had not yet clarified to myself how to accomplish this goal. I was a willful but modest young man who had acquired his incomplete and one-sided knowledge primarily through self-study. Greedy for deeper understanding but rather unreceptive and marred by poor memory, studying appeared by no means an easy task for me. With a feeling of well-justified insecurity, I registered for the admission examination to the engineering division. The exam painfully showed me the inadequacy of my preparation, though the examiners were patient and understanding. I thought it entirely justified that I flunked. But it was consoling that the physicist H. F. Weber let me know I could attend his course, were I to stay in Zurich. However, the director, Prof. Herzog, recommended me to the canton school in Aarau, where I passed the *Matura* after a year of study. This school has left an unforgettable impression on me through its liberal spirit and the modest seriousness of its teachers, who do not rely on any outside authority. In comparison with six years of school at a German authoritarian-led Gymnasium, I became acutely aware of how an education for free action and responsibility for oneself is greatly superior to that of relying on drill, external authority, and ambition. Genuine democracy is no mere delusion.

During this year in Aarau, the following conundrum occurred to me: When one runs after a light wave at the speed of light, one would have a time-independent field of waves before one. But such a thing does not seem to exist! This was the first childish thought-experiment related to the special theory of relativity. Invention is not the work of logical thinking, though the end product is linked to the logical form.

From 1896 to 1900, I studied in the specialized teachers' division of the Federal Polytechnic. I soon noticed that I had to content myself with being an average student. To be a good student, one needs the ease of comprehension, the willingness to concentrate one's energies on what is presented in lectures, and enough love of orderliness to take notes on the lectures and work through them afterward. All these qualities, I regretfully realized, were thoroughly missing in me. Thus, I gradually learned to live in peace with a somewhat guilty conscience and arrange my studies in a way corresponding to my intellectual appetites and interests. I followed some lectures with great interest, but otherwise skipped classes often, and studied the masters of theoretical physics with great zeal at home. This was in itself good and also served the purpose of weakening the bad conscience to the extent that the emotional equilibrium was not noticeably disturbed. This extended private studying was simply the continuation of an earlier habit; a Serbian student, Mileva Maric, whom I later married, participated in it. But I did work assiduously and passionately in H. F. Weber's physical laboratory. I was also fascinated by Professor Geiser's lectures on infinitesimal geometry, which were true masterpieces of pedagogical art and, later, of much help in the struggle with the general theory of relativity. Otherwise, however, higher mathematics interested me little during the years of study. It seemed to me, erroneously, that the field was so diverse that one could easily waste away one's entire energy in one of its distant provinces. In my innocence, I also thought it would suffice for a physicist to clearly grasp the elementary mathematical concepts and have them ready for application, that the rest consisted of unfruitful subtleties for the physicist—a regrettable error I recognized only later.

The mathematical skill was evidently insufficient to allow me to distinguish the central and fundamental aspects from the peripheral, basically unimportant ones.

During these years of study, a real friendship developed with a fellow student, Marcel Grossman. Once every week, I went with him to celebrate in the Café Metropol on the Limmatquai and talked to him not only about studies, but about all things that might interest open-minded young people. He was not a vagabond or loner like me, but rather someone who was rooted in the Swiss milieu without the loss of his inner independence. In addition, he had in great abundance those gifts I lacked: quick perceptiveness and any sort of orderliness. He not only attended all the lectures that concerned us, but worked through them in such an exemplary fashion that one could easily have printed and published his notebooks. He lent me these notebooks, which were a saving anchor for me in preparing for the exams; I'd rather not speculate on how I would have fared without them.

Despite this invaluable assistance, and even though all subjects were intrinsically interesting, I nevertheless had to strive hard to thoroughly study all these things. For people of my kind, with pensive interests, university studies are not necessarily beneficial. Forced to eat so many good things, one could persistently spoil one's appetite and stomach. The flicker of holy curiosity may then be lastingly extinguished. Fortunately, this intellectual depression lasted only one year after the successful completion of my studies.

The greatest thing that Marcel Grossmann did for me as a friend was the following: About one year after completing my studies, he recommended me to the director (Haller) of the Swiss Patent Office (which was then called "Office for Intellectual Property"), with help from his father. Following a thorough oral examination, Mr. Haller offered me employment there from 1902 to 1909, thereby freeing me from existential worries during the most productively creative years. In addition, work on the final formulation of technical patents was a true blessing for me. It forced me to think broadly and offered much stimulation for thinking about physics. In fact, a practical profession is generally a blessing for people like me, since a young person in an academic career is hard-pressed to produce a significant amount of scientific papers, a temptation into superficiality which only strong characters manage to withstand. Most practical professions are also of such a sort that a person of normal aptitude can deliver what is expected of him. In his civil life, he is not dependent on special insights, and if he has deeper scientific interests, he can immerse himself in his favorite problems alongside his required work. Fear that his efforts might remain fruitless need not oppress him. I owe my being in such a lucky position to Marcel Grossmann.

I mention only one of the scientific experiences brought by those happy Bern years—namely, an idea that proved the most fruitful of my life. The special theory of relativity was already a year old. Was the principle of relativity limited to inertial systems, i.e., to coordinate systems that move uniformly with respect to each other (linear coordinate transformations)? Formal instinct says, "Probably not." However, the foundation of all mechanics until then—the principle of inertia—appeared to exclude any extension of the relativity principle since, if one introduces a coordinate system that is accelerated (relative to an inertial system), then an "isolated" mass point relative to it will no longer move rectilinearly and uniformly. A spirit freed from inhibiting thought habits would then have asked: Does this behavior provide a means of distinguishing between an inertial and a non-inertial system? They would then have had to conclude (at least in the case of a rectilinear uniform acceleration) that this is not the case, as they could also interpret the mechanical behavior of bodies relative to such an accelerated coordinate system as the action of a gravitational field; this is made possible by the empirical fact that the acceleration of bodies, too, is always the same in a gravitational field, independent of their nature. This insight (principle of equivalence) made it not only probable that the laws of nature have to be invariant with regard to a larger group of transformations than the group of Lorentz transformations (extension of the principle of relativity), but also that this extension would lead to a deeper understanding of the theory of the gravitational field. I did not doubt that this idea was, in principle, correct, but the difficulties in its execution seemed almost insurmountable. At first, elementary considerations indicated that the transition to a broader transformation group would be incompatible with a direct physical interpretation of the space-time coordinates that had smoothed the path to the special theory of relativity. Moreover, it was at first unclear how to choose the expanded transformation group. In truth, I reached this equivalence principle by a detour, the description of which does not belong here.

From 1909 to 1912, while I had to teach theoretical physics at the Zurich and the Prague Universities, I constantly mulled over the problem. I had come significantly closer to its solution when I was called to the Zurich Polytechnic in 1912. H. Minkowski's analysis of the formal foundation of special relativity theory proved significant. It can be summarized in the sentence: Four-dimensional space has an (invariant) pseudo-Euclidean metric; this determines the experimentally observable metric properties of space, as well as the principle of inertia and, moreover, the form of the Lorentz-invariant equation systems. In this space, there are privileged—namely, quasi-Cartesian—coordinate systems, which are the only "natural" ones (inertial systems).

The principle of equivalence prompts us to introduce non-linear coordinate transformations into such a space—that is, non-Cartesian (curvilinear) coordinates. The pseudo-Euclidean metric then takes the general form $ds^2 = \Sigma g_{ik} dx_i dx_k$ (summed over the indices i and k [from 1 to 4]). These g_{ik} are, then, functions of the four coordinates which, according to the equivalence principle, describe not only the metric, but also the "gravitational field." The latter, of course, is of a very special kind here, since it can be brought through transformations into the special form $-dx_1^2 - dx_2^2 - dx_3^2 + dx_4^2$, i.e., into a form in which the g_{ik} are independent of the coordinates. In this case, the gravitational field described by g_{ik} can be "transformed away." In the latter special form, the inertial behavior of isolated bodies is expressed through a (timelike) straight line. In the general form, this corresponds to the "geodesic line."

This formulation still referred to the case of the pseudo-Euclidean space in all cases, but it clearly showed how a transition to general gravitational fields could be accomplished. Here, too, the gravitational field is to be described by a sort of metric, i.e., by a symmetrical tensor field g_{ik}. The generalization now simply consists of dropping the assumption that the field can be transformed into a pseudo-Euclidean one simply through a coordinate transformation.

The problem of gravitation was now reduced to a purely mathematical one. Are there differential equations for the g_{ik} that are invariant with non-linear coordinate transformations? Such, and *only* such, differential equations were to be considered as field equations of the gravitational field. The equation of the geodesic line would then give the law of motion for material points.

In 1912, with this task in mind, I looked for my old university friend, Marcel Grossmann, who had meanwhile become a professor of mathematics at the Federal Polytechnic. He immediately became fired up with enthusiasm, though, as a true mathematician, he had a somewhat skeptical attitude toward physics. When we were both still students and habitually exchanged our ideas over coffee, he once made such a pretty and typical remark, which I cannot refrain from reproducing here:

> I confess that I nevertheless gained something fundamental from the study of physics. Earlier, when sitting down in a chair, I would feel the heat of my "pre-sitter" seep through and shudder a bit. This has completely passed, because physics has taught me that heat is something completely impersonal.

That's how it so happened that he was willing to collaborate on the problem, but with the understanding that he would not take over responsibility for any physical statements or interpretations. He examined the literature and soon discovered that the suggested mathematical problem had already been solved, particularly by Riemann, Ricci, and Levi-Civita. This entire development was connected to the Gaussian theory of curved surfaces, which made systematic use of generalized coordinates for the first time. Riemann's accomplishment was the greatest. He showed how tensors of second differential order can be formed from the field of the g_{ik} tensors. One could thus see what the field equations of gravitation would look like, were one to demand the invariance of the group with regard to all continuous coordinate transformations. However, it was not so easy to realize if such a demand was justified, as I seemed to have found reasons against it. In fact, these incorrect reservations were the reason the theory appeared in its final form only in 1916.

While I was assiduously collaborating with my old friend, neither of us thought an insidious malady would wrest away this exceptional man. The desire to express my gratitude to Marcel Grossmann at least once in my life gave me the courage to send in this somewhat colorful autobiographical sketch as a contribution to this commemorative volume.

Forty years have now passed since the completion of the theory of gravitation. They were devoted almost exclusively to the effort of obtaining, through a generalization of the theory of gravitation, a field theory that would form a foundation for all physics. Many were working toward the same goal. I subsequently abandoned a number of papers that seemed promising. However, the last ten years finally led to a theory that seemed natural and hopeful to me. The fact I currently cannot convince myself whether I should consider this theory physically valuable is, for now, caused by insurmountable mathematical difficulties, the same being encountered for its application by any other non-linear field theory. Moreover, it appears quite doubtful whether a field theory can account for the atomistic structure of matter and radiation and for the quantum phenomena. Most physicists will answer, without hesitation, with a decided "no," since they believe that the quantum problem has been solved principally *in a different way*. Regardless, we still have Lessing's comforting words: The striving for truth is more delicious than its assured possession.[a]

Editorial Note

[a] Gotthold Ephraim Lessing, revered German Enlightenment writer, philosopher, and critic, wrote in 1778: "It is not the truth in whose possession any man is or thinks he is, but the sincere effort he has made to get at the truth, that constitutes the worth of man."

Bibliography

Abraham, M. (1902). "Dynamik des Elektrons." *Königliche Gesellschaft der Wissenschaften zu Göttingen. Mathematisch-physikalische Klasse. Nachrichten*, pp. 20–41.
Bayer A. von, et al. (1919). "To the Civilized World." *The North American Review*, 210, no. 765, pp. 284–287.
Cohn, E. (1913). *Physikalisches über Raum und Zeit*. 2nd rev. ed. Leipzig: Teubner.
Compton, A. H. (1923). "A Quantum Theory of the Scattering of X-Rays by Light Elements." *Physical Review*, 21, pp. 483–502.
Coudenhove-Kalergi, R. N. von (1923). *Pan-Europa*. Vienna: Pan-Europa-Verlag.
Eddington, A. S. (1921). "A Generalization of Weyl's Theory of Electromagnetic and Gravitational Fields." *Proceedings of the Royal Society of London*, A99, pp. 104–122.
Einstein, A. (1905a). "Über einen die Erzeugung und Verwandlung des Lichtes betreffenden heuristischen Gesichtspunkt." *Annalen der Physik*, 17, pp. 132–148.
— (1905b). "Zur Elektrodynamik bewegter Körper." *Annalen der Physik*, 17, pp. 891–921.
— (1907). "Die Plancksche Theorie der Strahlung und die Theorie der spezifischen Wärme." *Annalen der Physik*, 22, pp. 180–190.
— (1914a). "Antrittsrede des Hrn. Einstein." *Sitzungsberichte der Königlich Preussischen Akademie der Wissenschaften. Öffentliche Sitzung zur Feier des Leibnizischen Jahrestages vom 2. Juli*. Vol. 28, Berlin: Sitzungsberichte, pp. 739–742.
— (1914b) "Relativitätsprinzip." *Vossische Zeitung* (26 April): *Morgen-Ausgabe*, pp. 1–2.
— (1916a) "Ernst Mach." *Physikalische Zeitschrift*, 17, no. 7, pp. 101–104.
— (1916b). "Meine Meinung über den Krieg." In: Berliner Goethebund. *Das Land Goethes 1914–1916: Ein vaterländisches Gedenkbuch*. Ed. by I. Landau and E. Zabel. Stuttgart and Berlin: Deutsche Verlags-Anstalt, p. 30.
— (1917). "Der Angst-Traum." *Berliner Tageblatt* (25 December): *Morgen-Ausgabe*, p. 1.
— (1918a). "Besprechungen. Weyl, Hermann, Raum-Zeit-Materie: Vorlesungen über allgemeine Relativitätstheorie." *Die Naturwissenschaften* 6, p. 373.
— (1918b) "Motive des Forschen." In: E. Warburg, et al. *Zu Max Plancks sechzigstem Geburtstag. Ansprachen, gehalten am 26. April 1918 in der Deutschen Physikalischen Gesellschaft*. Karlsruhe: C. F. Müllersche Hofbuchhandlung, pp. 29–32.
— (1919a). "Die Zuwanderung aus dem Osten." *Berliner Tageblatt* (30 December): *Morgen-Ausgabe*, p. 2.
— (1919b). "Induktion und Deduktion in der Physik." *Berliner Tageblatt* (25 December): *Morgen-Ausgabe*, p. 4.
— (1919c). "Time, Space, and Gravitation." *The Times*. London (28 November), pp. 13–14.
— (1920a). "[A Confession]." *Israelitisches Wochenblatt für die Schweiz* (24 September), p. 10.
— (1920b). "[On the Contribution of Intellectuals to International Reconciliation]." German Social & Scientific Society of New York. *Thoughts on Reconciliation*. New York: Neumann Brothers, pp. 10–11.
— (1920c). *Äther und Relativitätstheorie*. Berlin: Verlag von Julius Springer.
— (1920d). "Die freie Vereinigung für technische Volksbildung." *Neue Freie Presse* (24 July): *Morgen-Ausgabe*, p. 8.
— (1920e). "Die Urteile der Deutschen Gelehrten." *Berliner Tageblatt* (25 July): *Morgen-Ausgabe*, p. 4.
— (1920f). "Meine Antwort: Ueber die anti-relativitätstheoretische G.m.b.H." *Berliner Tageblatt* (27 August): *Morgen-Ausgabe*, pp. 1–2.
— (1920g). "Tumultszenen bei einer Einstein—Vorlesung." *8-Uhr Abendblatt*. Berlin 73, no. 38 (13 February), pp. 2-3.

— (1921a). "[On a Jewish Palestine]." *Jüdische Rundschau* (1 July), p. 371.
— (1921b). "Das Gemeinsame am künstlerischen und wissenschaftlichen Erleben." *Menschen. Zeitschrift neuer Kunst* 4, p. 19.
— (1921c). "Die Not der deutschen Wissenschaft. Eine Gefahr für die Nation." *Neue Freie Presse* (25 December): *Morgen-Ausgabe*, p. 1.
— (1921d). "Geometrie und Erfahrung." *Sitzungsberichte der Preussischen Akademie der Wissenschaften. Öffentliche Sitzung zur Feier des Jahrestages König Friedrichs II, 27 Januar 1921.* Berlin: Verlag der Akademie der Wissenschaften, pp. 123-130.
— (1921e). "Kulturelle, politische, wirtschaftliche: Wissenschaft." *Die Friedensbewegung: Ein Handbuch der Weltfriedenströmungen der Gegenwart, unter Mitarbeit von 64 hervorragenden in-und ausländischen Vertretern des Pazifismus.* Ed. by K. Lenz and W. Fabian. Berlin: Schwetschke (1922), pp. 78-79.
— (1921f). "Wie ich Zionist wurde." *Jüdische Rundschau* No. 49 (21 June), pp. 351-352.
— (1922a). "In Memoriam Walther Rathenau." *Neue Rundschau* 33, pp. 815-816.
— (1922b). "Über die gegenwärtige Krise der theoretischen Physik." *Kaizo* 4, no. 12 (December), pp. 1-8.
— (1922c). "Vorwort." In: B. Russell. *Politische Ideale.* Berlin: Deutsche Verlagsgesellschaft für Politik und Geschichte, p. 5.
— (1923a). "My Impressions of Palestine." *New Palestine* 4, p. 341.
— (1923b). "Plauderei über meine Eindrücke in Japan." *Kaizo* 5, pp. 338-343.
— (1923c). "Prof. Einstein über seine Eindrücke in Palästina." *Jüdische Rundschau* 33, pp. 195-196.
— ed. (1923d). *Scripta universitatis atque bibliothecae hierosolymitanarum. Mathematica et physica 1.*
— (1924a). "Das Komptonsche Experiment: ist die Wissenschaft um ihrer selbst willen da?" *Berliner Tageblatt* (20 April): *Beiblatt*, p. 1.
— (1924b). "Einstein über den Völkerbund." *Frankfurter Zeitung* 69, no. 646 (29 August), p. 1.
— (1924c). "Elsbach's Buch: Kant und Einstein." *Deutsche Literaturzeitung* 1, cols. 1685-1692.
— (1924d). "Philosophie. J. Winternitz [Dr. phil], Relativitätstheorie und Erkenntnislehre." *Deutsche Literaturzeitung* 1, pp. 20-22.
— (1925a). "Botschaft." *Jüdische Rundschau*, 30, no. 14 (17 February), pp. 129-130.
— (1925b). "De los ideales." *La Prensa* (28 April), p. 10.
— (1925c). "Ein Wort auf den Weg." *Jüdische Rundschau*, 30, no. 27/28 (3 April), p. 244.
— (1925d). "Letters by Celebrated Individuals." *Letters from Russian Prisons, Consisting of Reprints of Documents by Political Prisoners in Soviet Prisons, Prison Camps and Exile, and Reprints of Affidavits Concerning Political Persecution in Soviet Russia, Official Statements by Soviet Authorities, Excerpts from Soviet Laws Pertaining to Civil Liberties, and Other Documents; With Introductory Letters by Twenty-Two Well- Known European and American Authors.* Ed. by A. Berkman. New York: A. & C. Boni, p. 7.
— (1925e). "Nichteuklidische Geometrie und Physik." *Die Neue Rundschau* 36, no. 1, pp. 16-20.
— (1925f) "Pan-Europa." *Das Junge Japan: Deutsch-Japanische Monatsschrift* 1, no. 10, pp. 369-372.
— (1926a). "Erklärung des Herrn Professor Dr. Albert Einstein." In: Keren Hayesod. *Das Palästinawerk: Eine Kundgebung deutscher Juden im ehemaligen Herrenhaus zu Berlin am 4. März 1926. Stenographischer Bericht.* Berlin: Keren Hayesod, p. 18.
— (1926b). "Space-time." In: *Encyclopædia Britannica.* Ed. by J. L. Garvin. 13th ed. Supplementary Vol. 3. New York: Encyclopædia Britannica, pp. 608-611.
— (1927a). "Neue Experimente über den Einfluß der Erdbewegung auf die Lichtgeschwindigkeit relativ zur Erde." *Forschungen und Fortschritte* 3, p. 36.
— (1927b). "Newtons Mechanik und ihr Einfluß auf die Gestaltung der theoretischen Physik." *Die Naturwissenschaften*, 15, no. 12, pp. 273-276.
— (1928). "A propos de «la déduction relativiste» de M. Émile Meyerson." *Revue Philosophique de la France et de l'Étranger*, 105, nos. 3-4 (January-June), pp. 161-166.
— (1929a). "The New Field Theory." *New York Times*, 4 February 1929, reprinted in *The Observatory* 52, pp. 82-87 and 114-118.
— (1929b). "An einen jungen Gelehrten." *Berliner Tageblatt* (25 December), p. 1.

— (1929c). "The Palestine Troubles." *Manchester Guardian Weekly* 21 (12 October), p. 314.
— (1930a). "Religion and Science." *New York Times Magazine* (9 November), p. 1.
— (1930b). "What I Believe–Living Philosophies XIII." *Forum and Century*, 84, no. 4 (October), pp. 193–194.
— (1931a). "Militant Pacifism." *World Tomorrow* 14, no. 1 (January), p. 9.
— (1931b). "The 1932 Disarmament Conference." *The Nation* 199, no. 3455 (23 September), p. 300.
— (1932a). "Is There a Jewish View of Life?" *Opinion* 11 (26 September), p. 7.
— (1932b). "To American Negroes." *Crisis* 41, no. 2, p. 45.
— (1933a). "Lettre à M. Nahon." *La Patrie Humaine* (18 August), 81: 1.
— (1933b). *On the Method of Theoretical Physics. The Herbert Spencer Lecture Delivered at Oxford 10 June 1933*. Oxford: Clarendon Press.
— (1933c). *Warum Krieg? Ein Briefwechsel, Albert Einstein und Sigmund Freud. Numerierte Auflage von nur 2,000 Ex.* Völkerbund: Internationales Institut für Geistige Zusammenarbeit.
— (1934). "Einiges über meine Eindrücke in Amerika." In: *Mein Weltbild*. Amsterdam: Querido Verlag, pp. 54–60.
— (1944). "Bemerkungen zu Bertrand Russells Erkenntnis-Theorie." *The Philosophy of Bertrand Russell*. Ed. by P. A. Schlipp. Evanston, Ill.: Northwestern University Press, pp. 278–291.
— (1955). "Erinnerungen—Souvenirs." *Schweizerische Hochschulzeitung* 28 (Sonderheft 100 Jahre ETH), pp. 145–153.
— (1960). "[Manuscript on Germany and Hitler]." In: *Einstein on Peace*. Ed. by O. Nathan and H. Norden. New York: Simon & Schuster, pp. 263–264.
Einstein, A., G. F. Nicolai, and F. W. Föster. (1917). "Aufruf an die Europäer." In: G. F. Nicolai. *Die Biologie des Krieges: Betrachtungen eines deutschen Naturforschers*. Zurich: Füssli, pp. 9–11.
Einstein, A., et al. (1931). *Living Philosophies—A Series of Intimate Credos*. New York: Simon and Schuster.
Elsbach, A. C. (1924). *Kant und Einstein. Untersuchungen über das Verhältnis der modernen Erkenntnistheorie zur Relativitätstheorie*. Berlin and Leipzig: W. de Gruyter & Co.
Epstein, P. S. (1916). "Zur Theorie des Starkeffektes." *Annalen der Physik* 50, pp. 489–520.
Fulda, L., et al. (1914). "Aufruf an die Kulturwelt." *Wilamowitz nach fünfzig Jahren*, p. 718.
Gehrcke, E. (1916). "Zur Kritik und Geschichte der neueren Gravitationstheorien." *Annalen der Physik* 51, pp. 119–124.
— (1920). *Die Relativitätstheorie: Eine wissenschaftliche Massensuggestion, gemeinverständlich dargestellt*. Berlin: Arbeitsgemeinschaft deutscher Naturforscher zur Erhaltung reiner Wissenschaft e. V.
Gerber, P. (1898). "Die räumliche und zeitliche Ausbreitung der Gravitation." *Zeitschrift für Mathematik und Physik* 43, pp. 93–104.
— (1902). "Die Fortpflanzungsgeschwindigkeit der Gravitation." Reprinted in *Annalen der Physik* 52 (1917), pp. 415–444.
Goenner, H. (1993). "The Reaction to Relativity Theory I: The Anti-Einstein Campaign in Germany in 1920." *Science in Context* 6, pp. 107–133.
Goethe, J. W. von (1952). *Werke*, vol 2. Hamburg: Christian Wegner.
Grebe, L., and A. Bachem (1920). "Über die Einsteinverschiebung im Gravitationsfeld der Sonne." *Zeitschrift für Physik* 1, pp. 51–54.
Gutfreund, H., and J. Renn (2020). *Einstein on Einstein.: Autobiographical and Scientific Reflections*. Princeton, N.J.: Princeton University Press.
Hastedt, G. (2004). *Encyclopedia of American Foreign Policy*. New York: Facts on File.
Helmholtz, H. von (1868). "Über die Thatsachen, die der Geometrie zum Grunde liegen." *Königliche Gesellschaft der Wissenschaften und Georg-Augusta-Universität zu Göttingen. Nachrichten*, pp. 193–221.
— (1884). "Über den Ursprung und die Bedeutung der geometrischen Axiome." *Vorträge und Reden*. Vol. 2. Braunschweig: Vieweg, pp. 1–34.
Herneck, F. (1966). "Zwei Tondokumente Einsteins zur Relativitätstheorie." *Forschungen und Fortschritte* 40, pp. 133–135.

Hermann, A., K. Von Meyenn, and Viktor F. Weisskopf, eds. (1979). *Wolfgang Pauli. Wissenschaftlicher Briefwechsel mit Bohr, Einstein, Heisenberg u.a.* Vol. 1, 1919-1929. New York: Springer.
Hertz, H. (1889). "Die Kräfte elektrischer Schwingungen behandelt nach der Maxwell'schen Theorie." *Annalen der Physik und Chemie* 36, pp. 1-22.
— (1892). *Untersuchungen über die Ausbreitung der elektrischen Kraft.* Leipzig: Barth.
Hilbert, D. (1899). *Grundlagen der Geometrie.* Leipzig: B. G. Teubner.
Kant, I. (1795). *Zum ewigen Frieden: Ein philosophischer Entwurf.* Königsberg: Friedrich Ricolovius.
Kennedy, R. J. (1926). "A Refinement of the Michelson-Morley Experiment." *National Academy of Sciences: Publications* 12, pp. 621-629.
Kleinert, A. (1993) "Paul Weyland, der Berliner Einstein-Töter." In *Naturwissenschaft und Technik in der Geschichte. 25 Jahre Lehrstuhl für Geschichte der Naturwissenschaften und Technik am Historischen Institut der Universität Stuttgart.* Ed. by Helmuth Albrecht. Stuttgart: Verlag für Geschichte der Naturwissenschaften und der Technik, pp. 198-232.
Lenard, P. (1918). "Über Relativitätsprinzip, Äther, Gravitation." *Jahrbuch der Radioaktivität und Elektronik* 15, pp. 117-136.
— (1920). *Über Relativitätsprinzip, Äther, Gravitation.* 2nd ed. Leipzig: Hirzel.
Levi-Civita, T. (1917). "Nozione di parallelismo in una varietá qualunque e conseguente specificazione geometrica della curvatura riemanniana." *Circolo Matematico di Palermo. Rendiconti* 42, pp. 173-205.
Lief, A., ed. (1933). *The Fight Against War.* John Day Pamphlets No. 20. New York: John Day.
Mach, E. (1897). *Die Mechanik in ihrer Entwickelung. Historisch-kritisch dargestellt.* 3rd rev. and enl. ed. Leipzig: Brockhaus.
— (1898). "On Some Phenomena Attending the Flight of Projectiles." *Popular Scientific Lectures.* 3rd rev. ed. La Salle, Ill: Open Court, pp. 309-337.
— (1903). "Über Erscheinungen an fliegenden Projektilen." *Populärwissenschaftliche Vorlesungen.* 3rd rev. ed. Leipzig: Barth, pp. 356-383.
Michelson, A., and E. W. Morley (1887). "On the Relative Motion of the Earth and Luminiferous Ether." *American Journal of Science* 34, pp. 333-345.
Miller, D. C. (1925a). "Ether-Drift Experiments at Mount Wilson." *National Academy of Sciences: Proceedings* 11, pp. 306-314.
— (1925b). "Ether-Drift Experiments at Mount Wilson." *Nature* 116, pp. 49-50.
— (1925c). "Experiments at Mount Wilson." *Science* 61, pp. 617-621.
Minkowski, H. (1908) "Die Grundgleichungen für die elektromagnetischen Vorgänge in bewegten Körpern." *Nachrichten von der Gesellschaft der Wissenschaften zu Göttingen, Mathematisch-Physikalische Klasse.* pp. 53-111.
Newton, I. (1846). *Newton's Principa: The Mathematical Principles of Natural Philosophy.* Ed. by N. W. Chittenden. Trans. by Andrew Motte. First American edition. New York: Daniel Adee, pp. 80-81.
Nietzsche, F. (1886). *Jenseits von Gut und Böse: Vorspiel einer Philosophie der Zukunft.* Leipzig: C. G. Naumann. pp. 240-256.
— (2003). *Morgenröthe. Gedanken über die moralischen Vorurtheile. Werke,* Part 5, Vol. 1. Berlin: De Gruyter.
Pauli. W. (1921). "Relativitätstheorie." In: *Encyklopädie der mathematischen Wissenschaften, mit Einschluss ihrer Anwendungen.* Vol. 5, Physik, part 2, pp. 539-775. Ed. by Arnold Sommerfeld. Leipzig: B. G. Teubner, 1904-1922. Issued 15 November 1921.
— (1958). *Theory of Relativity.* Translated by G. Field. London: Pergamon.
Petzolt, J. (1914). "Die Relativitätstheorie der Physik." *Zeitschrift für positivistische Philosophie* 2, pp. 1-56.
Piccard, A., and E. Stahel (1927). "Neue Resultate des Michelson-Experimentes." *Die Naturwissenschaften* 15, 140.
Planck, M. (1900). "Zur Theorie des Gesetzes der Energieverteilung im Normalspectrum." *Deutsche Physikalische Gesellschaft. Verhandlungen* 2, pp. 237-245.
— (1914). "Erwiderung." In *Königlich Preußische Akademie der Wissenschaften,* Berlin: Sitzungberichte, pp. 742-744.

Reves, E. (1945). *The Anatomy of Peace*. New York and London: Harper & Brothers.

Riemann, B. (1867). "Über die Hypothesen, welche der Geometrie zu Grunde liegen: Habilitationsvortrag Göttingen." *Abhandlungen der Königlichen Gesellschaft zu Göttingen* 13, pp. 133–150.

Rowe, D. E., and Robert Schulmann, eds. (2007). *Einstein on Politics: His Private Thoughts and Public Stands on Nationalism, Zionism, War, Peace, and the Bomb*. Princeton, N.J.: Princeton University Press.

Rutherford, E. (1919a). "Collision of α-Particles with Light Atoms." *Nature* 103, pp. 415–418.

— (1919b). "Collision of α-Particles with Light Atoms, I–IV." *Philosophical Magazine* 37, pp. 538–580.

Saint-Pierre, C. I. C. de (1713). *Projet pour rendre la paix perpétuelle en Europe*. Utrecht: A. Schouten.

Schlick, M. (1918). *Allgemeine Erkenntnislehre*. Berlin: Springer.

Spinoza, B. de (1677). *Ethica, ordine geometrico demonstrata*. Trans. by R. H. M. Elwes. Auckland, New Zealand: The Floating Press.

Thomson, J. J. (1881). "On the Electric and Magnetic Effects Produced by the Motion of Electrified Bodies." *Philosophical Magazine* 11, pp. 229–249.

Vaz Ferreira, Carlos (1914). *Le pragmatisme: Exposition et Critique*. Montevideo: Barreiro & Ramos.

Weyl, H. (1918a). "Gravitation und Elektrizität." *Preußische Akademie der Wissenschaften* (Berlin): *Sitzungsberichte*, pp. 465–478.

— (1918b). *Raum, Zeit, Materie: Vorlesungen über allgemeine Relativitätstheorie*. Berlin: Springer.

— (1919). "Eine neue Erweiterung der Relativitätstheorie." *Annalen der Physik* 59, pp. 101–133.

— (1923). "Zur allgemeinen Relativitätstheorie." *Physikalische Zeitschrift* 24, pp. 230–232.

Weyland, P. (1920). "Einsteins Relativitätstheorie: eine wissenschaftliche Massensuggestion." *Tägliche Rundschau* (2 August): *Abend-Ausgabe, Unterhaltungsbeilage*.

Winternitz, J. (1923). *Relativitätstheorie und Erkenntnislehre: Eine Untersuchung über die erkenntnistheoretischen Grundlagen der Einsteinschen Theorie und die Bedeutung ihrer Ergebnisse für die allgemeinen Probleme des Naturerkennen*. Leipzig and Berlin: Teubner.

Index

Aarau canton school, Switzerland, 362–63
Abraham, Max, 97
absolute motion, 13
absolute space, 52, 85, 221
academic youth, and antisemitism, 111–12
action-at-a-distance, 46, 47, 52, 97, 167, 176, 324, 338
Adler, Felix, 197
African Americans, 208–9, 283–84
Ahad Ha'am (A.Z.H. Ginsberg), 108
alpha-particles, 58–59
American Council of Judaism, 274–75
Americans. *See* African Americans; United States of America
analytical geometry, 317
anarchy, economic and political, 298–99
Anti-Relativity Company, Berlin, 60–63
antisemitism, 41, 44–45, 60, 78–80, 111–12, 238–45, 247
Arabs and Jews in Palestine, 185, 186–87, 188–90, 246
Aristotelian theory of concepts, 124
art and science, 65
Asian ideal, 144
atomic bomb, 251–52, 269–74, 277–80, 289; in the shadow of the, 287–89
atomic composition, 100–101
atomic energy, 59, 252, 269, 272–74; outlawing as a means of war, 289–91
atomic theory, 119, 320, 321, 333
atomistic structure of matter and energy, 223–24
Austria, plight of science in, 90–91
autobiographical notes, 312–52
axiomatic method, 66, 67–69, 77, 222
axioms, 39–40, 66–67, 130–31, 220; of parallels, 132

Bachem, Albert, 63
basic laws. *See* natural laws
Belgium, 218
beliefs
    Einstein's, 193–197, 247–50
    Jewish, 213–16
Benedict, Ruth, *Patterns of Culture*, 305
Berlin
    Eastern European Jews and Russians in, 41–42, 79–80
    Jewish people in, 88
    speech at Keren Ha-Yesod rally, 160–61
    University, 43–44, 58
the Bible, 313
black market, 42
Bohm, David, 360
Bohr, Niels, 100, 101, 116, 119, 272, 331
Bolsheviks, 41, 42
Boltzmann, Ludwig, 165, 320, 328, 330, 331
Boltzmann's entropy-probability-relation, 334
Born, Max, 61, 346, 355, 358–61
Brahe, Tycho, 162
Brownian motion, 331–32
Buddhism, 144

capitalism/capitalist society, 273, 297–98, 310–12
Cassirer, Ernst, 124, 126, 128
causality, 162, 163, 168, 282
centrifugal force, 37
Christianity, 144
civilization
    and science, 225–28
    and war in Europe, 7–9
classical mechanics, 6, 14, 29, 37, 62, 84, 99, 101, 320, 331
clocks, 156, 334, 336, 337
    and gravitational fields, 27–28, 30, 38, 84
    and relativity, 25–26, 27–28, 30, 55, 61, 69, 84, 152, 154, 155
    See also space-time
Cohen, Hermann, 127
Compton experiment, 116–17, 119–20
concepts and propositions, 316–17
conservation of mass-energy, 3, 37
conservation of momentum and energy, 338
coordinate systems, 15, 24, 26–29, 30–33, 35–36, 50, 151–52, 323–24, 341
    accelerated, 115–16
    and general relativity, 27, 29, 33, 37, 84–85, 341, 342
    inertial systems, 36, 37, 84, 152–53, 324, 334–35, 339, 341, 342–43, 365
Coudenhove-Kalergi, Count Richard von, 140
Cox, Rev. Ignatius W., 197

375

De Broglie, Louis, 168, 223, 360; wave and wavelength, 357, 358
Debye, Peter, 61
deductive method, 39–40
denationalization of military power, and security, 279–80
Descartes's theory of physical events, 171, 172, 179
Diaspora, 81, 82
differential geometry, 317
differential laws, 162–63, 168
Dirac, Paul, 223, 346
Disarmament Conference, Geneva, 1932, 207–8
dual particle and wave nature of light, 117–20, 177, 322

Earth's motion, effect on light velocity, 157–59
Eastern European Jews, 41–43, 80–81
economy
    capitalist society, 310–12
    planned, 311–12
    and socialism, 305–6, 311–12
Eddington, Arthur S., 61, 63, 169, 172
education, 103–4, 227, 232–36, 311
    final secondary school exam, 20
    higher/university, 142, 233
    mathematical, 192–93, 317, 363
    school, 233–36, 362–63
    technical, 57–58
electric charge, 51
electric masses, 322
electrodynamics, 40, 323, 327, 330;
    Maxwell's theory, 6, 14, 36–37, 48–50, 83, 99–100, 152, 164, 167, 323
electromagnetic equations, 97, 98
electromagnetic fields, 6, 97–98, 118, 165, 166, 176, 177, 323, 334, 343, 350
    and the ether, 48–51, 53–54, 97
    and gravitational fields, 181–83
electromagnetic forces, 323
electromagnetic inertia, 97, 328
electromagnetic radiation, 50, 120
electromagnetic theory, 97, 118, 165, 166, 177, 321, 326
electromagnetic waves, 176
electromagnetism, 176, 321, 323, 326
elliptical geometry, 77
Elsbach, Alfred C., *Kant and Einstein*, review, 124–29
energy
    from uranium, 251
    new sources of, 58–59
    potential, 325

English Jews, 79
epistemology, 10–11, 12, 321
    of Neo-Kantians, 123–29
    and relativity theory (Winternitz), 112–14
Epstein, Paul, 101
ether, 46–55, 158, 320
    and electromagnetic fields, 48–51, 53–54, 97, 327
    ether hypothesis, 47–48, 51–52
    and theory of relativity, 50, 51, 52–53, 54
    and light, 3, 6, 39, 47–48, 175, 320, 322
    Lorentzian ether, 49, 53
    Mach's view of, 52
    and matter, 48–49, 53–54, 322
    mechanical properties of, 48, 49
Euclidean geometry, 37, 53, 68–69, 75–77, 84, 97, 116, 127, 131–33, 148–51, 182–83, 220–21, 257, 316, 334
    application to pre-relativistic mechanics, 151–52
    continuum, 181, 182
    pseudo-Euclidean space, 366
Euler, Leonhard, 320
Europeanism, 144
Europe
    Americanization of, 144–45
    changes during the 1930s, 247–50
European community, 139–40
European ideal, 143–45
Europeans
    comparison to Americans, 204–6
    intellectual life, 138–39, 249
    Manifesto to (1914), 7–9
    as pacifists, 139
Exner, Wilhelm, 57
experience and reality, 123–25
exports, United States, 301

family ties, Japan, 104
Faraday, Michael, 54, 83, 97, 118, 164, 178, 323, 328
Faraday-Maxwellian field theory, 98, 325–26
Fascist states, 249
Fermi, Enrico, 251
field theory, 98, 166–68, 190, 223, 367–68.
    See also unified field theory
final secondary school exam, 20
finite universe, 71, 72–74, 85
Fizeau, Hippolyte, 48
forces acting at a distance, 97, 164, 166, 167, 176, 325
Förster, Friedrich Wilhelm, 7–9, 49

four-dimensional continuum/space, 70, 84, 115, 152–53, 157, 178, 179–81, 223–24, 336–37, 340, 349–50, 365–66
France, 218
free particles (quantum mechanics), 291–93
Freud, Sigmund, 210

Galilei, Galileo, 6, 35, 36, 39, 163, 220, 220, 259, 264, 326
Gauss, Carl Friedrich, 85, 97, 367
Gehrcke, Ernst, 24, 60–63
General Association for Popular Technical Education, 57–58
general theory of relativity, 14, 30, 35, 37–38, 84, 85, 98, 116, 222, 223, 341
  and coordinate systems, 27, 29, 33, 37, 84–85, 341, 342–43
  and ether, 50, 52–53, 54
  field-structure and group, 343–46, 349–50
  geometry and physical space, 65–77
  mathematical constructions, 179–81
  motion of perihelion of Mercury, 61, 62
  and Newtonian mechanics, 167
  path to, 342–47
  principle, 341
  reducing physics to space-time geometry, 169–73
  and space-time, 154, 155, 179–83
  and theory of gravitation, 115–16, 167, 178–79, 343–44, 345, 349, 351–52
  two fields as one, 181–83
geodesic line, equation of, 366
geometry
  analytical, 317
  axiomatic, 67–68
  differential, 317
  elliptical, 77
  Euclidean. *See* Euclidean geometry
  foundations of, 151
  four-dimensional, 152–53
  integral, 317
  non-Euclidean, 127, 128, 133–34
  and physical space in the light of general relativity, 65–77
  and physics, 130–34
  practical, 68, 70
  Riemann's, 70, 71, 133–34, 154–55
  and space, 127–28
  space-time, and general relativity, 169–70
  spherical, 75, 76
Gerber, Paul, 62
German Jews, 78–79, 81
  contribution to German civilization, 228–29

development, 86–87
  Hitler's hatred of, 231
German National People's Party, 41
German Reich, 18, 121
Germany, 58, 286, 296
  antisemitism, 41, 44–45, 78–80, 238–45
  contribution of German Jews to German civilization, 228–29
  criticism of relativity, 60–63
  Eastern European Jews in, 41–43, 80–81
  economy, 42–43
  entry into League of Nations, 121–22, 139
  extermination of Jews, 268
  final secondary school exam, 20
  Hitler's power, 215–16, 218, 230–31, 268, 297
  immigration from the East, 41–43
  losing the World War blamed on Jews, 239
  Nazi regime, 231, 243
  plight of science in, 90–91
  political and social intolerance, 216, 217
  rearmament, 218
  Russians in, 41
  schooling, 234
  social justice/social responsibility, 195
  Weimar Republic, 230, 231
Ginzberg (Ginossar), Shlomo, 108
God, 183–84, 197, 198–200, 213–16
Goethe, Johann W. von, 9, 139, 145
Goethebund, 17
gold, 300
Goldstein, Rabbi Herbert S., 183–84
gravitation, 323, 325
  field-theory of, 343–44
  and general relativity, 115–16, 167, 178–79, 343–45, 345, 349, 351–52
  general theory of, 4, 7, 178–79, 367
  and inertia, 179
  laws of, 37, 115
  Newton's theory, 40, 47, 54, 72, 83, 97, 164, 167, 259–60, 343–44
  scalar theory, 339
  and special relativity, 339–40, 344, 365
gravitational ether, 53, 54
gravitational fields, 4, 27, 29, 32, 34, 53–54, 72, 84, 115, 153, 339, 343–44, 349, 365–66
  and clocks, 27–28, 30, 38
  and electromagnetic fields, 181–83
  and laws of motion, 345
  light ray deflection, 4, 38, 62, 179

Great Britain
   and atomic bomb, 270
   supports re-establishment of Jewish home in Palestine, 186, 187, 189–92
Grebe, Leonhard, 63
Grossman, Marcel, 364–65, 367

Haber, Fritz, 39
Hahn, Otto, 272
Haller, Friedrich, 363
Halpern, Lipmann, 110
Hasenclever, Walter, 65
heat radiation, theory, 98–99, 117–18, 328–30, 333
Hebrew University of Jerusalem, 88, 136, 141–43, 189
Hegel, Georg W. F., 169, 171, 179, 263
Heisenberg, Werner, 224, 346, 357
Hellenism, 144
Helmholtz, Hermann von, 132, 317
Herbert Spencer Lecture, University of Oxford, 218
Hertz, Heinrich, 48, 49, 166, 172, 176, 317, 320, 323, 325, 327
Herzl, Theodor, 160
higher education, 142, 233
Hiroshima, bombing, 250, 288
Hitler, Adolf, 215–16, 218, 230–31, 268, 297
human development, "predatory phase" of, 306
Hume, David, 264–65, 266–67, 334
Hurwitz, Gregg, 317
Huygens-Young-Fresnel wave theory of light, 175
hydrodynamics, 320
hydrogen atoms, 59

ideals, 143–45
individuals, relationship to society, 283, 286, 307–9, 353–54
inductive method, 39, 40
inertia
   electromagnetic, 97, 328
   generalized law of, 167
   and gravitation, 179, 339
   and interaction of masses, 324
inertial systems, 36, 37, 61, 84, 115, 152, 153, 156, 324, 334–35, 339, 339, 341, 342–43, 365
infinite universe, 71, 72–73
integral geometry, 317
integral laws, 162
intellectual freedom
   aspiration and accomplishment, 227, 228–29, 242, 243, 253
   and science, 252–55, 272–73

intellectuals, contribution to international reconciliation, 63–64
interference experiment (Michelson-Morley), 157–58
International Court of Justice, 277
international reconciliation, intellectuals' contribution to, 63–64
international security, 212, 249, 270–72, 273, 276. *See also* League of Nations; United Nations
international understanding, 289–91
Israel, Einstein declines presidency of, 353–54

Jacobi, Carl, 360
Japan
   bombing of Hiroshima, 288
   impressions of, 102–7
Japanese
   art and style connected to nature, 105
   music, 105–6
   painting and woodcarving, 106–7
   tradition not to express one's feelings, 104
Jewish immigrants in Germany, 41–43, 80–81
Jewish Labor Office, 42
Jewish Palestine, 81, 86–88, 108–10, 160–61, 188, 276
   British support for, 186, 187, 189, 190
   co-operation with Arabs, 188–89, 246
   reconstruction, 186, 187, 246–47
Jewish suffering, 187–88
Jews
   adaptation to European nations, 87
   alleged crimes over the course of history, 239–40
   antisemitism. *See* antisemitism
   and Arabs in Palestine, 185, 186–87, 188–90, 246
   calling of the, 236–37
   characterization, 241–42, 243
   extermination in death camps, 269
   in Germany, 78–80, 88
   heroes of the Battle of the Warsaw ghetto, 268
   indebted to Roosevelt, 276
   intellectual aspiration and accomplishment, 229, 242, 243
   nationalism, 78, 136, 142, 160–61, 188, 246
   philosophy and beliefs, 213–16
   where oppression is a stimulus, 243–45

See also German Jews; Zionism/Zionist
    movement
Judaism, 144, 214–15, 237, 242–43; owes a
    debt of gratitude to Zionism, 245–47.
    See also American Council of
    Judaism

Kaluza, Theodor, 350
Kant, Immanuel, 113, 123–29, 139, 265, 317
Kant and Einstein (Elsbach), review, 123–29
Kantianism, relationship to relativity theory,
    123–29
Kauffmann, Richard, 109
Kayser, Rudolf, *Spinoza: Portrait of a*
    Spiritual Hero, foreword to, 282–83
Kelvin, Lord (William Thomson), 165
Kennedy, Roy J., 158–59
Kepler, Johannes, 162, 163, 200, 220, 260
Keren Ha-Yesod in Berlin, speech at rally for,
    160–61
kinematics, 36, 37
kinetic energy, 325, 327
kinetic theory of gases, 35, 40, 164, 320,
    328–29, 331–32
kinetic theory of heat, 164, 321
Kirchhoff, Gustav, 317
knowledge, theory of, 262–67

La Déduction Relativiste (Meyerson), review,
    169–73
Langevin, Paul, 61
language of science, 255–58
Larmor, Joseph, 61
Laue, Max von, 61
League of Nations, 121, 139–40, 249, 277
    Germany's entry into, 121–22, 139
    International Committee on Intellectual
        Cooperation (ICIC), 121
    International Institute of Intellectual
        Cooperation, 210
Leibniz, Gottfried Wilhelm, 23, 163
Lenard, Philipp, 24, 31, 61, 62
Letters from Russian Prisons (Berkman),
    introductory letter, 135
Levi-Civita, Tullio, 61, 134, 367
light
    Compton's scattering experiment,
        116–17, 119–20
    corpuscular and emission theory, 100,
        117–18, 175
    duality of particle and wave, 117–20, 177,
        322
    emission theory of, 100, 117–18
    and ether, 39, 47–48, 320, 322

Michelson-Morley's interference
    experiment, 157–58
principle of the constancy of the velocity
    of speed of, 2, 3, 6, 14, 36, 83, 336
propagation and equivalence of inertial
    systems, 115, 156
propagation in empty space, 175, 176
properties and elastic waves, 47
quantum theory, 116–20, 334–35
Roentgen radiation (X-rays), 120
and theory of electromagnetic fields, 176,
    326
and unified field theory, 176
velocity, Earth's motion effect on, 157–59
wave theory, 46, 48, 53, 118–19, 164, 175,
    320, 322, 362–63
light-ether, 3, 6, 48, 158, 175, 320
light particles, 175
light rays, 158; deflection in gravitational
    field, 4, 38, 62, 179
Lorentz, Hendrik A., 33, 46, 49, 61, 63, 115,
    166, 176, 327, 328
Lorentz contraction, 68
Lorentz transformations, 98, 336, 337, 339,
    340, 341, 345, 365
Lorentzian ether, 49, 53
luminiferous ether. *See* light-ether

Mach, Ernst, 10–16, 23, 72, 333, 334, 339
    ether, 52
    History of Mechanics, 321
    on Newton's bucket experiment, 14,
        323–24
    on Newton's views of time, space, and
        motion, 12–14, 85, 322–24
    philosophical studies, 15
    science views, 11–12
macro-systems, in quantum mechanics, 357,
    359–61
malaria, 109
Mandatory Power for Palestine, 190–92
Marić, Mileva, 363
Marxism, 286
mass-density, negative, 72
mathematics and mathematical education,
    192–93, 317–18, 363
matter
    and energy, atomistic structure, 224
    and ether, 48–49, 53–54, 322
    quantum properties, 223–24
    and space, 54, 174–79, 327
Maxwell, James Clerk, 54, 97
Maxwell's electrodynamics, 6, 14, 36–37,
    48–50, 83, 99–100, 152, 164, 167, 323

Maxwell's equations, 37, 49, 50, 97–98, 172, 178, 181, 223, 323, 328, 330, 334, 339, 345
Maxwell's theory of electromagnetic fields, 97, 118, 164, 165–66, 177, 320, 325, 327–29, 330, 333, 339, 346
Maxwell's theory of electromagnetism and theory of light, 176, 326
measuring rods, 55, 77, 152, 154, 155, 336, 337
mechanics
　classical, 6, 14, 29, 37, 62, 84, 99, 101, 320
　critique of as basis of physics, 322–26
　inner simplicity, 324–25
　laws of, 36, 96, 97
　Mach's critique, 12–14, 85, 323–24
　pre-relativistic, 151–52
　and quantum theory, 101–2
　See also Newtonian mechanics
Meitner, Lise, 272
Mercury
　motion of perihelion of, 62
　and theory of gravitation, 72
metaphysics, 266, 267
Meyerson, Émile, *La Déduction Relativiste*, 179; review, 169–73
Michelson-Morley interference experiment, 157–58
militant pacifism, 201–2; reversal of views on, 218
military mentality, and control of science projects, 285–87
military power, denationalization of, 279–80
Milky Way, stars and their masses, 72–73
Miller, Dayton C., 157–59
Minkowski, Hermann, 51, 84, 317, 337, 366
motion
　absolute motion, 13
　laws of (Galileo and Newton), 6, 35, 36, 37, 85, 162–64, 166–67, 259–60, 320, 324–25, 346
　Mach's view of Newton's ideas, 13–14, 85, 323–24
　non-uniform motion, 4, 7, 31
　planetary motion, 162–63
　relative motion, 1–2, 13
　and special relativity, 31, 33, 36–38
　uniform motion, 3–4, 7
　waves on the surface of water, 51
motives for research, 21–23

Nagasaki, bombing, 250
naïve realism, 253
nationalism, 78, 136, 142, 160–61, 188, 246, 290. *See also* Zionism

nationalists, 60
Natorp, Paul, 127, 128
nature
　natural laws, 39–40, 101, 170, 220, 336, 337, 341
　"physical reality," 291–95, 346–47, 357–58
Nazi regime, 231, 243, 268
neo-Kantian epistemology, 123, 125
Nernst, Walther, 39
new field theory. *See* unified field theory
Newton, Isaac, 6, 35, 52, 100, 161–68, 174, 200, 221–22, 258–61, 325, 346
　absolute space, 52, 85, 221
　action at a distance, 46, 47, 97, 167, 176, 324, 338
　awareness of weaknesses in his theories, 165–66
　bucket experiment, 14, 323–24
　corpuscular theory and emission of light, 117–18
　laws of motion, 6, 35, 36, 37, 85, 163, 164, 166, 167, 259–60, 320, 325, 346
　logical objections against his theories, 166–67
　Mach's critique of, 12–14, 85, 323–24
　on planetary motion, 162, 163–64
　significance of achievements, 164–65
　sound transmission, 320
　theory of gravitation, 40, 47, 54, 72, 83, 97, 164, 167, 178, 260–61, 343–44
Newtonian mechanics, 6, 35, 49, 83, 99, 320, 323, 338
　influence on formation of theoretical physics, 161–68
　and theory of fields, 167–68
Nicolai, Georg F., 7–9
Nietzsche, Friedrich, 136, 139
nitrogen atoms, 58, 59
non-inertial systems, 153, 341
nuclear chain reactions, 251
nuclear research, military mentality in control of, 285–87
Nuremberg trials, 353

O'Connell, Cardinal William Henry, 184
Ostwald, Wilhelm, 333

pacifism
　Europe, 139
　militant, 201–2, 218
　science's impact on development of, 89–90
Palestine, 42, 81, 82, 86–88, 108–10, 142, 276
　economic development, 109–10

Great Britain support for
  re-establishment of Jewish home in,
    186, 187, 189–92
  impressions of, 107–10
  Jewish workers and weight of debt, 108,
    109
  Jews and Arabs in, 185, 186–87, 188–90,
    246
  malaria in, 109
  Mandatory Power, 190–92
  partition, 246
  reconstruction and development, 109,
    186–87, 188–89, 190, 191, 245–47
  See also Jewish Palestine
pan-European movement, 138–40
partial differential equations, 320
passive resistance, 92
Pauli, Wolfgang, 360; *The Theory of
  Relativity*, review, 93
philosophy
  and beliefs, 193–197, 213–16, 247–50,
    313–14
  Jewish philosophy, 213–16
  and science, 123–29
  and theory of knowledge, 262–67
photoelectric effect, 100, 330
"physical reality." *See under* Nature
physics
  19th century, 320
  critique of mechanics as basis of, 322–26
  and geometry, 130–34
  induction and deduction in, 39–40
  and non-Euclidean geometry, 133–34
  and philosophy, 124–29
  reduced by general relativity to
    space-time geometry, 169–73
Piccard, Auguste, 158, 159
Planck, Max, 6, 21–23, 39, 61, 99, 116, 328,
  329–31, 333
Planck-Bohr-Rutherford quantum theory,
  96
Planck's constant, 101
Planck's law of heat radiation, 6, 99, 118–19,
  330, 333
Planck's quanta, 333
planetary motion, laws of, 162, 163–64
planned economy, 311–12
Poincaré, Henri, 68, 69, 133
Poland, 42, 81, 88, 268, 283
ponderable matter, 174–75
Prague University, 366
principle, theories of, 35
principle of contiguity, 293
principle of equivalence, 116
principle of relativity, 2, 3, 6–7, 116, 171

principles, development, 5, 6
propositions and concepts, 316–17
Prussian Academy of Sciences, 4, 43, 65;
  resignation from, 216–17
Prussian State Library, sound recordings,
  115–16
pseudo-Euclidean space, 366
Pythagorean theorem, 316

quantum mechanics
  and Born's statistical interpretation of
    wave function, 355, 358, 359–61
  interpretation of foundations of, 355–61
  in macro-systems, 359–60, 361
  and physical reality, 291–95, 346–47,
    357–58, 358
  quantum hypothesis, 6
  special example, 357–58
quantum theory, 96, 98–101, 177, 223–24,
  329, 349
  and Compton experiment, 116–17,
    119–20
  and light, 116–20, 333–34
  and radiation, 98–100, 116, 339
  statistical, 346–48
"quasi-periodic" mechanical systems, 101

Rabinowitsch, Wolf Ze'ev, 110
racial prejudice, 208–9
radiation
  alpha-particles, 58, 59
  average energy of a
    quasi-monochromatic oscillator, 329,
    333–34
  emission and absorption, 322
  energy, 331, 333
  field theory, 100
  heat, 98–99, 118–19, 328–30, 333
  theory and quanta, 98–100, 116, 119–20,
    339
Rathenau, Walther, *in memoriam*, 94–95
red shift of spectral lines, 7, 38, 62
refraction, indices of, 322
Reichenbach, Hans, 113
relative motion, 1–2, 13
"relativism," 170–71, 304
relativity
  general theory. *See* general theory of
    relativity
  principle of relativity, 2, 3, 6–7, 116, 171
  special theory. *See* special theory of
    relativity
  theory. *See* theory of relativity
Relativity Theory and Epistemology
  (Winternitz), review, 112–14

religion, 144–45, 183–84, 196, 313–14
   and the Bible, 313
   and science, 197–201, 302–5
   See also God; Spinoza's God
research, motives for, 21–23
Reves, Emery, *The Anatomy of Peace*, 273
Ricci, Giovanni, 367
Riemann, Bernhard, 85, 97, 367
Riemannian metric, 181, 182, 223, 343
Riemannian space, 178
Riemann's geometry, 70, 71, 133–34, 154–55
rigid bodies, 67–68, 69–70, 71, 75, 84, 115, 148, 155. *See also* Euclidean geometry
rigid rods, 158, 335, 336
Roentgen rays (X-rays), 120
Roosevelt, Franklin D., 299
   commemorative words for, 276
   letter on atomic energy to, 251–52
Royal Albert Hall speech, 225
Russell, Bertrand, 286
   Political Ideals, preface to, 92–93
   theory of knowledge, remarks on, 263–67
Russia, 139, 196, 234, 239, 270, 296
   political prisoners, 135
   socialism, 297–98
Russian Jews, 43, 78
Russians
   in Berlin, 41
   criticism of Einstein's support for World Government, 296–302
Rutherford, Ernest, 58–59
Rutherford-Bohr theory, 100–101, 120

Saint-Pierre, Charles-Irénée Castel de, 139
scalar theory of gravitation, 339
scattering experiment (Compton), 116–17, 119–20
Schlick, Moritz, 67, 113
Schilpp, Paul Arthur, 312
school education, 233–36, 362–63; final secondary school exam, 20
Schopenhauer, Arthur, 22
Schrödinger, Erwin, 168, 223, 346, 360
Schrödinger equation, 358, 359, 360
Schwarzschild, Karl, 101
science
   and art, 65
   and civilization, 225–28
   common language of, 255–58
   impact on development of pacifism, 89–90
   induction and deduction in, 39–40
   and intellectual freedom, 252–55, 272–73
   as a mental construct, 113–14

   military mentality in control of, 285–87
   and philosophy, 123–29
   plight of in Germany and Austria, 90–91
   and religion, 197–201, 302–5
   for science's sake, 117
scientific method, 257
security, 142, 190, 198, 206, 281, 286
   and denationalization of military power, 279–80
   international, 212, 249, 270–72, 273, 276
   supranational, 278–80, 287
sensory experiences, 124
simultaneity, 2–3, 14, 115, 156, 335, 338
singularities, 346
social justice/social responsibility, 195, 205, 241
social policy, 92
socialism, 273, 297–98, 305–12
   directed toward a social-ethical end, 306–7
   and economic science, 306, 311
society
   capitalist, 273, 297–98, 310–12
   individual's relationship to, 283, 286, 307–9
Society for Social Responsibility in Science, statement to, 353–54
Sommerfeld, Arnold, 61, 101
sound recordings, Prussian State Library, 115–16
sound transmission, 320
Soviet Union, 270, 271. *See also* Russia
Soviet scientists, criticism of Einstein, 296–302
space, 147–55
   concept of, 147, 151–52151–52
   and Euclidean geometry, 148–51
   and geometry, 127–28
   and matter, 54, 174–79, 327
   in pre-scientific thought, 147–48
   in reference to the Earth, 148
   and time, 152
space-time, 50, 85, 115, 127, 152
   effect of relativity, special and general, 152–54
   Elsbach's summary of Cohen's and Natorp's views, 121–23
   Encyclopædia Britannica, Einstein's article in, 146–57
   four-dimensional continuum, 70, 84, 115, 152–53, 157, 178, 179–80, 181, 223, 336–37, 349, 365–66
   and general relativity, 169–73, 179–83
   and gravitation, 34–39

Newton's and Mach's views, 12–13
  structure, 179–83
Spanish Jews, 109
special theory of relativity, 25–26, 84, 153, 177–78, 335, 336–409, 341, 345, 363
  and the ether, 50, 51
  formal foundation, 365–66
  and gravitation, 339–40, 344, 365
  and inertial systems, 36, 37, 61, 84, 336, 365
  metric of the theory, 85
  and motion, 31, 33, 36–37
  and Newton's mechanics, 167
  universal principle, 336
specific heats of solids, 99
spectral lines, 70, 331; red shift of, 7, 38, 62
speed of light, 338; law of constancy of, 2, 3, 6, 14, 36, 83, 336
spherical geometry (space), 75, 76, 77
"spinors," 223
Spinoza, Baruch, 10, 183, 263, 282–83, 305
Spinoza: Portrait of a Spiritual Hero (Kayser), foreword to, 282–83
Spinoza's God, belief in, 183–85
Stahel, E., 158, 159
statistical mechanics, 331
statistical quantum theory, 346–48
Stefan-Boltzmann law, 331
supranational security, 278–80, 287
Swiss Patent Office, 192, 364
Switzerland, 318, 319, 361–68
system of coordinates. *See* coordinate systems
Szilard, Leo, 251

Taoism, 144
teacher-student relationship, 234–36
technical education, 57–58
theoretical method, development, 220–21
theoretical physics, 319
  development phase, 96–98
  and experience, 220–22
  influence of Newtonian mechanics on formation of, 161–69
  method of, 219–24
  present crisis, 96–102
  and philosophy, 125
theories
  deduction in development of, 39–40
  and empirical facts, 321
  "inner perfection," 322
  "naturalness" or "logical simplicity" of the premises, 321
  of principle, 35
  truth of, 40

theory of fields, 166–68
theory of knowledge (Russell), remarks on, 263–67
theory of relativity, 1–4, 12, 68, 69, 75, 151
  compared to systems of Hegel and Descartes, 171–72
  development and present position (1921), 83–85
  dialogue on objections to, 24–34
  and ether, 46–55
  and field theory, 166–68, 177–78
  historical sketch, 174–83
  relationship to Kantianism, 123–29
  response to Anti-Relativity Company objections, 60–63
  and space-time, 153–54
  supporters, 61
  See also general theory of relativity; special theory of relativity
Theory of Relativity (Pauli), Einstein's review, 93–94
thermodynamics, 35, 40, 98, 164, 320, 328–30. *See also* kinetic theory of gases
"thinking" and "wondering," 315–16
Thomson, J. J., 97
time, 335–36
  concept of, 155–56
  measurement of, 156
  See also clocks; space-time

Ukraine, 81
undulatory theory of light. *See* wave (undulatory) theory of light
unified field theory, 169, 174, 179, 181–83, 295
uniform motion, 3–4, 7
unitary field theory. *See* unified field theory
United Nations, 270, 271, 298, 299
United States of America, 196
  American ideal, 144–45
  antisemitism, 80
  and atomic bomb, 251–52, 270, 287–89
  comparison of Americans to Europeans, 204–5
  economic oligarchy, 299
  economic and political domination, 298–300, 301
  Einstein's message to, 283–85
  exports financed through loans to foreign countries, 301
  impressions of, 203–6
  influence on international affairs, 205–6
  influence on United Nations, 299
  Jewish people in, 88

United States of America (cont.)
   military mentality, 286–87
   prohibition law, 205
   Russian scientists criticism, 298–99
   social responsibility, 205
   superiority in things technical and organizational, 204–5
   uranium and atomic research, 251–52
   victorious against Germany, 276
   See also African Americans
universal constants, 338–39
universe, 71–74, 85, 167
University of Berlin, 58; Einstein's lectures at, 43–44
university education, 142, 233
University of Leyden, 46
uranium, 251, 252, 273

Veblen, Thorstein, 262, 306

Warburg, Emil, 100
wars and war prevention, 226
   1932 Disarmament Conference, Geneva, 207–8
   danger of annihilation supersedes all others, 301–2
   outlawing of atomic energy as an instrument of war, 289–91
   "Why War?" 210–13
   and World Government, 273, 296–302
wave functions, Born's statistical interpretation (quantum mechanics), 355, 358–60
wave (undulatory) theory of light, 46, 48, 53, 118–19, 164, 175, 320, 322, 362–63

waves on the surface of water, 51
Weber, Heinrich Friedrich, 362, 363
Weimar Republic, 230, 231
Weizmann, Chaim, 354
western civilization, and war in Europe, 7–9
Weyl, Hermann, 54, 169, 172
Weyland, Paul, 60–61
Wien, Wilhelm, 116
Wien's law, 331
Wilhelm II, Kaiser, 286
Winternitz, Joseph, *Relativity Theory and Epistemology*, review, 112–14
"wondering" and "thinking," 315–16
World Government, 270–72, 273; reply to Soviet Scientists criticism of Einstein's support for, 296–302
World War I
   and *Manifesto to the Europeans*, 7–9
   opinion on, 18–20
   post-war impact on feeling of security, 249
   and reconciliation of nations, 63–64
World War II
   America victorious against Germany, 276
   Battle of the Warsaw ghetto, 268
   extermination of Jews in death camps, 268

Zionism/Zionist movement, 41, 78–80, 81, 87–88, 136–37, 187–88
   American Council of Judaism opposition to, 274–75
   Judaism's debt of gratitude to, 245–47
Zürich Polytechnic, 318; Einstein's experience at, 361–68